Available Energy
and Environmental
Economics

Available Energy and Environmental Economics

Robert H. Edgerton
Oakland University

LexingtonBooks
D.C. Heath and Company
Lexington, Massachusetts
Toronto

Library of Congress Cataloging in Publication Data

Edgerton, Robert H.
 Available energy and environmental economics.

 Includes bibliographical references and index.
 1. Power resources—Environmental aspects. 2. Thermodynamics.
3. Power (Mechanics) I. Title.
TD195.E49E33   333.79   81-47624
ISBN 0-669-04699-x ⟋ AACR2

Published simultaneously in Canada

Printed in the United States of America

International Standard Book Number: 0-669-04699-x

Library of Congress Catalog Card Number: 81-47624

Contents

Contents vii

List of Figures

List of Tables

Preface and Acknowledgments

The tasks of engineers and people in technical management have been modified in important ways in the last decade. The design of new devices and systems traditionally has involved optimization procedures that revolved around economic criteria. These procedures are now being changed with the introduction of environmental and energy-resource limitation criteria. This book focuses on the development of new techniques for handling these situations.

The area of technology assessment has been developed to analyze the consequences to society of introducing new technology. In this book another aspect of technology assessment is introduced. It is the aspect of evaluating alternative technologies based on environmental, economic, and thermodynamic factors. This book is also addressed to the student of science or engineering who has a need to follow up introductory environmental texts with more advanced techniques of analysis. Selected techniques from thermodynamics, economics, and chemical engineering have been introduced in an environmental context.

The author has been strongly influenced by the work of Robert Evans and Myron Tribus in thermodynamics and thermoeconomics. Their information-theory approach to available energy (*essergy*) appears in the thermodynamic function formulations and in the discussion of solar energy. Several environmental-ecological ideas have been drawn from the work of Harold Morowitz and Howard Odum. The contributions of many others cannot be acknowledged adequately. It is hoped, however, that in the attempt to integrate ideas from diverse areas, the spirit of each remains.

1 Introduction

Information consists of differences that make a difference.
—G. Bateson, *Mind and Nature* (1979)

Introduction

The objective of this book is to provide engineers and people making technological decisions with new techniques for dealing with energy and environmental situations. Quantitative methods are emphasized as the basis for the development of these techniques. Techniques of combining thermodynamic design and economic analysis are emphasized throughout the development.

Many papers and books have appeared in recent years devoted to describing the energy, economic, and environmental status of the world [1–10]. These are of limited value for design purposes because of their very general nature. They do, however, provide the basis for improving the dialogue between technical people and the rest of society.

The value of an action in an industrial society is often measured in terms of the difference between the benefits that accrue and the costs incurred. In economic terms this measure is monetary. In technological terms it is often efficiency. In environmental terms, however, the measure is not as easily defined. Ecologists talk in such terms as diversity of species, biological productivity, and evolutionary progress [10–13]. This apparent complication in dealing with environmental problems is, of course, due in part to the unfamiliarity of industrial society with the nature of the environment. Economists' difficulties in determining costs and benefits are related to the difference in the value of money to different individuals or groups. The monetary system provides an excellent measure of value as long as the consequences of choices are clear to the society. This is particularly true when: (1) choices do not affect the environment appreciably, (2) people have the ability to change their environment by transporting themselves to an acceptable environment, or (3) people can use their resources to secure a satisfactory artificial environment.

When one reaches the point where maintaining one's personal environment in the future appears to be beyond one's own control or has detrimental consequences for society, a shift in values is required. This shift can to some extent be accommodated by a monetary accounting system including

1

penalties for polluting, tax incentives, and so forth. The development of the value of environmental change must enter into consideration at this point. In the same way that, as Norbert Wiener [14] has stated, "there is no wage at which a laborer with a hand shovel can compete with a steam shovel," so there is no price at which an individual can buy the right to poison his environment or his fellow inhabitants of the earth.

An accounting system is required, not for the purposes of balancing costs and benefits, but for choosing among alternative futures. In this book an attempt is made to provide the basics of a thermodynamic accounting system for evaluating alternatives. Thermodynamics provides some of the most powerful principles for evaluating alternative solutions creatively. The idea of accounting is partly wrapped up with the idea that transactions are an exchange of equal value to the participants. Conservation of mass, momentum, charge, and energy are basic principles that all students of science and engineering are expected to learn and apply.

Thermodynamics is important to the extent that it proposes that in energy exchanges there is a quality as well as a quantity that must be taken into account. Not only is this quality of a descriptive nature, but it is also quantitatively described in the entropy function. This entropy function is assumed in environmental and economic decisions, but only on a qualitative basis. Characteristic of this are the notions found in economic literature dealing with depreciation, deterioration, loss, obsolescence, and so forth. In environmental literature terms such as *waste, dispersion, evolution,* and *pollution* are attempts to describe these principles qualitatively.

This book attempts to integrate these qualitative descriptions into a quantitative measure defined as the *available energy*. Available energy specifically represents that aspect of the energy of a material, substance, or field relative to the datum state that is potentially convertible into a useful work form [15, 16]. In the chemical sense each substance not at equilibrium with its environment has the potential to be used to produce a useful work effect. Simple examples are the thermal energy stored in underground hot rock strata, the energy of chemical bonds of hydrocarbons in the earth or in the biomass of plants, the kinetic energy in the wind, the potential energy of the water in a reservoir, the electromagnetic energy in the solar spectrum, and the nuclear energy in radioactive elements. The available energy is a thermodynamic description of the energy state in each of these cases relative to the environment in which it is present.

The processes of energy conversion involve the transformation of energy in one form to a form more useful for a given purpose. In this process a loss is inevitably incurred. This loss is a function of several basic processes. One is the inherent conversion loss to friction or heat if the process is carried out at a finite rate. This rate loss is described by the theory of irreversible thermodynamics [17–21], a theory that is often assumed to be too advanced for undergraduate university students. Its consequences are therefore not widely utilized in the analysis of most technological or environmental decisions. In chapter 8 this rate loss is simply described and incorporated as a criterion for economic analysis. Without this perspective decisions are static ones, and the basis for the trade-off of capital and operating costs is often irrelevant in energy and environmental situations.

In chapter 7 the important concept of impedance matching for the maximum energy transfer between systems is examined. This concept, well known in electrical-circuit theory and used in developing electrical-power converters, is one of the crucial elements in the design of wind-power systems and in the general field of power transfer. In automotive-efficiency optimization, the proper transmission matching of the engine characteristics to the driving pattern, aerodynamic characteristics, and dynamics of the vehicle is essential.

Available energy is converted to useful forms only at a finite rate. In classical thermodynamics a reversible process is hypothesized as a process by which the maximum energy can be extracted by a machine. This useful concept is extended to irreversible systems to show how the efficiency or *effectiveness* of a machine depends on the rate of energy transfer. *Expected* and *actual* work processes are defined and the difference used to illustrate the entropy generation in a system.

The relationships between reversible processes and ideal economic systems are illustrated in chapter 9. The ideas of thermodynamics and economics are integrated to demonstrate the costs of approaching reversibility in thermodynamic systems. The engineering science of heat-exchanger design is examined from thermodynamic and economic points of view, using available energy concepts. It is shown that the loss of available energy in heat exchangers is often avoidable. The optimization of heat exchangers can be made with the objective of maximum available energy conversion per unit cost. This is more important than simply maximizing the heat-transfer coefficient. The cost of improved exchanger performance is examined from an optimization position with regard to the alternatives of size and operational costs.

Many studies have been completed utilizing data from the *Census of Manufacturers* to determine the energy required to produce a consumer product [22]. Notable are the estimation of energy required to produce an automobile [23] and the energy required to grow and process a certain quantity of food. Investment companies have even provided their customers with information on energy use per unit value of products in many industries to assist them in investment decisions.

Critical studies by Steinhart and Steinhart [24] and Pimentel et al. [25] have demonstrated, by careful analysis of the energy inputs to industrialized agriculture, the energy subsidy used to produce food energy. They have demonstrated that between 5 and 10 calories of energy are required to produce 1 calorie equivalent of food in highly industrialized systems. Table

1–1 and figure 1–1 illustrate these energy requirements for food production. Particularly important are the increased direct use of fossil fuel in food production and the energy required for commercial fertilizers. As Hayami and Ruttan have demonstrated [26], the energy inputs associated with mechanization have substituted fossil fuels for human or animal effort without any increase in productivity. The increases in productivity are primarily associated with the increased use of commercial fertilizers, irrigation, and genetically improved varieties of plants. Ruttan has also pointed out the fallacy in agriculture of an energy balance as an accounting of efficiency [27]. Using a marginal economic optimization, it would appear that as long as each additional calorie expended returns more than 1 food calorie, the additional energy should be expended. The problem revolves around the fact that more value is assigned to a calorie of food than to one of fuel. The reason for this is that food is a scarce commodity in a hungry world. In the presence of cheap or abundant fossil fuel without competing demands, this balance would be maintained. However, in the future the competing demands for fuels as basic materials for the chemical industry will increase the fuel costs. As long as energy as heat per se is the competing use, fuels can be evaluated reasonably well by the calorie heating value. When value is associated with the informational content of the fuel, then a new accounting system must be designed. In food it is not the calorie value that is important. The structural value of a protein is greater than that of a carbohydrate. In plants the carbohydrate parts are often the residue of stalks, leaves, roots, and so forth that are not harvested. Analysis has shown that the fertilizer value of crop residues is smaller than that of the equivalent fertilizers that could be manufactured using the crop residues as fuels. A simple analysis demonstrates, however, that economically it is to the farmer's advantage to apply the capital that would be required to harvest the residues to the planting and harvesting of a larger food-crop area. As long as excess cropland is available, as in the United States, or optimum yields are not obtained from present agricultural land, it will be more economical to invest energy or dollars in improved farming practices or expanded land use. Crop residues, as shown in figure 1–2, represent a temendously large potential for materials but not for fuels. The use of sugarcane residues for wallboard is a prime example of the possibilities for other residues such as cornstalks or wheat stubble. To convert these to fuels is possible only at a disadvantage economically and environmentally. The combustion of these residues is inherently wasteful because their available energy-to-heating-value ratio is much higher than that of fossil fuels.

Pyrolysis techniques are now being used in pilot plant operations to reduce cellulose in waste products to petroleum. Fermentation and bacterial action yield hydrocarbon gases in controlled biological operations. In these operations the product hydrocarbons have lower energy than the reactants.

Table 1-1
Energy Use in the U.S. Food System

Component	1940	1947	1950	1954	1958	1960	1964	1968	1970
On farm									
Fuel (direct use)	70.0	136.0	158.0	172.8	179.0	188.0	213.9	226.0	232.0
Electricity	0.7	32.0	32.9	40.0	44.0	46.1	50.0	57.3	63.8
Fertilizer	12.4	19.5	24.0	30.6	32.2	41.0	60.0	87.0	94.0
Agricultural steel	1.6	2.0	2.7	2.5	2.0	1.7	2.5	2.4	2.0
Farm machinery	9.0	34.7	30.0	29.5	50.2	52.0	60.0	75.0	80.0
Tractors	12.8	25.0	30.8	23.6	16.4	11.8	20.0	20.5	19.3
Irrigation	18.0	22.8	25.0	29.6	32.5	33.3	34.1	34.8	35.0
Subtotal	124.5	272.0	303.4	328.6	356.3	373.9	440.5	503.0	526.1
Processing industry									
Food-processing industry	147.0	177.5	192.0	211.5	212.6	224.0	249.0	295.0	308.0
Food-processing machinery	0.7	5.7	5.0	4.9	4.9	5.0	6.0	6.0	6.0
Paper packaging	8.5	14.8	17.0	20.0	26.0	28.0	31.0	35.7	38.0
Glass containers	14.0	25.7	26.0	27.0	30.2	31.0	34.0	41.9	47.0
Steel cans and aluminum	38.0	55.8	62.0	73.7	85.4	86.0	91.0	112.2	122.0
Transport (fuel)	49.6	86.1	102.0	122.3	140.2	153.3	184.0	226.6	246.9
Trucks and trailers (manufacture)	28.0	42.0	49.5	47.0	43.0	44.2	61.0	70.2	74.0
Subtotal	285.8	407.6	453.5	506.4	542.3	571.5	656.0	787.6	841.9
Commercial and home									
Commercial refrigeration and cooking	121.0	141.0	150.0	161.0	176.0	186.2	209.0	241.0	263.0
Refrigeration machinery (home and commercial)	10.0	24.0	25.0	27.5	29.4	32.0	40.0	56.0	61.0
Home refrigeration and cooking	144.2	184.0	202.3	228.0	257.0	276.6	345.0	433.9	480.0
Subtotal	275.2	349.0	377.3	416.5	462.4	494.8	594.0	730.9	804.0
Grand total	685.5	1,028.6	1,134.2	1,251.5	1,361.0	1,440.2	1,690.5	2,021.5	2,172.0

Source: Reprinted with permission from J. Steinhart and C. Steinhart, "Energy Use in the U.S. Food System," *Science* 154, (1974):309.

Note: All values are multiplied by 10^{12} kcal.

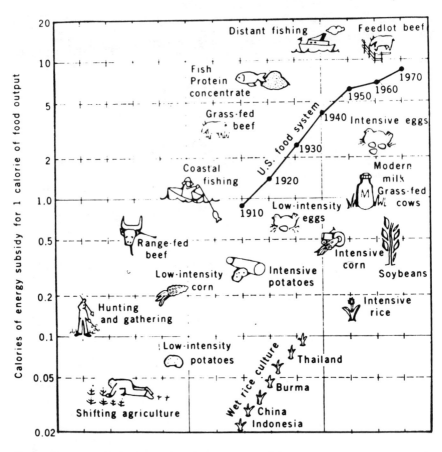

Source: Reprinted with permission from J. Steinhart and C. Steinhart, "Energy Use in the U.S. Food System," *Science* 154(1974):312.

Note: The energy history of the U.S. food system is shown for comparison.

Figure 1-1. Energy Subsidies for Various Food Crops

The products, however, are more easily adapted to present chemical-reactor or combustion technology.

    A question related to this problem is whether nonrenewable hydrocarbon resources should and can be managed on a sustained-yield basis as renewable resources can be managed? To answer this question, let us explore the meaning of *resources* and *sustained yield*. Resources imply a level of technology as well as of potential resource. As technology improves, resources are produced. As the efficiency of oil extraction increases, it becomes economic to extract poorer-quality oil reserves. Nonrenewable resources can then be utilized on a sustained-yield basis if the technology

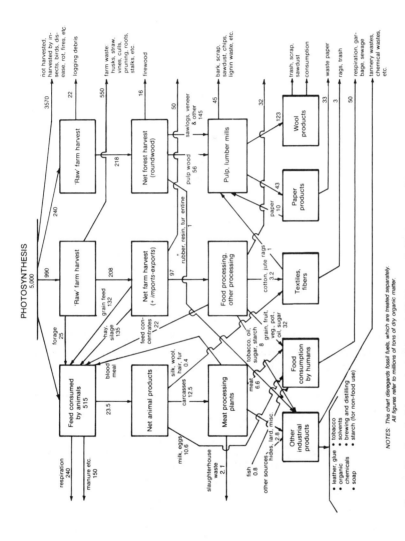

PHOTOSYNTHESIS
5,000

NOTES: *This chart disregards fossil fuels, which are treated separately.*
All figures refer to millions of tons of dry organic matter.

Source: Reprinted with permission from A.V. Kneese, R.U. Ayers, and R.C. D'Arge, *Economics and the Environment* (Baltimore, Md.: Resources for the Future, Johns Hopkins University Press, 1979).

Figure 1–2. Production and Disposal of Products of Photosynthesis

improves faster than the extraction (see chapter 11). A second way to extend nonrenewable resources is by conservation, or improved efficiency in the use of the resource. For example, if one calculates that the supply of oil available will last twenty years at present extraction rates, then one can extend this by instituting conservation methods that reduce the rate of extraction. This technique approaches sustained yield but does not achieve real sustained yield forever. It is not necessary to extend the resource forever if (1) the technology of utilizing a substitute resource is improved or (2) a new resource is provided by technology (for example, fusion power). The resource must be extended until an alternative can be developed, however. This means that oil resources need to be extended until coal-liquefication technology is improved to the point where its cost is less than the cost of oil. This time must include development of technology, construction of processing facilities, and capital accumulation.

Suppose the sustained-yield rate of extraction of natural gas is set to give a twenty-year supply. The rate of extraction can be adjusted continously to provide the twenty-year buffer supply. The extension can be made indefinitely provided the measures just indicated are successful. For example, suppose the efficiency of home furnaces is improved by 10 percent. It this efficiency extends the twenty-year buffer to twenty-two years, then gas companies would be allowed to increase their processing rate to bring it back to twenty years if they wished.

This proposal does two interesting things. First, it links conservation through technology to improved availability of resources for economic development. Second, it provides a transition mechanism between fuel sources without disruption of supply. The same might apply to the case of switching from gas- to coal-burning furnaces, or to solar heating. In the latter case, solar heating can be directly linked to economic development.

Thermodynamics and Economics: Resources

The concept of entropy in relation to economics was perhaps first introduced by Helm, who in 1887 [36] related money to the value of goods of low entropy. Georgescue-Roegen [37, 38] proposes that a necessary condition for something to have value is that it be in a condition of low entropy. He states that it is not a sufficient condition and illustrates it by an analogy to the relation between economic value and price. Unfortunately, Georgescue-Roegen associates value with low entropy in a manner such that low-entropy flows enter a process and are equated to high-entropy flows out (waste). In the biological context Lotka [39] expresses the idea that the earth receives a low-entropy stream of energy from the sun and utilizes this only in a delaying action, later discarding the energy at high entropy to outer space. In his view humans serve two functions. First, they utilize this low entropy stored over the past (for example, mining of coal). Second, they develop ways to intercept and store more of this low-entropy energy source (irrigation of land, fertilization of land, development of new crops, solar-energy utilization, tide-energy utilization, windmills, and so on).

The definition of resources must be introduced into decisions regarding their cost of recovery and availability. As Zimmerman [40] has pointed out, a resource implies a technology. It also implies the existence of that resource. In other words, the estimation of the quantity of a material or element for decisions regarding its use requires a measure of the uncertainty about its existence and also the uncertainty about the availability of technology to extract it economically. Estimates of the abundance of most materials present in the earth's crust are available. The economics of recovery of each are a function of the technology available. Reserves commonly include those materials that are economically recoverable in identified deposits. Resources, on the other hand, include in addition an estimate of deposits that are not yet discovered and of those that cannot be recovered at present. In some estimates such technical terms as *measured, indicated* and *inferred* might be more appropriate—or, in economic terms, *recoverable, paramarginal*, and *submarginal*.

Another measure might be appropriate. This is the measure in the case where the available energy required for extraction exceeds the available energy difference between the usable form and the deposit form. This may be artificial for situations in which no substitute material is available. It certainly is a level that is required for fuel resources in the sense that if the available energy utilized exceeds that recovered, then no resource exists. Figure 1–3 includes a further breakdown of resources and reserves, indicating an available energy influence, a technological level, and a geological effect. Note the addition of a seawater available-energy level, which is introduced to provide a recognition of the important future role of ocean resources.

A relationship between available energy and economic value is illustrated schematically in figure 1–4. The available energy of materials that are well dispersed in the environment is assumed to be zero. This implies a "Clarke" level or concentration of minerals equivalent to that in the earth's crust.[1] These mineral concentrations for the earth's crust and oceans are shown in tables 1–2 through 1–4. Either of these represents the datum from which minerals would need to be extracted in the far distant future. People take advantage of the nonuniform distribution of these minerals, which resulted from geological events in the past. Mining therefore extracts most minerals with higher available energy than the Clarke level. Some, like magnesium and sodium chloride, are from above the Clarke level by evaporation of seawater. Other materials, such as nitrogen, are extracted from the air to make ammonia for fertilizers (table 1–5 indicates atmospheric-component concentrations). Once the ores or compound have been processed through

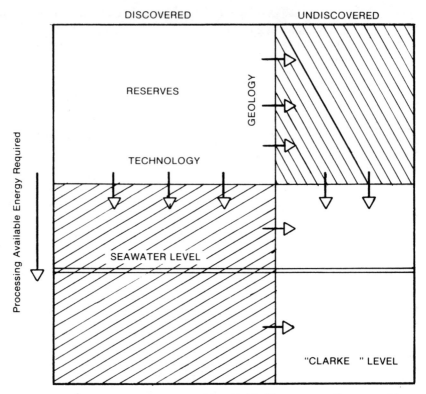

Figure 1-3. Diagram of Processing Available Energy as a Resource Factor with Technology and Geology as Variables

the application of energy, their available energy has increased, as has their value. Specialized alloys or compounds with higher available energy, such as high-strength aluminum or polymerized plastics, are then processed.

At this point the monatonically increasing relation between value and available energy is expected to have reached its limit. From this point the processing of materials into functional products involves the application of energy to increase the value. The value is in the structural information of parts, assemblies, systems, and so forth. These are improbable forms characterized by their interconnections and processing functions. In a broader perspective they represent the structure required for the processing of materials and information to meet human needs.

In any event, thermodynamics was not originally formulated to deal with this extended region, and the empirical methods of economic theory are conventionally applied. Incorporation of information theory into thermodynamic formulations is one way to deal with some of these interactions. In

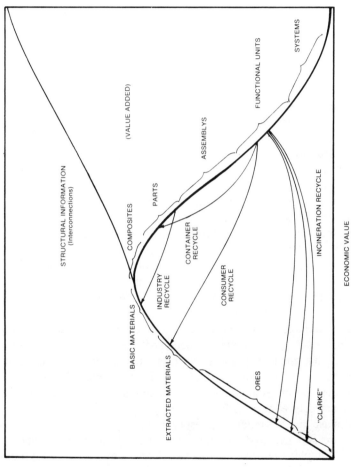

Note: A schematic indicating the relationship of available energy and economic value in resource processing from extraction to final use.

Figure 1–4. Available-Energy Value and Economic Value of Materials

Table 1–2
Abundance of Earth Elements

Element	Percentage
Oxygen	46.6
Silicon	27.7
Aluminum	8.1
Iron	5.0
Cadmium	3.6
Sodium	2.8
Potassium	2.6
Magnesium	2.1

ecological work, for example, information-theory views of the interactions of species and the structure of communities has proved valuable [10, 11, 12, 13, 31, 32].

In the discussion of recycling, reference to the available energy-value diagram is instructive. It helps to indicate that the purpose of recycle is to increase the available energy, but with the value reduced to a previous state in processing. Energy is applied in the recycling process.

Historically, in the progress of technological society, the first extraction of materials from the environment requires little energy because the ores are of high concentration. In this primitive economy the limitations are on the distribution of such easily worked metals as copper or lead, or perhaps that of birchbark. This phase is followed by the situation in which greater energy is required because the product has a higher available energy. Iron and alloy materials fall into this category in that the basic materials have higher structure and available energy relative to the ore. In this phase the technology of the production of "good" materials becomes crucial. Secret processes of smelting, alloying, quenching, and so forth provide advantageous economic value. It is the art, not the science, that provides economic advantage. This phase precedes the understanding of phase diagrams, transition temperatures, atomic structure, and the like.

A later phase that is similar with respect to energy is the scientific phase, in which more sophisticated structures with higher information value are synthesized. Simpler raw materials like coal are converted to a myriad of hydrocarbon derivatives. Polymerization and organic synthesis of new products proceeds. The value of the resource is small, and the added value is scientific. Energy requirements increase as the range of input resources expands and the synthesis becomes important.

The next two stages are substitution and recycle. Both arise from resource limitations rather than energy limitations. Energy requirements increase with substitution and decrease with recycle. If this energy addition does not result in a higher available-energy product, its effect is dissipative of both the

Table 1–3
Relative Abundance of Metals in the Earth

Metal	PPM	Metal	PPM
Silicon	277,200	Arsenic	5
Aluminum	83,000	Scandium	5
Iron	58,000	Dysprosium	4.5
Calcium	36,300	Hafnium	4.5
Sodium	28,300	Boron	3
Potassium	25,900	Ytterbium	2.7
Magnesium	20,900	Erbium	2.5
Titanium	6,400	Tantalum	2.1
Manganese	1,300	Uranium	1.7
Rubidium	310	Tin	1.7
Strontium	300	Molybdenum	1.3
Barium	250	Holmium	1.2
Zirconium	220	Tungsten	1.1
Vanadium	150	Europium	1.1
Chromium	110	Antimony	1.0
Zinc	94	Terbium	0.9
Nickel	89	Lutetium	0.8
Copper	63	Thallium	0.6
Cerium	46	Bismuth	0.2
Yttrium	28	Thulium	0.2
Cobalt	25	Cadmium	0.15
Columbium	24	Indium	0.1
Neodymium	24	Selenium	0.09
Lithium	21	Mercury	0.089
Lanthanium	18	Silver	0.075
Gallium	15	Palladium	0.01
Lead	12	Platinum	0.005
Thorium	12	Gold	0.0035
Cesium	7	Tellurium	0.002
Germanium	7	Iridium	0.001
Samarium	6.5	Osmium	0.001
Gadolinium	6.4	Rhenium	0.001
Beryllium	6	Rhodium	0.001
Praseodymium	5.5	Ruthenium	0.001

Sources: J. McHale, *The Ecological Context* (New York: Braziller, 1970); B. Mason, *Principles of Geochemistry* (New York: Wiley, 1958).

Note: Present in more than 0.0009 parts per million.

energy used and the available energy of the components. Incineration is the simplest illustration this dissipative process. It involves not only conversion of available energy to heat, but also dispersion of combustion products to the atmosphere and the earth. Combustion in an incinerator is at present only the acceleration of the rate of oxidation of the reactants. Communities pay from $10 to $25 per ton in incineration costs—in effect paying for a more rapid dissipation of energy, an entirely entropic process.

Note that the notion that the highest-quality resources are exploited first

Table 1–4
Relative Abundance of Metals in Seawater

Metal	PPM	Metal	PPM
Sodium	10,561	Barium	0.05
Magnesium	1,272	Arsenic	0.024
Calcium	400	Iron	0.02
Potassium	380	Zinc	0.014
Strontium	13	Manganese	0.01
Boron	4.6	Lead	0.005
Silicon	4.0	Selenium	0.004
Aluminum	1.9	Tin	0.003
Rubidium	0.2	Cesium	0.002
Lithium	0.1	Molybdenum	0.002
Copper	0.09	Uranium	0.0016

Note: Present in more than 0.0015 parts per million.

in a historical sequence is not without exception. Galbraith [33] notes that in the United States the best agricultural land was not utilized first because of the large trees in these areas. It was easier to clear the poorer-quality land where fewer and smaller trees were growing. Kakela [34] also notes that taconite iron ores are of higher quality than some of the iron ores smelted before the process for converting taconite was developed.

Incineration to produce heating or electrical power is an improvement, but the available energy of the refuse is high compared with the heating value or electrical energy that results. Separation-process recycling systems are an improvement both environmentally and in terms of energy. The costs are often higher because the facilities for utilizing the separated materials are not planned at the time the disposal facility is planned. This often leads to the value of the recycled materials being less than that of the original material. Other functions in recycling economics involve the uncertain content of waste materials and irregular input flow with respect to time.

Table 1–5
Atmospheric-Materials Abundance

Component	PPM
Nitrogen	780,900
Oxygen	209,400
Carbon dioxide	9,300
Argon	315
Neon	18
Helium	5
Methane	1
Nitrous oxide	0.5
Hydrogen	0.5

Source: Data from A.C. Stern, ed., *Air Pollution*, 3 vols. (New York: Academic Press, 1968).

Material processing and flows are examined in chapter 11 from an available-energy point of view as well as from a mass-flow basis. The consideration of available energy adds the basis for estimating future costs, as resources are depleted and additional waste materials are produced in their processing. Substitution and recycle generally involve energy considerations as well. Industry recycle, as shown in figure 1–4, requires little energy. For example, it represents energy to remelt defective castings or scrap cuttings. Within the automotive industry approximately 80 percent of the charge in an electric-arc furnace is recycled metal. In the aircraft industry identification of specially alloyed parts is made to facilitate their recycle. Substitution of aluminum for copper in electrical applications has been primarily a resource limitation enhanced by energy and resource-extraction technology problems. Economy-of-scale advantages in copper-ore extraction have been considerable but have limits when the concentration of the ore deposits is diminished.

Projections indicate that recycle from more disperse sources, such as consumers, will continue. The energy cost is higher because of the increased transportation and sorting requirements. If this recycle energy cost is less than that of the ore-recovery energy, it will lead to increased recycling. Economists predict that recycle materials will be the principal source in the future. In that case the limitation on throughput or recycle flow will be energy flows. Materials and energy are essentially conserved, but available energy is not. Available energy, then, provides the currency or measure of value in these situations.

Thermodynamics and Economics: Technology

Recently questions of economics have been placed in a different context by Kenneth Boulding [41]:

> By constrast, in the spaceman economy, through-put is by no means a desideratum, and is indeed to be regarded as something to be minimized rather than maximized. The essential measure of the success of the economy is not production and consumption at all, but the nature, extent, quality and complexity of the total capital stock, including in this the state of the human bodies and minds included in the system. In the spaceman economy, what we are primarily concerned with is stock maintenance, and any technological change which results in the maintenance of a given total stock with a lessened through-put (that is, less production and consumption) is clearly a gain.

This concept of maximizing the capital store has been recast by others in the form that says that a society tries or should try to maximize the information structure or diversity for the energy available to it. In ecological literature this corresponds to the idea of a climax community. An ecological community reaches the steady state of a climax after passing through the stages of succession. This climax stage of growth or development is the point at which theoretically energy is channeled into maintenance with no net growth. Another way to look at this situation is that the system has reached a point where its structure is maximized for the energy flow available to it.

In biological literature this climax community is reached only after a successional stage. In this successional stage the system maximizes the energy throughput or power flow. This principle is sometimes referred to as Lotka's principle [42]. Surviving species are not those that are the most efficient but those that can channel more energy flow through their systems.[2] The nature of the transition from survival of those species that are inefficient but can channel more energy through their systems to the climax community where efficiency is at a premium has yet to be resolved. In plants the energy is needed for growth, reproduction, and maintenance. In the climax stage only maintenance is required.

Morowitz [43] has postulated that ecologically and biologically another principle may also apply. This principle is that organisms tend to maximize their structure for a given throughput of energy. This is similar to Boulding's concept just quoted. For a fixed amount of energy available, these systems maximize their structure. Plants capture more energy from the sun's stream by maximizing their structure. Whether the structure or the energy flow is being maximized has not been resolved. Some evidence from measures of diversity as representative of structure indicate that diversity in human civilizations may increase during successional stages as energy use is increased, but may decrease after reaching a maximum. This is theoretically described in thermodynamic terms by Morowitz as characteristic of energy flow in open systems. The energy flow starts to break down the diversity of energy-storage forms as the thermal-energy levels approach the structural-energy levels and energy interchange between them is increased (this is discussed further in Chapter 17). Before leaving this discussion, Lotka's analogy of the earth as a bottleneck for energy flow should be discussed in the economic-structure context. Lotka observed that the biological system of the earth acted in its evolutionary way to intercept and delay the flow of energy from the sun to outer space. Evolution, in his view, involves increasing the time delay of this energy by storing more energy in biological forms or structure. Studies also indicate that in the process of building irrigation systems, the biological storage has been increased. Similarly, the storage of energy in hardware structures such as buildings or processing facilities increases this time delay in the flow. Other technological activities, however, particularly the use of fossil fuels, certainly have reduced the time delay of this solar energy that has been stored in organic forms.

In thermodynamic terms, as outlined in Chapter 4, the maximum-efficiency systems are not the maximum-energy-throughput systems. In Boulding's terms, the desideratum is the capital stock of the structure. Stored fossil fuels are also in part the structure built up in the past that is being utilized or converted to other structures. The conversion requires a through-put of available energy in which a portion is reduced to unavailable forms.[3]

In order to reformulate economic measures in terms of Boulding's capital stock, a measure of the value of these capital stocks must be supplied. It is hypothesized here that the available energy invested in the structure-building process could be one measure of the capital stock. One of the problems, of course, is the difficulty of establishing this measure in terms of present value. The available energy required changes with technological advancement, the type of structure, the construction materials, and the fabrication methods. In general, these changes have been in the direction of increased available-energy expenditure. For example, a house constructed of wood materials with hand tools has less available energy tied up than does an aluminum and steel structure built with power tools.

This analysis also needs to include the fact that both of these involve a capital-fuel-stock change (the materials) and a capital-stock utilization in the processing. The use of modern domed structures involves a reduction in the total materials needed to provide a function compared with that required by older structures. The use of air-supported or inflated structures involves the saving of the capital-energy expenditure at the cost of increased operational energy and maintenance cost. The net benefit or loss can be evaluated from an available-energy analysis.

In summary, the techniques outlined in this book are primarily applicable in industrial processes, where available energy and value have close correspondence. The available-energy concept is extended importantly to environmental considerations in an attempt to quantify environmental factors.

Practical Production Questions

If one is in the business of producing a product, whether it be a nail, a television set, or an electric-power plant, one is faced with questions relating to the selection of materials for the product and the processing machinery. Some of the selection criteria for the product's material are summarized in table 1–6. The first is a classification of economic questions. These deal primarily with the situation in which a known technology for processing the candidate materials is either in place or available on an economic basis. The questions then revolve on making a choice based on minimum cost. Minimum cost, however, implies a time in the sense that the minimization of cost is based on a minimum cost over a certain time scale. In the past, with an expected relatively certain continued supply of most materials and with decreasing costs of these materials, selection could often be made based on present costs. Also, as illustrated elsewhere, when faced with a choice with a

Table 1–6
Materials-Selection Criteria

1. Cost
   a. Material—resource
   b. Manufacturing—processing
2. Performance
   a. Physical properties
   b. Chemical properties
   c. Biological properties
3. Administrative
   a. Availability of supply
   b. Consistency of supply
   c. Substitution availability
   d. Future potential
   e. Marketability—acceptability
4. Health—environment
   a. Toxicity
   b. Safety in use
   c. Disposability
5. Legal—ethical
   a. Acceptability in use
   b. Allocation of resources

fairly uncertain continued supply, one could decide to maximize profits, or growth, or whatever, based on present values of materials. The soft criteria for selection are, then, the cost of the materials, the availability of future supplies, the ease of substitution of another material if future supplies are uncertain, and the marketability of the product with a certain material. As an example of the latter, the poor quality of early plastic materials often led to continued use of standard metallic or wood materials even where use of plastics might have been indicated. Overcoming consumer resistance built up by past plastic-product failures was a difficult marketing problem.

The second or hard criterion might be considered the technical criterion—the minimum requirements of the product to ensure its function. These include meeting the engineering specifications of durability, strength, reliability, safety, and so forth. They also include processing requirements for formability, including such requirements as machinability, moldability, weldability, and the like. Aluminum is a material that requires specialized welding equipment for joining that might preclude its use on the grounds that skilled workers needed for fabrication were not available or that the welding equipment was not reliable or consistent in its operation. Included also are consideration of the articulation of the material in technical decisions such as the possibility of the joining of dissimilar materials leading to corrosion problems or thermal-expansion difficulties, and articulation of the processing equipment with other substitution materials. A further consideration is future value. If future developments in processing of materials such as plastics lead to molding capabilities for products now machined, then investment in the facility for molding, and selection of more expensive molding materials might be made.

Environmental criteria are an area of recent emphasis in material-selection processes. Here the questions revolve around disposal or recycling possibilities. The importance of these has long been recognized by industry from an economic point of view. As an illustration, the use of sheet metal and metallic parts in the automotive industry is in part justifiable on the grounds that the scrap material from the forming operations is used in their foundries to mold other parts, such as engine blocks.

This criterion has been only informally examined in optimization of designs and rational material selection. The associated criterion of disposability of the product after it has lost its product value has been a recent concern. This is an important criterion and is at the center of the problem of recycling of consumer products. A proposal to include more aluminum in automobiles so that their scrap value would be enhanced is moving in this direction. Is this a reasonable proposal? How should it be evaluated?

The concept of conservation of matter is basic to the thinking of most economists and scientists. It seems obvious from an input-output view of a process that the material that enters must leave either as a product or as a waste; the number of atoms of each material is conserved. The inputs and outputs may far exceed the purchased or sold quantities. Figure 1–5, for example, shows the large, unaccounted flows in iron smelting.

From a strict conservationist point of view, the composition of "spaceship earth" is constant. The classical conservationist has always excoriated those who used up precious resources and called for economy in the use of the earth's store. The so-called cowboy economy of recent decades, which measured success by total production (GNP) or accumulation of goods, is clearly wasteful. The nature of materials problems was recently put in new perspective by the observation by environmentalists that the limitations on the progress of a consuming society may be on the output rather than the input side. In other words, some of the limits on what society can produce are determined by the earth's capacity to assimilate wastes. This assimilative capacity is unfortunately extremely difficult to estimate. In uncertainty terms, the assimilative capacity of the earth is much more uncertain than is its resource capacity. Over two hundred years of geological and geographical exploration have been aimed at mapping and estimating the distribution of mineral and agricultural resources. The earth's capacity for assimilation has hardly been explored.

Some have tried to simplify the waste problem by saying that pollution is only a resource out of place. However, this concept is erroneous in both substance and concept. Although it is true that recycling and waste industries exist to process and redistribute the waste materials of one production unit to make it acceptable for input resource for another, these industries also produce waste themselves and do not return materials for use without a

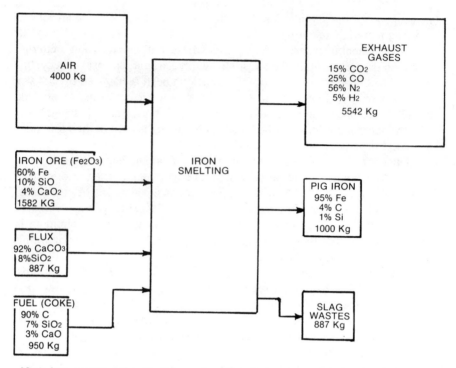

Note: Areas are proportional to the mass flow. Note the large mass of gas handled relative to the mass of iron produced.

Figure 1–5. Iron-Smelting Mass-Flow Diagram

material or energy cost. Materials and energy are required for the transportation and processing functions as well as for the capital machinery of these industries. It is clear that the costs to reprocess materials from waste must be less than from geological formations if recycling industries are to thrive in the present economic system.

Notes

1. The Clarke level was first introduced as a term by F.W. Clarke of the U.S. Geological Survey in 1924 [29].

2. Successional-stage plants include rapidly growing species like poplar trees, crabgrass, and horseweed. These commonly appear in the natural revegetation process in the northeastern United States.

3. Appendix 1A gives a surprising but simple account of the historical development of the steam engine in terms of economic, technical, and thermodynamic factors.

References

1. Clarke, W. *Energy for Survival: The Alternative to Extinction*. Garden City, N.Y.: Anchor Books, 1975.
2. Commoner, B. *The Poverty of Power*. New York: Knopf, 1976.
3. Dorf, R.C. *Energy, Resources and Policy*. Reading, Mass.: Addison-Wesley, 1978.
4. Masters, G.M. *Introduction to Environmental Science and Technology*. New York: Wiley, 1974.
5. Wagner, R.H. *Environment and Man*. New York: Norton, 1971.
6. Victor, P.A. *Pollution, Economy and Environment*. Toronto: University of Toronto Press, 1972.
7. Spiro, T.G., And Stigliani, W.M. *Environmental Science in Perspective*. Albany: State University of New York Press, 1980.
8. Baker, J.J.W., and Allen, G.E. *Matter, Energy and Life*. Reading, Mass.: Addison-Wesley, 1965.
9. Kneese, A.V.; Ayres, R.V.; and d'Arge, R.L. *Economics and the Environment*. Baltimore, Md.: Resources for the Future, Johns Hopkins University Press, 1970.
10. Levins, R. *Evolution in Changing Environments*. Princeton, N.J.: Princeton University Press, 1968.
11. Margalef, R. *Perspectives in Ecological Theory*. Chicago: University of Chicago Press, 1968.
12. Pielou, E.C. *Ecological Diversity*. New York: Wiley, 1975.
13. May, R.M. *Stability and Complexity in Model Ecosystems*. Princeton, N.J.: Princeton University Press, 1973.
14. Wiener, N. *The Human Use of Human Beings*. Boston: Houghton-Mifflin, 1954.
15. Keenan, J.H. "Availability and Irreversibility in Thermodynamics." *Britt. J. Appl. Phys.* 2(1951):183.
16. Gibbs, J.W. "On the Equilibrium of Heterogeneous Substances." 1878. *The Collected Works*. New Haven: Yale University Press, 1948.
17. Onsager, L. "Reciprocal Relations in Irreversible Processes," I. *Phys. Rev.* 37(1931):405.
18. Onsager, L. "Reciprocal Relations in Irreversible Processes," II. *Phys. Rev.* 38(1931):2265.
19. Denbigh, K.G. *Thermodynamics of the Steady State*. London: Methuen, 1950.
20. DeGroot, S.R. *Thermodynamics of Irreversible Processes*. Amsterdam: North-Holland, 1951.
21. Prigogine, I. *An Introduction to Thermodynamics of Irreversible Processes*, 3rd ed. New York: Interscience, 1968.
22. U.S. Bureau of the Census. *1967 Census of Manufacturers*. Washington, D.C.: U.S. Department of Commerce, 1971.

23. Berry, R.S. "Recycling Thermodynamics and Environmental Thrift." *Bull. At. Scient.* 28(May 1972):8.
24. Steinhart, J. and Steinhart, C. "Energy Use in the U.S. Food System." *Science* 184(1973):307.
25. Pimintel, D.; Hurd, L.E.; Bellotti, A.C.; Forster, M.J.; Oka, I.N.; Sholes, O.D.; and Whitman, R.J. "Food Production and the Energy Crisis." *Science* 182(1973):443.
26. Hayami, Y., and Ruttan, V.W. *Agricultural Development: An International Perspective.* Baltimore, Md.: Johns Hopkins University Press, 1971.
27. Ruttan, V.W. "Food Production and the Energy Crisis: A Comment." *Science* 187(1975):560.
28. Kneese, A.V., and Schultze, C.L. *Pollution, Prices and Public Policy.* Washington, D.C.: Brookings, 1975.
29. Clarke, F.W. "The Composition of the Earth's Crust." U.S. Geological Survey Paper no. 127, 1924.
30. McHale, J. *The Ecological Context.* New York: Braziller, 1970.
31. MacArthur, R.H., and Wilson, E.D. *The Theory of Island Biogeology.* Princeton, N.J.: Princeton University Press, 1963.
32. McArthur, R.H., and Connell, J.H. *The Biology of Populations.* New York: Wiley, 1966.
33. Galbraith, J.K. *Money.* Boston: Houghton-Mifflin, 1976.
34. Kakela, P.J. "Iron Ore: Energy, Labor and Capital Changes with Technology." *Science* 202(1978)1151.
35. Stern, A.C., ed. *Air Pollution.* 3 vols. New York: Academic Press, 1968.
36. Helm, G. *Die Lehre von der Energie.* Leipzig: A. Felix Publisher, 1887, p. 72.
37. Georgescu-Roegen, N. *The Entropy Law and the Economic Process.* Cambridge, Mass.: Harvard University Press, 1971.
38. Georgescu-Roegen, N. "Energy and Economic Myths." *Southern Economic J.* 41(1975):347.
39. Lotka, A.J. *Elements of Physical Biology.* Baltimore, Md.: Williams and Wilkins, 1922.
40. Zimmermann, E.W. *World Resources and Industries.* New York: Harper Brothers, 1933.
41. Boulding, K.E. "The Economics of the Coming Space-Ship Earth." In *Environmental Quality in a Growing Economy,* ed. H. Jarrett. Baltimore, Md.: Johns Hopkins University Press, 1966.
42. Hardin, G. *Nature and Man's Fate.* New York: Holt, Rinehart and Winston, 1959.
43. Morowitz, H. *Energy Flow in Biology.* New York: Academic Press, 1968.
44. Mason, B. *Principles of Geochemistry.* New York: Wiley, 1958.

Appendix 1A:
A Historical
Perspective on
Thermodynamics
and Economics

The concepts in thermodynamics of conservation of energy and the first law were elements related to studies in physics, although the equivalence of heat and work was associated with the boring of cannon, a military application. The conservation of energy as a hypothesis has, along with the conservation of matter, been a postulate of science and engineering that is accepted as a law (except in the relativity and quantum work in physics, where the equivalence theory $E = mc^2$ is made). Economics has long utilized the concept of material balances as a basic element of its theories. The conservation of energy has also been used, but not productively, because of the losses associated with real processes.

The economics of the industrial age began with the steam engine. Efficiency represented a new concept in energy-conversion processes. The concept that efficiency was important involved a judgment that some energy forms (work) were more useful than others (heat). The concept of *utility* was thus introduced. The history of the development of steam engines parallels the development of Carnot's thermodynamic reasoning.

It is generally believed that the improvements in steam engines, particularly in Watt's designs, were motivated by economic reasons. Certain facts are apparent from examination of this development: (1) the engines decreased in size or mass, and (2) the engine efficiency increased. The first implies a decrease in capital cost and the second a decrease in operating cost. In perspective, the second seems less a pressing need of the time since the early machines were primarily used for pumping out coal mines and a cheap source of fuel was available. It is thus unlikely that engine efficiency was a goal of the businessman of the day. A general principle is that the larger the machine, the more efficient it is expected to be. The need to reduce the size or mass was real, however, as iron was a scarce commodity at that time. The question remains: How did the men of the day evolve a design process that led to both smaller machines and improved efficiency?

The progression of the design indicates that the designers or inventors were seeking to get more power output from a given machine (that is, get more energy flow out in a given time). The invention of the condenser, the positive-pressure system, and the double-stroke systems all point in this direction with apparently little regard for the efficiency increase (more product flow for a given capital investment). In fact, thermodynamic principles were not used effectively until near the end of the nineteenth century in the design of these machines. The efficiency increased, of course, because the temperature and pressure of the entering steam were increased and the temperature and pressure of the water leaving were decreased. Pressure increase on the input side was a design change that produced a power increase. In producing this higher pressure, the temperature also increased. Similarly, on the exhaust stroke the steam was condensed to lower the pressure. The temperature was decreased to do this because the lower the temperature, the faster the steam condensed—not because the design involved any Carnot-cycle analysis.

The principal reason for the introduction of the condenser was the closing of the water cycle. This was extremely important for two reasons. First, corrosion and scale deposits on heat-exchanger surfaces were serious when water from coal mines that was high in dissolved solids and acidity was used in the system. Second, when water with a large fraction of air is used, it produces a lower-pressure steam with less available energy. By closing the cycle, air could be removed from the system, thereby improving the performance. In a system that relied on the condensing of the steam for the power stroke, as did the atmospheric engines, this was a crucial problem.

Machines became smaller because of several other factors than the capital cost of bulk iron. One was the difficulty in making large casting for high-pressure vessels, and a second was the limits to which the increased power could be used at one location. Power transmission was in a very primitive stage. The application required only a certain maximum power at a single location. Very large power machines were neither useful nor necessary.

References

1. Cardwell, D.S.L. *From Watt to Clausius*. Ithaca, N.Y.: Cornell University Press, 1971.
2. Storer, J.D. *A Simple History of the Steam Engine*. London: John Baker, 1969.

2

Net Energy and Available Energy

The concept of resources presupposes a person. Resources increase as man's knowledge increases. Uranium would not be considered a resource before 1900, for example. Neither would petroleum before 1800. Resources are created by man as well as discovered and exploited.
— Zimmerman, "World Resources and Industries" (1922)

Net-Energy and Available-Energy Analysis

Net-energy analysis is a method of energy accounting developed by several analysts [1,2,3,4,5] to evaluate the benefits of different technologies. Notable examples of this technique are: (1) the attempts to show that nuclear-power production is a net producer of energy after all the energy inputs of fuel processing, construction, machinery, and so forth are considered [10,11,12, 13,14] (see tables 2–1 and 2–2); (2) the work [8,16] to show the fossil-fuel subsidy required to produce food energy; and (3) the work of Berry and Fels [18,19] in delineating the energy inefficiency of the automobile as a transportation system. This methodology [15] is best illustrated by reference to figure 2–1.

The net energy output in a process is defined as the useful energy produced in the process minus the energy used both directly and indirectly to provide this energy output. In the case of resource utilization, such as a coal-fired power system, the energy value of the extracted coal resource is not included. This in effect assigns no energy value to the principal resource being processed, but gives an energy value to this resource if it is extracted for indirect use. The indirect energy used is determined from U.S. census data [21,22,23] as compiled into input-output form similar to economic input-output tables [1,5,6,7]. The results of net-energy analysis depend on the boundary on the system under consideration, as the selection of inputs to the processing system are to some extent arbitrary.

The first level of inputs includes the direct energy inputs to the process—fuels and electrical energy purchased directly. Fuel-energy values are considered as heating values, and electrical energy is multiplied by power-plant heat rates, typically 10,500–11,400 Btu/kW-hr. Several important considerations are not included in such an analysis. Natural gas has environmental and technical advantages over coal (as a fuel, for example) that do not appear in this analysis. In electrical-energy generation, no

Table 2–1
Energy Requirements for Light-Water-Reactor Nuclear-Fuel-Cycle Elements

Process	Equivalent Thermal Energy (trillion Btu)	Percentage of Total
Mining	2.935	2.1
Milling	3.037	2.1
Conversion	5.334	3.9
Enrichment	103.037	72.1
Fuel fabrication	4.096	2.9
Power-plant operation	23.401	16.4
Fuel storage	0.398	0.3
Waste storage	0.240	0.2
Transportation		
Natural U	0.061	0.1
Fuel	0.230	0.2
Totals	143[a]	100[a]

Sources: Development Sciences, Inc., "A Study to Develop Energy Estimates of Merit for Selected Fuel Technologies" (Prepared for U.S. Department of Interior, Office of Research and Development, September 1975); Institute for Energy Analysis, Oak Ridge Associated Universities, "Net Energy from Nuclear Power," IEA-75-3 (November 1975).
[a]Rounded.

Table 2–2
Comparison of Net-Energy Results for Nuclear-Power Systems

Investigator	Units of External Energy Input per 1,000 Units of Output
Development Sciences, Inc.	238
State of Oregon study	194[a]
University of Illinois, Center for Advanced Computation (Pilati and Richard)	210
Institute for Energy Analysis	248
ERDA-76-1	262

Sources: Development Sciences, Inc., "A Study to Develop Energy Estimates of Merit for Selected Fuel Technologies" (Prepared for U.S. Department of Interior, Office of Research and Development, September 1975); Institute for Energy Analysis, Oak Ridge Associated Universities, "Net Energy from Nuclear Power," IEA-75-3 (November 1975); University of Oklahoma, Science and Public Policy Program, "Energy Alternatives: A Comparative Analysis" (Washington, D.C.: U.S. Government Printing Office, 1975); State of Oregon, Office of the Governor, Office of Energy Research and Planning, "Transition" (January 1975); D.A. Pilati and R. Richard, "Total Energy Requirements for Nine Electricity Generating Systems," CAC Document no. 165, Center for Advanced Computation, University of Illinois, Urbana (1975).
[a]Adjusted to comparable basis.

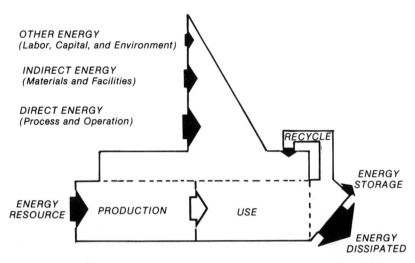

Figure 2–1. Diagram for Illustrating the Methodology of Net-Energy Analysis

consideration is given to the source of this energy, whether it be hydroelectic, nuclear, fossil fuel, wind, and so on. The crucial factor—whether the energy is from a renewable or a nonrenewable source—is thus not included. Energy purchased directly to operate pollution-control systems is also not evaluated separately.

The second-level inputs are indirect inputs of energy. These include energy required to manufacture machinery, construct buildings, and repair equipment; food energy used to feed workers on the site; and so forth. This indirect energy includes each item purchased directly by the processing entity that requires energy for its preparation and transportation to the site. This accounting is particularly difficult to evaluate because of the large variation in energy requirements for similar material goods. Food in uncooked cereal form would require less energy than the same cereal ground and baked in the form of bread. This indirect energy input also has the inherent difficulty of depending to a large degree on the energy costs of transportation.

The third level of inputs, which includes such things as labor, ecosystem, and organization energy, is even more nebulous. This level may include under ecosystem the energy required to process waste materials, energy used for reforestation, energy for draining marshes, and the like. Labor energy includes, in some analyses, the energy costs to transport labor to and from the processing system, as well as manual energy supplied within a processing system in the movement of goods. Organizational costs may include cost of a utility's communication system, energy costs for securing licensing, computer energy costs, and energy to maintain general administrative functions. This third level of energy inputs is often omitted on the basis that (1) it would not differ significantly regardless of the alternative process under consideration, (2) it contains large uncertainties, and (3) it involves vague comparisons—"apples and oranges."

Berry and Fels [19] have an improved approach to net-energy analysis that differentiates most energy qualities, as illustrated by figure 2–2. This figure depicts the free-energy requirements for the manufacture of an automobile. The numbers in the triangular boxes represent the free-energy requirements for each of the processes. This approach considers free energy as the commodity of value. In the case of different fuel inputs, the free energy of the fuel rather than its heating value is utilized. In the case of electrical-energy inputs, the electrical energy is regarded as totally free energy and no account is taken of the free energy of the fuel used to generate the electrical energy. This approach using free energy allows consideration of substitute fuels on a better basis than heating values, but does not allow for the consideration of the flow character of the processing functions. It allows (although it does not include in the present analysis) the consideration of energy inputs other than fuels. It could allow energy inputs in the form of such materials as oxygen.

In consideration of the efficiency of real versus ideal processes, Berry and Fels [19] examine basic processes from a thermodynamic point of view. In this examination they calculate the ideal free-energy change of the basic chemical reactions in the processing of iron, aluminum, copper, and zinc (see table 2–3). They note that in certain processes the ideal (thermodynamic) free-energy requirement is negative. In these cases the free energy of the actual product is less than the free energy of the reactants. These processes thermodynamically should be energy producers, not consumers, according to this method of analysis. For example, this comes about in the smelting of copper. The overall reaction of copper sulfide ore with oxygen to give metallic copper and sulfur dioxide is:

$$Cu_2S + O_2 \rightleftharpoons 2Cu + SO_2 \qquad \Delta H = -51.76 \text{ cal/gm-mole}$$
$$\Delta G = -51.19 \text{ cal/gm-mole}.$$

The free energy (G) of the products is less than that of the reactants. The reaction, if it proceeds, will lose this free energy. In this case the free energy would be lost as heat (H) to the environment. Converting this free energy to a ton of copper-product basis gives $\Delta G = -425$ kW-hr/ton. If the usual heating value were used, this would be $\Delta H \simeq -430$ kW-hr/ton. The actual smelting operation requires $\Delta E_{act} = 11,730$ kW-hr/ton. This difference,

$$\Delta E_{act} - (\Delta G) = 11,730 - (-425) = 12,155 \text{ kW-hr/ton},$$

is the energy that is "wasted" in this process. This wasted energy provides several functions: (1) melting of the reactants to provide mixing and uniformity, (2) reasonable reaction rates, and (3) separation of impurities from the ore. All of these involve economic and technical factors that are controllable.

Of more importance, however, is the lost energy that could be recovered or reused. In this case one major improvement would be the use of the molten copper to preheat the incoming ore rather than allow it to be lost. In the copper case the reactants leave the process region at a temperature of approximately 1,500 K. At this state both the liquid copper and gaseous sulfur dioxide have available energy that could be used for the production of work. An example might be the use of the hot gases for a heat-exchanger boiler system to produce steam from the copper to drive a turbine. Since copper is the only useful product of this reaction as considered, let us examine the possibilities of the SO_2. This may be used as input to a reactor for the production of sulfuric acid. In most smelter operations the SO_2 is a pollutant that must be removed from the exhaust. This requires both energy and capital investment.

If SO_2 is not an acceptable output product, then the reaction considered is not the ideal. The ideal must end with all reactants acceptable in the environment.

In the case of sulfuric acid as output product, the overall reaction could then be:

$$Cu_2S + 3/2O_2 + H_2O \rightleftharpoons 2Cu + H_2SO_4 \qquad \Delta G = -0.18 \text{ MJ/mole Cu.}$$

Equivalently, the ideal free energy would be -735 kW-hr/ton of copper. The waste would then be

$$E_{act} - \Delta G = 11,730 - (-735) = 12,465 \text{ kW-hr/ton Cu,}$$

if no extra energy was required for the pollution control. In this situation the reaction of water and sulfur dioxide to produce sulfuric acid has a net free-energy output. The difference between this net waste and the waste without SO_2 removed is certainly not a "waste," since it should be required to produce acceptable products.

The water input to a scrubber cools the emission gases as well as producing a loss in energy that could be used otherwise for preheating. The scrubbing process thus reduces the amount of energy that can be extracted from the hot product stream. Sulfuric acid, however, has thermodynamic value as an oxidant for use in many chemical processes. Its production is thus not a loss but must be accounted as an available-energy output of the system.

This example illustrates several important points relative to energy

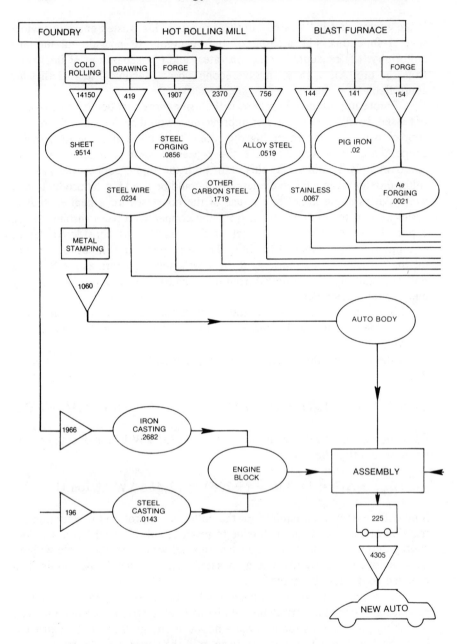

Note: Free-energy values indicated in triangular boxes are Gibbs free energy values,

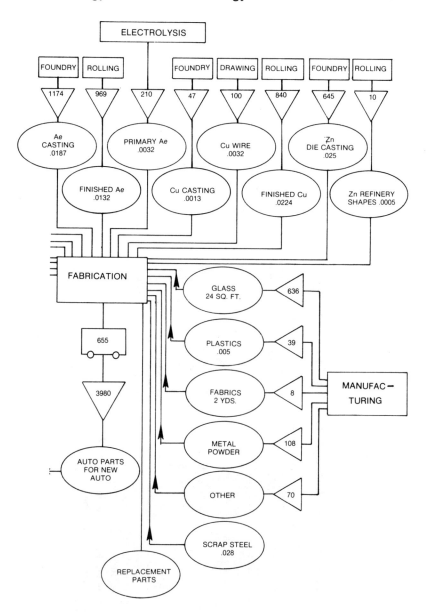

$G = H - TS$ of fuel inputs. Numbers in triangles and rectangles are the Gibbs free energy in kilowatt hours required for the process. Numbers in ellipses are tons of material processed.

Figure 2–2. Materials and Processes, with Corresponding Free Energies, for the Manufacture of an Automobile from Primary Materials

Table 2–3
Comparison of Ideal Free-Energy Cost and Real Free Energy Consumed for Various Processes and for the Total Automobile

Process	Ideal Free-Energy Cost[a] (kwh)	Real Free Energy Consumed (kwh)	Free-Energy Waste (kwh)	Waste Factor[b]	To Produce
Coke oven	−105	785	890	1.13	1 ton coke
Blast furnace	565	5,925	5,360	0.90	1 ton iron
Steel furnace	−260	1,375	1,635	1.19	1 ton steel
Refining of bauxite	0	3,220	3,220	1.00	Alumina for 1 ton aluminum
Smelting aluminum	4,610	60,970	56,360	0.92	1 ton aluminum
Smelting copper	−425	11,730	12,155	1.04	1 ton copper
Smelting zinc	−390	16,115	16,505	1.02	1 ton zinc
Total process	1,035	37,275	36,240	0.97	1 automobile

Source: R.S. Berry and M. Fels, "The Energy Cost of Automobiles," *Science and Public Affairs (Bull. At. Scient.)* 29(1973):13. Reprinted by permission of *The Bulletin of the Atomic Scientists*, a magazine of science and public affairs. Copyright © 1973 by the Educational Foundation for Nuclear Science, Chicago, Ill. 60637.

[a]Based on the reactions of the dominant technology now in use.

[b]The ratio of waste to real total consumption.

analysis. First, the ideal chemical process often has products that have less free energy than the reactants. Second, the principal product (copper metal) has greater free energy when separated than when combined as the ore (with sulfur). Energy is therefore required to separate the metal from the sulfide. Theoretically, this would be supplied by the oxygen reacting with the sulfur, with the resultant SO_2 having a lower free energy. In this reaction excess energy is released, which is lost to the environment. This excess energy could theoretically be used to heat the materials decreasing the reaction time. The energy release, however, is small compared with the energy required to maintain the material at the molten-copper point. In effect, it is the excess heat loss from the materials that must be added to maintain the processing rate. If the materials were formed in an insulated batch reactor, no additional energy would theoretically be necessary. In practice, the metal is poured out of the reactor, and the thermal energy is lost.

Two immediate mechanical solutions to the problem of reducing the free energy requirement are:

1. Insulate the reactor.
2. Use the heat rejected to heat incoming material (regenerative heating).

Two chemical solutions are:

3. Increase the reaction rate at low temperature by the use of a catalyst or solvent.
4. Change the reactant that separates the sulfur from the copper.

The first three solutions require an increase in capital investment and accompanying energy investment in material. The fourth solution may require a reactant that requires processing energy. In the illustration, if the oxygen is assumed available from the air, then its energy cost is only that required to move it into the reactor.

The third possibility noted involves research into the mechanisms of kinetic reactions that can produce high reaction rates without high-temperature product-energy losses. Industrial chemistry is heavily directed at the search for better catalysts for increasing reaction rates. Achievement of the objective of increased reaction rates even at high temperature will reduce the losses per unit product if heat recovery is used. The primary reason for the emphasis on catalyst research is the reduction of the size of reactors; with higher reaction rates more product can be produced from a smaller reactor, with obvious economic advantages. Catalytic combustion—the combustion of fuels at lower temperatures—is an interesting development in this context. The combustion of fuels to provide heat often has large exhaust-gas energy losses. In a home furnace, for example, approximately 35 percent of the heating value of the fuel is lost up the chimney. One way to reduce this loss is through catalytic combustion, whereby the products of combustion are at a lower temperature. Such a system in a house would also allow individualized heating of separate rooms, reducing duct and piping losses. With present flame temperatures, in the order of 2,000 K, used for space heating where temperatures of the order of 400 K are adequate, a large loss of available energy is incurred. In lowered-temperature combustion the available energy losses would be less, since less heat would be lost through the chimney. Another alternative would be the combustion of hydrogen, whereby the combustion products, H_2O, could be exhausted to the interior of the house, providing humidification as well as heating.

Net-energy analysis is most easily interpreted in those cases where the product is energy used to provide heating or a work effect. In the cases of products that have other than thermodynamic value, the analysis can be made but the efficiency or effectiveness is only relative. The prime example of this is the analysis of Pimentel et al. [8], who used net energy analysis to show the energy-input requirement per unit of corn production in the United States. They showed that approximately 1 kW-hr of fossil-fuel energy input is required to produce 2 kW-hr energy-value corn. The energy value of the corn product is certainly not a very appropriate measure of a food-product value. Hybridization that increases the energy value of corn at the expense of protein content can actually give a net loss in food value.

Input-output analysis that associates energy consumption per unit value of product does not clarify the situation with respect to technological factors. In the corn situation energy analysis shows that the increased yields have been primarily the result of intensive use of chemical fertilizers and plant breeding, not increased mechanization. The corn situation is further confused by the fact that the total energy value of the biomass produced per unit of land has increased very little. The result of hybridization has been to increase the seed biomass partly at the expense of the stalk and cob biomass. The stalk and cob biomass, of course, have energy value that is roughly equivalent to the seed energy value. Based on a net-energy analysis, the energy value of this crop residue should be included.

The usual breakdown of inputs in net-energy analysis is done on the basis that the efficiency of conversion of fuel energy to electrical energy is three fuel-energy units per electrical-energy unit. Since most fossil fuels have available-energy and heating values within 5 percent (see table 2–4), this is a good approximation for most analyses.

In the case of nuclear-energy inputs, where identifiable, this method is also utilized except that a lower efficiency of conversion is used in line with the lower efficiencies of light-water fission reactors. The concept of available energy in the case of nuclear-energy conversion requires some discussion. Nuclear energy is used primarily for the production of electrical power,

Table 2-4
Energies Associated with Combustion of Various Fuels

Energy Terms	Hydrogen H_2	Carbon C (to CO_2)	Carbon Monoxide	Methane CH_4	Ethane C_2H_6	Propane C_3H_8	Ethylene C_2H_4	Liquid Octane C_8H_{18}
(1) $E - E_f$	57.5	94.1	67.3	191.8	341.5	489.1	316.2	1216
(2) $P_0(V - V_f)$ (a)	+0.3	0	+0.3	0	−0.3	−0.6	0	−3
(3) $-T_0(S - S_f)$ (b)	−3.2	+0.2	−6.2	−0.4	+3.3	+7.2	−2.1	+35
(b)	+0.4	0	+0.4	+2.0	+3.3	+4.5	+2.5	+11
(4) Heat of combustion, $-\Delta H = (1) + (2)$	57.8	94.1	67.6	191.8	341.2	488.5	316.2	1213
(5) Available work without diffusion, $B = (1) + (2) + (3)$	55.0	94.3	61.8	193.4	347.8	500.2	316.6	1259
(6) Percentage change from $|\Delta H|$ to B	−4.8%	+0.2%	−8.6%	+0.8%	+1.9%	+2.4%	+0.1%	+3.8%
(7) Additional diffusion contribution B_d	1.9 (3.5%)	3.9 (4%)	4.0 (6%)	6.5 (3%)	11.3 (3%)	16.2 (3%)	9.8 (3%)	41 (3%)
(5) Available work in other units[a]								
kcal/gm	27.3	7.86	2.21	12.06	11.58	11.35	11.29	11.03
MJ/kg	114	32.9	9.23	50.5	48.4	47.5	47.3	46.1
kBtu/lb	49.1	14.1	3.97	21.7	20.8	20.4	20.3	19.9
kBtu/lb-mole	99.0	170	111	348	626	900	570	2266
eV/amu	1.18	0.341	0.096	0.523	0.502	0.492	0.490	0.478

Source: American Institute of Physics, *Efficient Use of Energy*, part I-A, *Physics Perspective*, ed. K.W. Ford, G. I. Rochlin, and R.H. Socolow (New York, 1975). Reprinted with permission of American Institute of Physics, 335 East 45th St. New York, NY 10017.

Note: Calculations are for combustion in air with gaseous H_2O product ("low heating value"). Energies in the upper part of the table are in kcal/mole of fuel. Available work is given in other units in the lower part of the table.

[a]Available work in other units does not include the diffusion contribution B_d.

although Dow Chemical Corporation, in cooperation with Consumers Power, is constructing a plant in Midland, Michigan, to provide process steam (see chapter 15). The available energy of nuclear fuel is technologically limited by the reactor materials and the more conservative design requirements on nuclear reactors. The available energy of nuclear-reactor waste products is more difficult to evaluate, since energy is required to process them for either storage or recycling. The net-energy analysis for nuclear energy is particularly sensitive to where one draws the boundary on energy inputs to the system. Under contention at the present is whether the large amount of energy expended in research and development and the long-term energy required for storage processing and monitoring should be charged to the nuclear-energy system. Without this inclusion the nuclear technology is clearly a net energy producer, as indicated by many independent analyses, summarized in table 2–2. Costs of reprocessing energy need to be included but are uncertain at this time. Real feasibility of nuclear energy requires reprocessing to reduce the radioactive-storage problem (the criterion is that the products be environmentally acceptable).

Summary

Net-energy analysis attempts to put into input-output format the accounting of energy utilized to produce goods and services in an economic system. Both direct and indirect energy use is thus included in determining such data as energy use per dollar value of output of different economic sectors. Stockbrokers can utilize such tables to let investors know which investment opportunities are most affected by changing energy supplies or prices. Economists can utilize this information to project changes in employment and capital investment due to shortages, taxation, or energy-cost changes.

What is the possible use by engineering managers or industrial-process designers? If shortages are reflected in higher prices of energy alone, this is adequately represented by economic input-output tables. The engineer can alter processes to reduce costs of fuel or change the materials used based on this information to good advantage. What then does energy accounting add? The first addition to decision making is with respect to the choice among alternative "fuels" for a task. Energy accounting delineates the energy received per dollar spent for alternative energy sources. This allows planning to modify processes to take advantage of supply price changes.

More importantly, if one is aware of (1) specific future energy-supply shortages, (2) alternative energy-processing technologies, or (3) renewable-resource possibilities, then one can utilize thermodynamic-energy information supplied by net-energy analysis for making energy-use decisions [20].

Net energy, then, supplies a future predictive value that is not available from economic input-output analysis. This should be an advantage in industrial and manufacturing enterprises, where the investment in equipment is on a long-term basis. Where short-term decisions are involved, the economic input-output analysis is more useful.

The combination of net-energy analysis utilizing available energy and effectiveness (second-law analysis) provides an additional advantage in the assessment of possible technological improvements. Without a measure of technological-change possibilities, net-energy analysis can prove misleading. Energy accounting must include consideration of technological changes that make older equipment obsolete. For example:

1. In the commercial production of ammonia for fertilizer, the energy required per pound of ammonia produced does not represent the true production cost. Natural gas is the principal input reactant, together with nitrogen from the air. The theoretical fuel energy to convert using propane is approximately 1.1 times that required for natural gas; the actual energy using propane is closer to 2.5 times that required with natural gas.
2. The extraction of minerals and fuels from the earth's crust will require increased energy costs in the future. The available-energy decrease of the minerals due to dispersion in the earth's crust, and changes in technology of extraction, are two factors that influence the choices, as shown in chapter 13.
3. The quantity of energy required for production depends on the rate or intensity of the process flow. The energy-dissipation dependence on the processing rate requires consideration of available-energy factors not included in energy analysis alone (see chapter 8).
4. The basic element in many heating systems, the heat exchanger, requires available-energy accounting to demonstrate the acceptable efficiency within economic constraints (see chapter 9).
5. Analysis of solar-energy systems requires consideration of the available energy of the sunlight, not its total energy. The capability for increasing the concentration of sunlight through mirror or lens systems provides additional value (see chapter 14).

In order to examine in detail the methods available to resolve these problems, we will introduce in chapters 3 and 4 the appropriate economic and thermodynamic basis.

References

1. Ballard, C., and Herendeen, R. "Energy Costs of Goods and Services, 1963 and 1967." Document 140, Center for Advanced Computation., University of Illinois, Urbana, March 1975.

2. Makhijani, A., and Lichtenberg, A. "Energy and Well-Being." *Environment* 14(1972):10.
3. Hirst, E. "How Much Overall Energy Does the Automobile Require?" *Automotive Engineering* 80(1972):35.
4. McGowan, J., and Kirchoff, R. "How Much Energy is Needed to Produce an Automobile?" *Automotive Engineering* 80(1972):39.
5. Herendeen, R. "An Energy Input-Output Matrix for the United States, 1963." User's Guide Document no. 89, Center for Advanced Computation, University of Illinois, Urbana, March 1973.
6. Herendeen, R. "Use of Input-Output Analysis to Determine the Energy Cost of Goods and Services." In *Energy: Demand, Conservation and Institutional Problems*, ed., M.S. Macrakis. Cambridge, Mass.: MIT Press, 1973.
7. Just, J. "Impacts of New Energy Technology Using Generalized Input-Output Analysis." In *Energy: Demand, Conservation and Institutional Problems*, ed. M.S. Macrakis. Cambridge, Mass.: MIT Press, 1973.
8. Pimentel, D.; Hurd, L.E.; Bellotti, A.C.; Forster, M.J.; Oka, I.N.; Sholes, O.D.; and Whitman, R.J. "Food Production and the Energy Crisis." *Science* 182(1973):443.
9. Hannon, B. "System Energy and Recycling, A Study of the Soft Drink Industry." *Environment* 14(1972):11.
10. Development Sciences, Inc. "A Study to Develop Energy Estimates of Merit for Selected Fuel Technologies." Prepared for U.S. Department of Interior, Office of Research and Development, September 1975.
11. Institute for Energy Analysis, Oak Ridge Associated Universities. "Net Energy from Nuclear Power." IEA-75-3, November 1975.
12. University of Oklahoma, Science and Public Policy Program. "Energy Alternatives: A Comparative Analysis." Washington, D.C.: U.S. Government Printing Office, 1975.
13. State of Oregon, Office of the Governor, Office of EnergyResearch and Planning. "Transition." January 1975.
14. Pilati, D.A., and Richard, R. "Total Energy requirements for Nine Electricity Generating Systems." CAC Document no. 165, Center for Advanced Computation, University of Illinois, Urbana, 1975.
15. "A National Plan for Energy Research, Development and Demonstration: Creating Energy Choices for the Future." ERDA Report 76-1, 1976.
16. Steinhart, J., and Steinhart, C. "Energy Use in the U.S. Food System." *Science* 184(1973):307.
17. Berry, R.S. "Recycling Thermodynamics and Environmental Thrift." *Bull. At. Scient.* 28(May 1972):8.
18. Berry, R.S.; Fels, M.; and Makino, H. "A Thermodynamic Valuation of Resource Use: Making Automobiles and Other Processes." In *Energy:*

Demand, Conservation and Institutional Problems, ed. M.S. Macrakis. Cambridge, Mass.: MIT Press, 1973.

19. Berry, R.S., and Fels, M. "The Energy Cost of Automobiles." *Science and Public Affairs (Bull. At. Scient.)* 29(1973):13.

20. Stanford Research Institute. "Patterns of Energy Consumption in the United States." Report to the Office of Science and Technology, Executive Office of the President, Washington, D.C., 1972.

21. U.S. Bureau of the Census, *1967 Census of Manufacturers*. Washington, D.C.: U.S. Department of Commerce, 1971.

22. U.S. Bureau of the Census. *1967 Census of Mineral Industries*. Washington, D.C.: U.S. Department of Commerce, 1971.

23. U.S. Bureau of the Census. *1967 Census of Transportation*. Washington, D.C.: U.S. Department of Commerce, 1971.

24. American Institute of Physics. *Efficient Use of Energy*, part I-A, *Physics Perspective*, ed. K.W. Ford, G.I. Rochlin, and R.H. Socolow. New York, 1975.

3 Economic Preliminaries

Economy is a distributive virtue, and consists not in saving but selecting.
—Edmund Burke, *Letter to a Noble Lord* (1796)

Economic Declaration

The introduction of thermodynamic analysis into the social science of economics requires that one's perspective be defined. Most authors have personal views representing an industrial, governmental, or consumer bias. Thus it is hazardous to try to synthesize the energy, economics, and environmental viewpoints without seeming naive. During the investigative hearings following the spectacular Tacoma Narrows Bridge failure, the story goes, experts were called in to determine the cause of the failure. Each identified his special interest: a legal counsel for the state, a suspension-cable expert, a vibrations expert, a foundations expert, the construction engineers, the construction corporation, an expert on plate construction, an insurance expert, and so forth. The last expert called on to identify the specialty he represented was Professor Theodore Von Karman, an internationally known aerodynamics expert. His response was that he "represented the *wind*" [1]. Similarly, contemporary environmentalists claim to represent the earth. This position is attractive but ambiguous in the sense that it must include people as well. "Environment," as noted by R. Buckminster Fuller "is everything but *me.*"

Therefore, I have taken the position that a long-range view of economics must deal with the economics of "spaceship earth." This position is both idealistic and realistic, idealistic in the sense that few people now view economics in this light, but realistic in that it is one of the directions future-oriented economists appear to be taking [2].

The reader should test the view of economics presented in this book with the question: Is it consistent with spaceship economy? It is important to consider a spaceship economy as dynamic, not static. Economic changes and opportunities are expected to be just as important in the future as they are today. An introduction to the dynamics of economics might well begin with a reading of Defoe's *Robinson Crusoe* [3].

Economic Criteria

This chapter will not attempt to provide economics for everybody. The criteria for selection of topics will be as follows:

1. Are these topics directly applicable to energy-environment decisions?
2. Since energy-technology decisions involve choices among alternative methods, what is the technological decision that can be made from the analysis?
3. Does the economics analysis include recycling?
4. Are transportation processes included if, indeed, pollution is a resource out of place?
5. Can the accounting system be used in a physcial sense that includes conservation of mass and energy?
6. Can the accounting system include nonconservative systems, where the quality as well as the quantity of inputs and outputs can be handled?
7. Is the system expandable to dynamic-time-dependent processes?
8. Is the economics of value included in the design process?
9. Does the economics attempt to predict as well as describe?
10. Is it operational for an engineer or manager in the sense that inputs to the models are obtainable?

An economics for use implies a user. Let us discuss three categories of user and the expected requirements for each.

Consumer

The consumer's task, economically, is selection based on his or her value system and resources. It is probably not realistic to assume that consumers' choices will change the products available to them or that they will make their choices among alternatives on the basis of a hope their choices will influence manufacturers to change their products. (Although this is the basis of macroeconomic theory, it is an integrative effect that probably has little influence on individual consumer behavior.) A consumer selects a small car not because he wants manufacturers to produce more small cars but because he wants to save money or energy or space. For consumers economic decisions will require an understanding of financing, including the methods of determining total cost, capital costs, and operating costs: (1) the methods for including interest rates, depreciation, amortization; the relation between use and operating, and the influence of inflation on decisions; (2) the difference between fixed and variable costs; and (3) how to average over the expected life of an item to minimize cost.

It is widely believed that people tend to buy such items as automobiles, appliances, and houses on the basis of lowest initial cost within the constraints of their value system. If one is uncertain about the operating or variable cost in a system, this a rational basis. A system must then be developed to provide consumers with variable- or operating-cost estimates that enable them to make decisions based on total costs rather than initial cost. The idea that manufacturers should label appliances with expected energy use and automobiles with expected gas mileage is a step in this direction. The methods of analysis required are similar to standard engineering-economy methods except for some integration of cost information. It is this method of integration of dollar cost and energy cost with environmental cost that is attempted in this book.

Information and a method alone are not sufficient. One also needs an awareness of the resources one is using and the environmental consequences of one's decisons. Without these there is little basis for a rational decison about what is being optimized and why.

The intent here is to provide a simple methodology for accounting for cost, energy, and environment. The emphasis or weighting for each will depend on the economic constraints on the consumer, the resource and disposal constraints of the environment, and the technological and ethical constraints of society.

Industrial Management

The objective with respect to the producing sector of the economy is to provide methods to deal with optimization within environmental constraints. This is the sector that requires engineering decisions. Engineering economics has been based on profit maximization through technical optimization. Engineering has appropriately developed technical alternatives and refined them to increase the profit from a given operation. Classically, however, the engineer has considered the constraints to be on the resource side. Except for a few environmental engineers concerned with residual handling, technology has generally been developed to produce a product at least cost in terms of resources used. Disposal costs are rarely included in the economics of manufacture. The engineering preoccupation with efficiency is based on maximizing output for a given input. In espousing this attitude of efficiency, the manager has always expressed it in terms of dollars, not energy or materials. To the manager, *efficiency* has meant minimum cost for a given product value. This is why engineers often misinterpret managerial economy. The purpose of the analysis of economics in this section is to summarize what managers already know and then to provide a method of extending the analysis to energy and environmental situations.

Input-out analysis is included for two reasons. The first is that in a historical period in which both the absolute and the relative prices of goods and services are changing rapidly, the engineer or manager must be able to estimate the overall impact on his or her designs or processes. If the cost of oil increases by 20 percent, what will be the expected cost increase of nylon relative to that of polyethylene or aluminum (that is, the indirect effect on the cost of materials that are to be used in a process)? What will be the expected change in the cost of electrical energy as the result of changes in SO_2 emission standards? How will the costs of disposal of the waste from a manager's facilities be affected by environmental limitations? How will the value of the waste from a process be changed by new recycling methods or new product requirements?

Rate considerations are introduced to make the engineer or manager aware that it is the process rate that is important and that the change in rate and scale are interconnected in basic physical processes. The value of material and energy in different forms must also be included in the economics. Decisions must be made about the cost of making them suitable for a process or of reducing them to valuable products.

Macroscale

The macroeconomics of large-scale decisions at governmental or large-industrial levels requires a method of assessing new technology and identifying new areas toward which resources should be directed. A plethora of recycling schemes and energy-saving methods has been developed by engineers. Now sound methods of analysis to assess the limitations and opportunities of these methods must be developed. These require greater emphasis on technical evaluation. An economic analysis can only compare costs, but a technical evaluation can project feasibilities. At this level the interactions of different pollution-control strategies must be accounted for. The imposition of water-quality standards that increase air-pollution level is a typical example. The method of analysis must include interactions among many elements to ascertain the optimum strategies for obtaining the maximum societal benefits at the least societal costs.

Input-output techniques can be extended beyond mass conservation to available-energy accounting. For this purpose a summary of input-output analysis is provided with an extension to the externality problem.

Input-Output Method of Analysis

The first method of economic analysis to be discussed is the input-output technique initiated by Leontief [4] to assist in analyzing the interactions of the economic sectors of the United States. The original analysis divided the U.S. economy into eighty-one sectors and analyzed the inputs and outputs in terms of the dollar value of each. More recently, extensions to the environmental and energy sectors have been attempted to assist in developing environmental-control strategies or energy policies [5–8]. These methods are discussed because they will assist in formulating the problems in available-energy terms in a later chapter. The method is also applicable at the several levels noted. An engineer engaged in designing new systems or devices must have a way to analyze the effect of the change in prices for input materials or energy on the cost of his product. The recent dramatic rise in energy costs affected material costs as well. Knowing how these costs affect substitute materials can help one redesign products to allow use of reduced-cost materials. For example, if the filter to be designed can use alternative materials of nylon, polyethylene, vinyl, and so forth, how will the choice be affected by an increase in the cost of crude oil or in the pollution-control requirements on refineries? Similar projections by consumers can anticipate price increases on consumer goods. The initial example that is expanded on in this section is taken from a basic paper by Leontief [4].

Input-output analysis is based on the idea that there is a given set of final demands for products by consumers. The usual question, then, is how these demands are met by outputs of different industries or economic sectors. In a sense it is a production-oriented model that attempts to determine the outputs required of industries to meet demands.

Figure 3–1 shows a block diagram of the flows of goods from two basic economic sectors, industry and agriculture. In the figure, X_j and X_k represent the total production of the agricultural and industrial sectors. They are determined by the demand sector and the other sectors in the model, as shown in the following equations:

$$X_j = Y_j + C_{jj}X_j + C_{jk}X_k, \qquad (3.1)$$

$$X_k = Y_k + C_{kk}X_k + C_{kj}X_j. \qquad (3.2)$$

In this form $C_{jk}X_k$ represents the output of the jth sector required by the kth sector. This is the characteristic of a demand economy in which demands or sales are the limiting factor in production decisions. This is a sort of nonenvironmental way of looking at economics. From an environmental viewpoint the resource inputs required from the environment would be more appropriate. Similarly, in underdeveloped countries it is resources, not demand, that limit production. The factor C_{jk}, then, represents the fraction of the output of the kth sector that is required from the jth sector to meet the final demand of the kth sector. These equations may be written in the general form

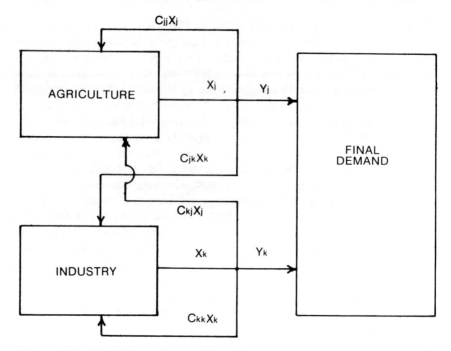

Figure 3–1. Basic Input-Output Block Diagram for a Two-Sector Producing Economy

$$X_i = Y_i + \sum_n C_{in} X_n, \tag{3.3}$$

where n is summed over all sectors. The inverse problem of knowing all the X_is and C_{in}s and requiring the final demands (Y_is) obviously can be written as

$$Y_i = X_i - \sum_n C_{in} X_n. \tag{3.4}$$

Leontief systematized the presentation of the coefficients in these equations by his input-output tables. For example, suppose X_j, the total output of the agricultural sector, is 1,000 tons of corn, and X_k, the industrial output in a system, is 500 barrels of paint. These are further divided in such a way that of the total agricultural output, 200 tons are needed for the farm operation and 250 tons are needed by the industrial sector, leaving the final demand from the agricultural sector as

$$Y_j = X_j - C_{jj} X_j - C_{jk} X_k,$$

$$Y_j = 1{,}000 - 200 - 250 = 550 \text{ tons of corn.}$$

Similarly, 500 barrels of paint, the total output of industry, are required to meet a final demand of 300 barrels, a demand of agriculture of 100 barrels, and a self-demand of industry of 100 barrels.

$$Y_k = X_k - C_{kk}X_k - C_{kj}X_j,$$

$$Y_k = 500 - 100 - 100 = 300 \text{ barrels of paint.}$$

Note that the units of production and demand of each sector are different— tons of corn and gallons of paint. Thus they cannot be added directly, and the coefficients are ratios that have different units.

The coefficients are then found as

$$C_{jj} = \frac{200}{X_j} = \frac{200}{1,000} = 0.20,$$

$$C_{jk} = \frac{250}{X_k} = \frac{250}{500} = 0.50 \text{ ton/barrel,}$$

$$C_{kk} = \frac{100}{X_k} = \frac{100}{500} = 0.20,$$

$$C_{kj} = \frac{100}{X_j} = \frac{100}{1000} = 0.10 \text{ barrel/ton.}$$

These coefficients represent inputs per unit output and are tabulated in an input-output format in table 3–1. If the total outputs of each sector are known, then table 3–1 can be used with equation 3.4 to find the final demand.

A more useful formulation from the demand point of view is the tabulation of inverse coefficients. These would be coefficients such that if the final demands Y_j and Y_k were known, the total outputs of each sector could be calculated.

Table 3–1
Table of Input-Output Coefficients for a Two-Sector Economy

From Sector	Into Sector	
	j Agriculture	k Industry
j—Agriculture k—Industry	$C_{jj} = 0.20$ $C_{kj} = 0.10$	$C_{jk} = 0.50$ $C_{kk} = 0.20$

If equation 3.4 is written as a matrix

$$\begin{bmatrix} Y_j \\ Y_k \end{bmatrix} = \begin{bmatrix} (1 - C_{jj}) & - C_{jk} \\ - C_{kj}(1 - C_{kk}) \end{bmatrix} \begin{bmatrix} X_j \\ X_k \end{bmatrix},$$

or $Y_i = [I - C]X_n$, where I is the identity matrix and C is the matrix of the input-output coefficients.

The inverse would then be

$$X_i = [I - C]^{-1} Y_n = A_{in} Y_n. \tag{3.5}$$

The coefficients of the inverse matrix would be required to calculate the total outputs of each sector to meet the final demand. The inverse coefficients can then be calculated either from the input-output coefficients or directly from the initial data, as follows. In this case they may be interpreted as the sum of the final demand and the direct demand of a sector divided by the final demand of that sector.

For the agricultural sector, from the agricultural sector, it is

$$\frac{200 + 550}{550} = 1.36.$$

For the industrial sector, from the industrial sector, it is

$$\frac{100 + 300}{300} = 1.33.$$

For the agricultural from the industrial it is

$$\frac{100}{550} = 0.18.$$

For the industrial from the agricultural it is

$$\frac{250}{300} = 0.83.$$

The inverse matrix is then

$$[A] = [I - C]^{-1} = \begin{bmatrix} 1 - C_{kk} & C_{jk} \\ C_{kj} & 1 - C_{kk} \end{bmatrix}^{-1} = \begin{bmatrix} 1.36 & 0.83 \\ 0.18 & 1.33 \end{bmatrix}.$$

In table form, inverse coefficients would appear as shown in table 3–2. From this table the total output of each sector can be calculated as follows, knowing the final demand of each as:

$$X_j = A_{jj} Y_j + A_{jk} Y_k,$$
$$X_j = 1.36(550) + 0.83(300) = 1,000.$$
$$X_k = A_{kj} Y_j + A_{kk} Y_k,$$
$$X_k = 0.18(550) + 1.33(300) = 500.$$

The inverse coefficients are particularly useful for examining the effects of changes in demand on the suppliers to major final-demand producers. For example, the effect of the decrease in demand for automobiles on the steel industry can be estimated from tables of this type. Similarly, the effect of additional final demand for environmental cleanup systems on the rest of the economy can be estimated.

This interpretation of the inverse coefficients is limited to a two-sector economy. A much more important interpretation is that the inverse coefficient A_{in} represents the input, both direct and indirect, required by sector n from sector i per unit of final demand Y_n. As an example, consider the three-sector economy of figure 3–2: the inverse coefficient A_{ej} represents the input both directly and indirectly required from the environmental-industry sector per unit of final demand Y_j. The indirect input in this case is the input required from the environmental industry by the industrial sector to produce the industrial input to agriculture.

This property of the inverse coefficients allows one to readily estimate the impact on an industry of the increase in energy costs on all purchases. It also allows an engineer to estimate the relative changes in the prices of materials or products due to basic-resource cost changes. In an environmental context it can be used to estimate the pass-through costs of environmental regulations that require new technology for environmental improvement.

In input-output matrices there is usually a conversion to dollar values to simplify calculating procedures by eliminating the need to consider different units associated with the coefficients. Several modifications to this type of

Table 3–2
Inverse Coefficients for a Two-Sector Economy

From Sector	Into Sector	
	j	k
j	1.36	0.83
k	0.18	1.33

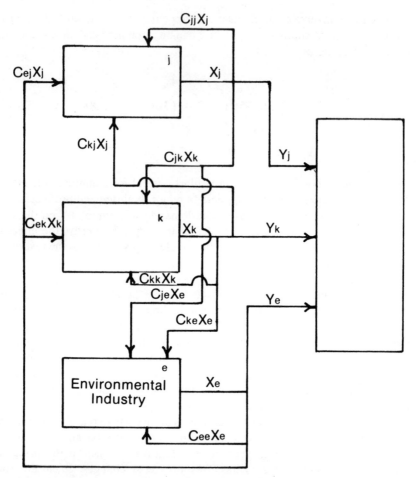

Figure 3–2. Addition of an Environmental-Industry Sector to the Basic Input-Output Diagram

analysis have been proposed to take environmental considerations into account. The simplest modification noted by Leontief consists of hypothesizing a third sector—a sector required by the society to reduce the pollutant flow to the environment. In terms of the addition of the line flows shown in figure 3–2, there is a new set of relations between demands and the outputs of each sector. In general, the output of each sector must be increased if part of the output is required by the environmental sector. Another way to look at it is that if the total output of a sector is limited, then the final demand must be reduced by the amount of the output required for pollution control. In this case,

$$Y_j = X_j - C_{jj}X_j - C_{jk}X_k - C_{je}X_e,$$
$$Y_k = X_k - C_{kk}X_k - C_{kj}X_j - C_{ke}X_e, \qquad (3.6)$$
$$Y_e = X_e - C_{ee}X_e - C_{ej}X_j - C_{ek}X_k,$$

where

Y_e is the consumer demand for a certain environmental quality, which may be so many gallons of "good" water, or the like;

X_e is then the total good water that must be produced to meet the consumer demand;

$C_{ej}X_j$ is the demand by the agricultural sector for good water;

$C_{ek}X_k$ is the demand by industry for good water.

Figure 3–2 and equation 3.6 immediately indicate that if the final demands Y_j and Y_k remain the same, then the additional demand for Y_e, a certain-quality environment, increases the total outputs required from each industry or agricultural sector. This in turn increases the demand by these sectors for more processing water in the illustration given. The environmental industry needs chemicals, filters, and so forth from the output of other sectors. The environmental industry itself may require water for processing as well.

The principal difficulty with the inclusion of the environmental industry in this way is that there is still a factor missing, namely, that the demand on the environment produced by the addition of the environmental-industry sector is increased because of the increased requirements of industry and agriculture for resources. Hence a greater waste flow from each also results. This is the next consideration that must be included in an input-output analysis. A simple way to do this is indicated in figure 3–3, where an *environment sector*—not an environmental industry—is introduced.

With the environment considered as a separate sector, the input-output analysis takes on a different aspect, as shown by the revised figure 3–3 and the following equations:

$$Y_j = X_j - C_{jj}X_j - C_{jk}X_k,$$
$$Y_k = X_k - C_{kk}X_k - C_{kj}X_j,$$
$$Y_e = X_e - C_{ej}X_j - C_{ek}X_k - C_{ke}X_k - C_{je}X_j. \qquad (3.7)$$

The last equation needs a new interpretation.

The terms $C_{ej}X_j$ and $C_{ek}X_k$ represent resource utilization by the respective sectors *from* the environment. The terms $C_{je}X_j$ and $C_{ke}X_k$ represent waste-assimilation requirements—waste flows *to* the environment.

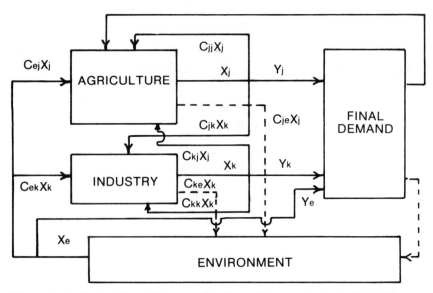

Figure 3–3. Basic Flow Diagram for Input-Output Analysis Including an Environmental Sector

All these are demands by the production sectors on the environment. Y_e represents the consumers' demands directly on the environment, and X_e represents the total demand on the environment by all sectors. This formulation is also useful if the flows are considered as mass flows. In that context, conservation of mass requires that a flow from the final demand to the environment be introduced. The flows can then be examined analytically. The results of such an analysis usually only indicate changes over time of the flows without any indication of the environmental effect. The typical requirement of conservation of mass specifies that the total mass flow into each sector be equal to the amount flowing out. In this assumption storage is assumed negligible, a reasonable assumption for most macroscopic problem analysis. This formulation however, neglect the important factor of capital stock, discussed in chapter 1.

Further refinements include the introduction of recycle from the final-demand sectors to the agricultural and industrial sectors, as well as an output from the final-demand sector to the environment. This latter would represent the disposal of waste materials from consumers. Such inclusions are usually required if the conservation-of-mass principle is applied. The assumption of a static environment is also required if an environmental sector is included in the analysis.

The inadequacy of a constraint on mass conservation is easily demonstrated by the fact that the net mass flow of pollutants to the environment is increased by the addition of a pollution-control system if the final-demand inputs are to be maintained. The explanation is simply that greater mass-flow inputs are required for each sector if part of the sector's output is needed to supply the pollution-control sector. Any pollution-control system, then, increases the total mass flow of materials both *to* and *from* the environment. The objective of a pollution-control system is hence not the reduction of the total mass flow to the environment, but a change in the quality of these inputs to the environment.

In some cases the reduction of the pollutant flow to water sytems is chosen at the expense of extra air-pollutant flows. Of course, pollution-control systems require fuel resources from the environment and produce combustion-product flows to the atmosphere.

Recycle systems are basic to the reduction of the waste flows to the environment. They are discussed in the section on chemical processing and integrated into the anlysis of available-energy flows later.

The coefficients in the input-output matrices are the factors that are of primary concern to the engineer. These coefficients depend on many elements but, in particular, (1) technology, (2) alternative supplies, (3) geography, and (4) labor. Control of these coefficients is the main mechanism for improving a system's performance. If the input-output tables are in dollar-value terms, then prices of goods are also included (a value-system factor).

Improved technology reduces the coefficients by reducing the input requirements of each sector and improving the efficiencies. Alternative supplies are reflected in different coefficients for different materials. Geographical factors influence the choice of location of the operations as well as the transportation system needed. Labor enters economic input-output analysis in several ways. Its consideration with respect to the coefficients is that the skill of the engineers, operators, and workers can reduce the resource requirements and the waste outputs. Coefficients vary from one industrial plant to another depending on these factors. In standard input-output analysis services are included in the tables, and a value-added category is also included to account for the difference between the dollar cost of inputs to a sector and the value of the products delivered. The sample input-output table (table 3–3) indicates the forward type of information available. These coefficients are in terms of dollars input per dollar output. The selling price of the output of a sector can be determined by summing the product of the price of the output of each sector times the coefficient of that sector (this is the sum of the column value for each sector input), plus the value added by that sector. The price of the agricultural output would be, for the following system,

$$P_1 = 0.25P_1 + 0.14P_2 + 0.15P_3 + 0.05P_4 + V_1,$$

Table 3–3
Input-Output Coefficients for an Economy with an Environmental-Industry Sector

Sector	1	2	3	4
1 Agriculture	0.25	0.4	0.133	0.1
2 Industry	0.14	0.12	0.133	0.1
3 Energy	0.15	0.4	0.133	0.1
4 Environmental Industry	0.05	0.2	0.133	0.1
Value added V	V_1	V_2	V_3	V_4

where V_1 represents the value added in the production system for agriculture. This is sometimes attributed to labor but obviously includes capital-investment return as well.

In general terms,

$$P_i = \sum_n C_{in} P_n + V_i = [C]P_i + V_i, \qquad (3.8)$$

or, inverted,

$$V_i = [I - C]^{-1} P_j,$$

where I represents the diagonal identity matrix. The ratio of total input to total output for specific sectors can also be calculated from these. Typical input/output (I/O) ratios for segments of the economy are given in table 3–4.

There are several ways to interpret these ratios. In one light, a low I/O ratio indicates a labor-intensive industry, since labor is not included in the input costs in this calculation. On the other hand, a high ratio can represent a high degree of automation and capital investment. Remember that the I/O ratio may be written as

$$I/O = \frac{I}{I + V},$$

where V is the value added. A low ratio represents a sector that adds greater value to the throughput than does a high-ratio sector like retail trade. Most pollution-control-systems costs are assumed to be capital costs, not labor and operating costs. This idea is in part the result of the publicity accorded to the high initial costs of pollution-control systems. Most effective control systems depend on automated operation, so that it is expected that the addition of pollution-control systems to an industry will raise the input/output ratio.

The price format of input-output is particularly useful for assessing the change in the cost of producing a product attributable to price increases

Table 3–4
Input Cost/Output Sales Ratios for a Sample of U.S. Economic Sectors

Sector	I/O
Food and kindred products	0.14
Textile-mill products	0.16
Lumber and wood products	0.20
Paper products	0.24
Chemical products	0.27
Plastics	0.28
Petroleum refining	0.31
Leather	0.33
Stone and clay products	0.36
Primary iron and steel	0.37
Primary nonferrous metals	0.38
Motor vehicles and equipment	0.59
Aircraft and parts	0.60
Electric, gas, water, and sanitary services	0.68
Wholesale and retail trade	0.69

Source: Data from R.U. Ayres and A.V. Kneese, "Production, Consumption and Externalities," *Amer. Econ. Rev.* 59(1969):282.

elsewhere in the economy. A particularly relevant example is the expected change in the cost of production of food due to an energy-cost increase that appears indirectly in the higher price of fertilizers.

The price equation

$$P_i = \sum_j C_{ij} P_j + V_i$$

becomes, in expanded form,

$$P_1 = C_{11} P_1 + C_{12} P_2 + C_{13} P_3 \cdots + V_1,$$
$$P_2 = C_{21} P_1 + C_{22} P_2 + C_{23} P_3 \cdots + V_2,$$
$$P_n = C_{n1} P_1 + C_{n2} P_2 \cdots C_{nn} P_n + V_n.$$

Suppose the price of gasoline P_2 is increased by Δ, with all coefficients and value-added factors held constant. The change in P_1 will increase directly by the factor $C_{12} \Delta$, but also indirectly since all other Ps will increase by a factor proportional to their use of energy C_{ij}s and Δ. This formulation can then be used to evaluate the expected total price of inputs to a given sector. Similarly, the effect of increased cost of production of a sector due to environmental-control costs on any other sector can also be estimated.

Resource-use or waste-generation matrices can also be constructed in an input-output format. One way to do this is to establish resource use or waste generation for each sector per unit dollar-value output by that sector. This has been done for water use by McGauhey et al. for nine Bay Area counties around San Francisco [10] (see table 3–5) and for energy use by R.E. Moor of A.G. Becker Co. [11] for national economic sectors.

The first step is to determine the use of per total dollar output and put it in the format of a diagonal $n \times n$ matrix where E_{11} is the use per total dollar output of sector 1, and so forth. This is often easier than finding the use per dollar demand directly, because you can determine technically the use for each unit of product produced.

$$E = \begin{bmatrix} E_{11} & 0 & 0 & 0 & 0 & 0 \\ 0 & E_{22} & 0 & 0 & 0 & 0 \\ 0 & 0 & E_{33} & 0 & 0 & 0 \\ 0 & 0 & 0 & & 0 & 0 \\ 0 & 0 & 0 & 0 & & 0 \\ 0 & 0 & 0 & 0 & & E_{nn} \end{bmatrix}$$

If the inverse input-output matrix is $[I - C]^{-1}$ from before in terms of dollar outputs, then an $n \times n$ use matrix can be found as

$$E_D = [I - C]^{-1\,T} E, \tag{3.9}$$

Table 3–5
Water Demands by Economic Sector
(Acre-feet/million dollars of output)

Sector	E_{ii}
1. Agriculture, forestry, and fisheries	4,345
2. Mining	950
3. Construction	13
4. Food and kindred products	16
5. Paper and allied products	92
6. Chemical and chemical products	26
7. Petroleum refining and related industries	16
8. Stone and clay products	26
9. Fabricated-metal products	2
10. Transportation, communication, gas, electric	160
11. Wholesale and retail trade	8
12. Finance, insurance, and real estate	13
13. Services and government enterprises	300
14. Manufacturing	15

Source: Data from D.H. McGauhey et al., "Final Report, Economic Evaluation of Water," Sanitary Engineering Research Laboratory, University of California (November 1969).

where T is the transpose of the inverse matrix found by replacing columns by rows in the matrix.

This E_D matrix can then be used to determine the resource use R_D required for each sector if the demand Y_D changes by the matrix operation

$$R_D = [E_D]Y_D. \qquad (3.10)$$

Table 3–6 indicates approximate energy use per dollar output for energy sectors as determined by the A.G. Becker study [11]. This, when combined with the inverse coefficients, can then be used to determine how the energy requirements of the economy would change with a change in demand. Future energy reductions can also be examined in terms of their effect on final outputs and demands by proper manipulation of the data in this matrix format.

The linear nature of the input-output format presents difficulties in predicting future expectations accurately. The nonlinear nature of the economy is manifest in several ways. One is the economy of scale in which an increase in production-unit size in industry or agriculture reduces the cost per unit product. Similar factors regarding pollution control and energy use are nonlinear to the extent that large industries, through their research efforts, are likely to institute new, more efficient processes or processes producing less waste per unit output at an earlier date. Examples are new electrolytic aluminum-processing methods [12], with lower electrical-energy requirements, and new paper-processing methods requiring less water through recycling systems [13]. Other factors include the shift of input materials to a process when prices of inputs change, as has occurred in the shift from copper to aluminum electrical wiring and the replacement of various wood and steel articles by plastic and aluminum substitutes. Changes due to resource depletion or technological advances are not incorporated in input-output analysis in a form useful for predicting future developments if a shift in inputs from one kind to another occurs.

These input-output tables are practical for short-term decisions because they reflect well the indirect effects of changes in an economy. For long-term predictions they are not expected to prove very valuable because of their static and simplistically linear nature. However, they can help the engineer project the overall impact of a new technology under development. In environmental situations this is most important.

Energy Transactions

The energy transactions in the analysis of the A.G. Becker report are in terms of how energy is distributed to the consuming sectors from the primary energy suppliers. A more complete analysis requires investigation of

Table 3-6
Energy Demands by Industry
(*Millions of Btu per dollar of final demand*)

Sector	E_{dd}
Agricultural, forestry, and fisheries	
Poultry and eggs	0.08
Cotton	0.06
Oil-bearing crops	0.06
Vegetables	0.04
Construction	
Highways	0.10
Residential	0.06
Energy	
Coal mining	0.10
Crude petroleum and natural gas	0.11
Petroleum refining	0.20
Asphalt products	0.45
Manufacturing	
Meat products	0.06
Fluid milk	0.06
Sugar	0.08
Manufactured ice	0.09
Alcoholic beverages	0.04
Textiles	0.04
Wood products	0.05
Furniture (wood)	0.05
Furniture (metal)	0.09
Paper	0.20
Plastics	0.22
Fertilizers	0.18
Synthetic rubber	0.27
Paint	0.14
Cement	0.42
Tires	0.13
Minerals	
Steel (basic blast furnace)	0.26
Iron-foundry products	0.10
Primary copper	0.12
Copper rolling and drawing	0.10
Primary aluminum	0.38
Aluminum rolling and drawing	0.23
Transportation and communication	
Railroads	0.08
Water transportation	0.14
Air transportation	0.15
Pipeline transportation	0.49
Radio and TV broadcasting	0.02
Utilities	
Electric utilities	0.10
Gas utilities	0.16
Water and sewer	0.12
Services	
Motion pictures	0.03
Hospitals	0.04
Education	0.05
Post office	0.02

Source: Data from R.E. Moor, and G.E. Garrison, "The Economics of Enervated Energy," *Market Economics* (Chicago: A.G. Becker, 1973).

intermediate secondary energy transactions between all sectors—for example, the indirect energy value of fertilizers, machinery, and so forth to agriculture from industrial supplies; the transportation energy required to make these transfers; the waste-output energy value; and the energy value of the product. The first two of these four elements are partially accounted for in the energy input-output analysis discussed in chapter 2 with respect to fossil fuels and published in extensive tables (see Boastead and Hancock [18]). The latter two have not yet been incorporated in an analysis.

We are faced with several decisions with respect to this accounting. The first is the misuse of the energy value of products. The second is the distribution of capital energy utilized to produce a product. Third, there is the association of energy-loss values with waste products. Several possible trade-offs can be made with respect to energy-intensive versus capital-intensive production. The accounting must include a measure of the interconvertibility of energy forms. In chapter 6 on available energy, a measure is developed for comparing energy values of products. An attempt is also made in chapter 13 to connect this measure with environmental quality. In this chapter the accounting principles for applying this measure to decisions are developed. A framework has been outlined in the input-output method. Basic economics of operating cost versus initial capital costs are introduced in the next section to provide background for energy accounting.

Recently attempts have been made to use input-output techniques to analyze an economy from Boulding's point of view [2]. In these formulations [22–24], capital stock rather than the gross national product is taken as the measure of economic welfare. This means that personal consumption (final demand) and governmental expenditures are internal transactions. The output of the economy is then measured by (1) gross capital formation, (2) net inventory change, and (3) net exports. To these might be added expenditures for improved education, personal capital formation like housing, new governmental service facilities, and so forth.

The current search for a better economic-objective function than the present GNP is encouraging from an environmental viewpoint. Consumption per se is often wasteful. An objective function that measures progress by environmental inprovements, reduction of waste, and better resource utilization would be welcomed.

The input-output methods described here are extended in chapter 13 to factors of energy, economics, and environment in extensive interacting resource-processing systems.

Economics of Cost

The conventional cost analyses used to determine the operating point for a system often break down the cost into two categories: *Fixed Cost* and *variable cost*. The two are differentiated in that the variable cost changes with output, but fixed cost does not. This does not mean that the fixed cost does not change with time. The fixed cost may in fact change with time as taxes, rent, or interest change. The variable costs, however, change as we change the output desired from the system. These costs may include costs of raw materials, energy, labor, and so forth. The differentiation is only on the basis of output changes and not by category of material or labor input. The cost of electricity, maintenance, or heating, which do not depend on the output, are considered as fixed costs. It is important to note that fixed costs are not actually fixed, since they change with time, but are fixed in the sense that changing the output does not change the fixed costs. Mathematically, this may be simply expressed as the total costs C_T. $C_T = C_F + C_V$, with the fixed and variable costs differentiated by the relation to the output Y. Thus fixed costs are costs for which a change in the output Y produces no change in the total cost. Now,

$$\frac{dC_T}{dY} = \frac{dC_F}{dY} + \frac{dC_V}{dY}.$$

Then

$$\frac{dC_F}{dY} = 0,$$

and

$$\frac{dC_T}{dY} = \frac{dC_V}{dY}.$$

This change in cost with output is often termed the *marginal cost*. This marginal cost is used as a measure of the extent to which production should be increased. (An interesting discussion of marginal cost in the case of energy use to produce food is presented by Ruttan [14].) This division into fixed and variable costs must usually be discussed in terms of the cost per unit output. This may be expressed as

$$\bar{C}_T = \frac{C_T}{Y},$$

where Y is the total units of the output.

In a production process it is often assumed that an objective is to reduce the average cost of the product by altering the production rate until a minimum is reached. A representative situation is shown in figure 3–4. The

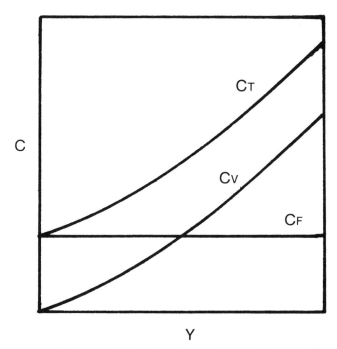

Figure 3–4. Classical Economic Cost (C) versus Output (Y) Behavior Defining Fixed Cost (C_F), Variable Cost (C_V), and Total Cost (C_T)

average cost or cost per unit of production for each is then approximately as shown in figure 3–5. This figure illustrates the classical Kelvin law behavior characteristic of situations wherein a system is limited in capacity. An optimum production rate requires that the average variable costs increase with output. This is characteristic of situations wherein an optimum is designed into the original system. As shown in the available-energy case in chapter 6, this is expected in situations in which operating energy costs are significant. It is important to note that this behavior does not represent the idea of *economy of scale*. To consider economy of scale, let us compare situations in which two alternative systems are available: one with high fixed cost and low operating cost, and the other with low fixed cost but high operating cost.

Cost comparison of these two alternative production systems is shown in figures 3–6 and 3–7. System A has a lower fixed cost than system B, but a higher variable cost. Initially, with a low production rate, the manufacturer who selects system A has an advantage over the one who selects B. Up to a production Y_e, choice A is preferable; for a B system in competition, the production required by demand may preclude B from gaining an advantage

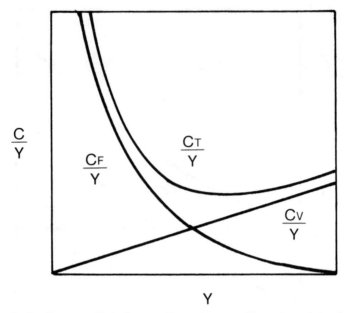

Figure 3–5. Cost per Unit Output Behavior as a Function of the Output in Classical Terms

before losing out in the competition. In a demand-limited economy, then, the choice of low-initial-investment, high-operating-cost systems is preferred. This can lead to a wasteful economy—for example, if the variable cost is primarily the cost of processing energy or if a resource cost is high because the processing system is inefficient, producing more waste. The choice of system A or B then depends on the output required. In this case, if the demand is too low system B may be overdesigned, with excess capacity. Similarly, system A may be underdesigned if demand increases with time above the minimum-cost design point.

An alternative sometimes available is to duplicate system A if demand increases. This is advantageous only if the demand increases in increments such that operation at minimum points for each is possible. Usually the proper first design will produce a minimum cost since the economy of scale is not applicable to duplicate production facilities and the minimum average of two A systems will not be lower than that of system A in any case, whereas the minimum of system B is expected to be lower.

In order to discuss the choice of system A or B, one must examine other factors, including the risk or uncertainty about:

1. demand and how it will change with time;
2. operation and cost of the actual system;

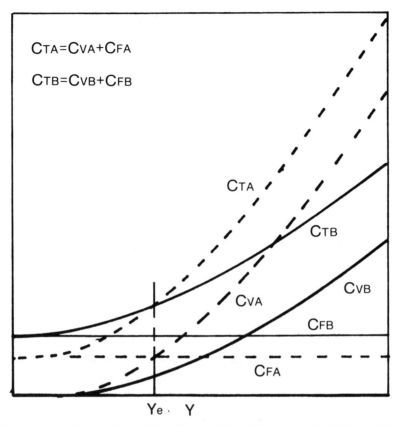

$$C_{TA} = C_{VA} + C_{FA}$$

$$C_{TB} = C_{VB} + C_{FB}$$

Figure 3–6. Comparison of the Costs of Two Systems with Different Fixed and Variable Costs

3. availability and cost of money as a function of time;
4. obsolescence of the system;
5. alternative investment possibilities;
6. competitor behavior.

All these factors must be included in an analysis before any decision is made. Since all require prediction of the future, all decisions are dynamic in nature. Static analysis is becoming increasingly obsolete. Some static-analysis techniques are useful, however, in estimating the coupling of market and cost factors and hence the inclusion of input-output techniques in the decision process.

Time is the independent variable that enters into most economic decision making. Consideration of future costs is the basis for most decisions that are

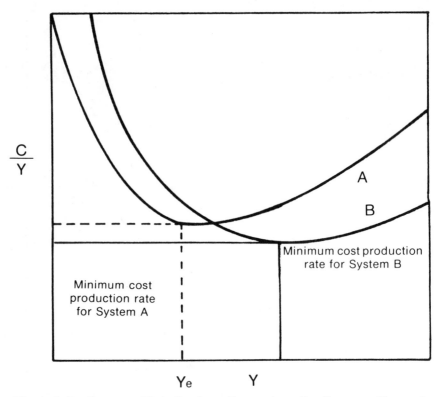

$\dfrac{C}{Y}$

A

B

Minimum cost production
rate for System B

Minimum cost
production rate
for System A

Y_e Y

Figure 3–7. Cost per Unit Product Comparison for Systems Shown in
Figure 3–6, Illustrating the Optimization in Terms of Fixed and
Variable Cost

termed rational. Neglecting to allow for future costs is, however, the most
prevalant error of the lay person's economic decisions. The reasons for this
may be summarized by the fact that money is a scarce commodity and that
most consumers have a multitude of needs or demands. Rather than attempt
to analyze consumer behavior as it is, let us try to project how it can be
improved through better economic analysis. Consumer behavior will be
characterized here as rational in the sense that, in the absence of information
on operating cost of a product or system, the consumer selects the product
with least first cost.

     Recently the U.S. Department of Commerce proposed that electrical
appliances be labeled as to the operating energy requirements and that
automobiles be labeled with their expected gasoline-consumption rate. The
purchaser is still left with the choice between higher-energy-use items and
least initial cost, with little real basis for comparison between kilowatts and

dollars. At equal price the decision can be made on a least-energy-consumption basis. But what is the decision in the case of a lower initial cost, but a higher energy or operating cost? To allow rational decisions, more than simply operating energy costs must be considered, including the expected use, the lifetime, the replacement cost, the inflation rate, the obsolescence rate, and so forth.

Economic analysis of luxury buying is not of concern in this chapter. We are attempting to analyze not luxury economics, but conservation and environmental economics. From the point of view of a consumer and a producer, we will seek least-cost solutions within the environmental constraints.

Time is related to cost in several ways. The first is the determination of the expected useful life of any new equipment to be purchased or old equipment currently in use. The capital cost must be distributed over a time horizon in order to compare the costs of several alternatives. Expected life is a point of uncertainty. The useful life of a system is also determined by some of the following factors:

1. deterioration;
2. changing cost of operation of the system:
   a. maintenance,
   b. resource-price change,
   c. efficient-disposal cost change;
3. obsolescence:
   1. competitive methods—technology,
   b. change in demands—sociology,
   c. new opportunities for money—economic,
   c. competitive advantage to moving operations or expansion;
4. inflation;
5. intensity of use;
6. availability of funds for replacement.

Most industries dealing with this problem use statistical techniques to determine the expected life of machines, based on repair costs with deterioration or use, and replace or repair machines on regular schedules. Similar decisions based on obsolescence and change to a new system are made in response to changing conditions and, in effect, usually shorten the expected life used for calculation procedures. If the expected life is shortened, this increases the capital cost compared with the operating cost. This often lends more justification to decisions weighted heavily toward least capital cost.

In a mobile society, where home occupancy by a single family averages three to five years, it makes sense economically to forgo the expense of extra insulation, storm or insulated windows, or more efficient heating systems.

Each successive family gains more economically by higher operating costs than if any one family were to install these improvements. Proper accounting by the purchasers at each stage could save both energy and money. But without this, all lose because each minimizes his separate cost. One may argue that if the value is there, it will appear in the price of the home. This is theoretically correct but inoperative in practice. The value of a house often depends more on its location, availability, and other intangible factors. Several factors are operating at present that could lead to a change in this situation. Computer listing of houses could be expanded to include energy-saving factors, and computer programs could be developed to allow buyers to evaluate a new house on a total cost basis including initial and operating costs. This could also be used by sellers to help them determine the value of their homes more realistically. Records are available from appraisals by tax agencies and could easily be expanded to include the required information.

An important question in economic thought, which appears to be little discussed in relation to production, is the cost of increasing production rate per capital cost with fixed capital investment, as opposed to the cost of decreased production rate per capital cost but with increased capital investment. Similarly, the cost of continuous twenty-four-hour operation at moderate rates versus the cost of high production rates for short times is an economic question that must be resolved. In energy-intensive industries this decision should be based on the available-energy analysis. In general, doubling a production rate, for example, is not accomplished by doubling the energy resources. Energy cost is not linearly related to production rate but must be examined through consideration of available-energy-dissipation processes.

Efficiency of Size: Economy of Scale

Economy of scale is a well-established economic concept indicating that it is less costly to produce a large quantity of goods in a single plant than to produce a similar quantity in several smaller plants. In process-equipment cost estimation, an often-used algorithm is the "six-tenths factor" used to estimate the cost of a projected unit when a unit cost of a different size is known [15]. The cost is related to the capacity as

$$Cost = (Capacity)^{0.6}.$$

The classical example of economy of scale is the spherical tank whose cost is proportional to the surface area times the thickness, and whose capacity is the volume. Then,

$$Cost = C*4\pi r^2 t,$$

and

$$\text{Capacity} = V = 4/3\pi r^3.$$

Then the cost in terms of the volume or capacity is

$$\text{Cost} = C^* \frac{4\pi t}{[(4/3)\pi]^{2/3}} V^{2/3} = KV^{2/3}.$$

This is not strictly correct, as the thickness must be increased with capacity to maintain structural integrity, and the construction cost per unit capacity decreases as the size increases. It is a good approximation, however; and it is generally less expensive to purchase one large tank than two small tanks with the same total capacity.

In the area of transportation, the frictional forces on a vehicle are proportional to the surface area, whereas the capacity is proportional to the volume. Thus the most efficient vehicle, in terms of cost per unit capacity, is a large one. This reasoning makes apparent the advantage of large oil tankers, long trains, and large aircraft. If we require these larger vehicles to travel at high speeds, this economy may be lost in the higher operating cost associated with speed.

In structural analysis there is a similar constraint. The weight of similar structures is proportional to the cube of their linear dimensions, but their strength is proportional to the structure's cross-sectional area or the square of its linear dimension. In a larger structure a greater percentage of its weight must be devoted to structure, and the less to the payload. This relationship is valid only for similar geometries and materials. The advance of metallurgical and structural technology has increased the strength-to-weight ratio in new construction. Examples are steel-cable bridges, filament-wound containers, and high-strength aluminum alloys used in composite structures.

If the available energy required to produce structural materials is proportional to the weight or volume, but the cost is proportional to the volume to the two-thirds power, then the cost per unit available energy is proportional to $V^{-(1/3)}$. For equivalent-strength structures the cost-to-strength ratio does not change with volume. Structures, however, are not built to equivalent strength; small structures are generally overdesigned. Hence the economy of large structures is still a good approximation, but not on analytic grounds. The cost to build a single large structure in terms of both energy and dollars is less than that for several small structures.

In heating buildings the energy required is proportional to the surface area. The cost to heat a building per unit volume is then inversely proportional to the volume to the one-third power. Large apartment buildings are thus more efficient from the point of view of energy cost than equivalent-volume individual homes. Hence economy of scale is operative in both the construction and the heating of buildings.

In the generation of electricity, the economy of scale in terms of weight per unit of power-output capacity is shown in table 3–7 [16]. If capital available energy is proportional to the weight, then the energy savings are clear. Economy of scale is applicable to power plants in both dollar and available-energy terms.

An important factor in the trend toward larger units is the longer time required for construction and the reduced availability in case of downtime maintenance compared with small units. With interests rates over 15 percent and a five-year construction time for a large electric plant, the interest can account for half of the total capital cost before operation begins. Thus high interest rates can mediate for the construction of smaller, less efficient units with short construction times.

The operating-energy economy of large scale may be summarized by the fact that both the frictional loss and the heat-transfer losses are proportional to the area, whereas the power output is proportional to the volume of the system. Magnetohydrodynamic-generation advocates have argued for years that a prototype full-scale operation is required to demonstrate the improved efficiency over that obtained in small-scale research units.

In nuclear-fission-power reactors, neutron-absorption control is crucial to the sustained chain reaction. Neutrons released in a fission must be moderated before reaction with other fissionable material. The probability of reacting with other material is proportional to the volume of the reactor. Neutrons escape from the reactor in proportion to the surface area. Although graphite or other materials may be used to partially reflect neutrons back into the reactor, the net result is that the larger the reactor, the more efficient it is

Table 3–7
Variation in Specific Weight of Turbogenerators and Some Major Components with Unit Size
(*Specific weight, tons/MW*)

	Unit Size (MW)	
Component	*200*	*500*
1. High-pressure cylinder	0.23	0.13
2. Generator stator	0.8	0.4
3. Complete generator	1.2	0.87
4. Complete turbogenerator	4.0	3.1

Source: J.L. Gray, "Economies of Scale—Generation of Electric Power," in *Economics of Technical Change*, ed. E.M. Hugh-Jones (New York: A.M. Kelly, 1969), p. 102. Reprinted with permission.

in converting nuclear fuel to thermal energy. Again, an increased efficiency is achieved with an increase in size.

The economy of scale in industrial production systems does not seem to be predictable in terms of energy use per unit product. One reason for this is the small fraction of the cost of many products that is due to energy use. Replacement of hand labor by automatic equipment will increase energy use per unit product, but if the energy use is low, the change may be insignificant. The availability of inexpensive energy has been an incentive for the use of more energy in many processes. Automation with increased energy use is often used to improve control of product quality as well. Many energy-conversion devices have energy economy-of-scale attributes, but initial cost

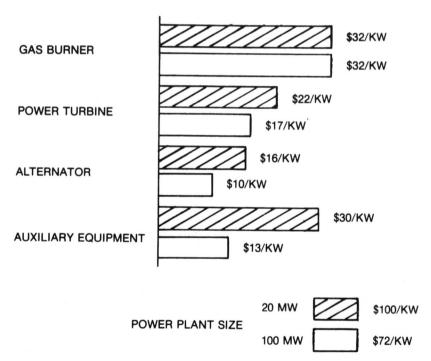

Source: From J. Verheus, "Energy from Petroleum Fuels: Technical and Economic Aspects of Pollution Control," in *Proceedings of the Ninth World Energy Conference*, sec. 3.1-9 (1974).

Figure 3–8. Specific Costs in Dollars per Kilowatt

factors often militate against their use. With appropriate accounting methods and increased energy costs, this may change in the future.

In gas-turbine-power peaking systems the economy of scale [21] is illustrated in figure 3–8, which compares the relative costs as a function of unit size. As this breakdown shows, the major savings obtained are not in the capital cost of the principal chemical-thermal-mechanical conversion unit. The scale savings occur principally in the supporting services, including building and control costs and such factors as operation and management activities. Dollar savings are incurred because the support services and software are nearly equivalent, whereas the generating unit's specific costs remain nearly the same. The scale saving is not due to technological factors. An available-energy analysis supports this decision for increased scale. Two crucial factors are not delineated in this figure. The first is the saving due to increased fuel efficiency and the second the delivery cost incurred with larger units, which would have higher transmission costs attributable to their supplying a larger geographical area.

Economy of scale is also manifest, though in a slightly different manner, in the chemical and materials-processing industries. The purity of a product can be improved by using a larger reactor vessel. Impurities and waste products are often present in the reactor or piping in proportion to the surface area. This may be because impurities are associated with the properties of the surfaces as a result of corrosion or other reactions, or because of separation processes such as settling, grinding, filtering, and so forth. The area-to-volume ratio in a large reactor is smaller than in a small reactor. Large reactors are thus more economical than smaller ones on a physical basis. Typical examples are polymer-chemical processing, drug manufacture, and aluminum smelting. In the polymer industry it is sometimes difficult to obtain a satisfactory product in pilot operations. Scaling up to production operation can result in a better product in these cases.

The major emphasis on small-scale technology, as outlined by Lovins [19] and Commoner [20], is that there is a limit to the economic size of units. The argument is based on the fact that as size increases, the costs of the following increase to limit the economic size of production units:

1. transportation costs;
2. downtime repair or reconditioning costs;
3. control costs;
4. administrative and political costs;
5. environmental costs;
6. cost of a failure;
7. social and psychological costs.

In addition to these arguments, the argument is made that the scale and type of supply system should be matched to those of the use. This has been termed *appropriate technology* [19].

References

1. Von Karman, T., with Edson, L. *The Wind and Beyond.* Boston: Little, Brown, 1967. Emphasis added.
2. Boulding, K.E. "The Economics of the Coming Space-Ship Earth." In *Sixth Resources for the Future Forum*, ed. H. Jarrett. Baltimore, Md.: Johns Hopkins University Press, 1966.
3. Defoe, D. *Robinson Crusoe.* New York: Grosset and Dunlap, 1944.
4. Leontief, W. "The Structure of the U.S. Economy." *Sci. Amer.* 212(1965):4.
5. Victor, P. *Pollution: Economy and Environment.* Toronto: University of Toronto Press, 1972.
6. Berry, R.S., and Fels, M. "The Energy Cost of Automobiles." *Science and Public Affairs* 29(1973):11.
7. Ballard, C., and Herendeen, R. "Energy Costs of Goods and Services." Document no. 140, Center for Advanced Computation, University of Illinois, Urbana, 1975.
8. Leontief, W. "Environmental Repercussions and the Economic Structure: An Input-Output Approach." *Rev. Econ. Stat.* 52(1970):262.
9. Ayres, R.U., and Kneese, A.V. "Production, Consumption and Externalities." *Amer. Econ. Rev.* 59(1969):282.
10. McGauhey, D.H., et al. "Final Report, Economic Evaluation of Water." Sanitary Engineering Research Laboratory, University of California, November 1969.
11. Moor, R.E., and Garrison, G.E. "The Economics of Enervated Energy." *Market Economics.* Chicago: A.G Becker, 1973.
12. Bureau of Mines Report of Investigations. "The Electrowining of Aluminum from Aluminum Chloride." Report no. 7353, U.S. Department of Interior, 1970.
13. Environmental Protection Agency. "Summary Report on the Paper and Allied Products Industry." Industrial Waste Studies Program, 1972.
14. Ruttan, V.W. "Food Production and the Energy Crisis: A Comment." *Science* 187(1975):560.
15. Williams, R. "Six-Tenths Factor Aids in Approximating Costs." In *Cost Engineering in the Process Industries*, ed. C.H. Chilton. New York: McGraw-Hill, 1960.

16. Gray, J.L. "Economies of Scale—Generation of Electric Power." In *Economics of Technical Change*, ed. E.M. Hugh-Jones. New York: A.M. Kelly, 1969.

17. Grainger, L. "Conversion of Solid Fuels into Other Energy Forms." In *Proceedings of the Ninth World Energy Conference*, sec. 3.1–9, 1974.

18. Boastead, I., and Hancock, G.F. "Handbook of Industrial Energy Analysis." New York: Wiley, 1979.

19. Lovins, A.B. *Soft Energy Paths*. New York: Harper and Row, 1977.

20. Commoner, B. *The Poverty of Power*. New York: Knopf, 1976.

21. Verheus, J. "Energy from Petroleum Fuels: Technical and Economic Aspects of Pollution Control." In *Proceedings of the Ninth World Energy Conference*, sec. 3.1–9, 1974.

22. Daley, H.E. *Steady State Economics*. San Francisco: W.H. Freeman, 1977.

23. Kendrick, J.W. *The Formation and Stocks of Total Capital*. New York: National Bureau of Economic Research, 1976.

24. Costanza, R. "Embodied Energy and Economic Valuation." *Science* 210(1980):1291.

4 Thermodynamic Preliminaries

In the whole domain of science, no field of cultivation is poorer in such labor-saving devices than that of human history, yet Man as a form of energy is in most need of getting a firm footing on the law of thermodynamics.
—Henry Adams, *Letter to American Teachers of History* (1910),
The Degradation of the Democratic Dogma

Thermodynamic Introduction

This chapter outlines the principles of chemical thermodynamics required for understanding the basics of available energy. The important relations are derived to allow one to calculate thermodynamic properties of a system. Since the inherent nature of the problems to be confronted involves changes in the composition of systems, the analysis will begin with open systems; closed systems will be considered as a special case. The concepts of energy, entropy, equilibrium, state, and process will be discussed in detail. The basic relations between state variables and composition will be examined. The classification of heat and work in terms of quality of energy will be specified. Emphasis will be placed on the operational methods for determining properties from tabulated data. The use of the concept of available energy in the context of Gibbs free energy is outlined.

The concept of energy is basic to the science of thermodynamics [1-3]. Energy must first be referred to as the energy of *something*. It is a property of something. We commonly refer to the kinetic energy of a particle, or the potential energy of a rock on a cliff, or the chemical energy of the battery, or the energy in an electrical field. From this point of view energy is a property of a portion of space or of a particle or collection of particles. In thermodynamics it is convenient to talk about a thermodynamic system that represents a collection of particles occupying a portion of space at some time. The energy of these particles represents their energy relative to some reference state. We will take energy to mean the sum of kinetic and interaction energies of the particles of a system.

The concept of conservation of energy is useful for solving practical problems just as are the concepts of conservation of mass, momentum, or charge. Often this is useful as an accounting mechanism based on an idea that the inputs to a system must equal its outputs. However, the importance of balance must be tempered with the reality that transformations occur that change the character of energy. This is the basis of thermodynamics, which is the science that deals with the change in the character of energy during transactions. Common descriptions—energy dissipation, energy loss, heat generation, friction, and so forth—express this change. The idea that energy is more useful in certain forms than in others is clear from the economic value of energy. There are two particular values that are easily described. First, high-energy-density systems have higher value than dilute-energy systems; that is, energy per unit mass is an important and useful measure of its usefulness (see table 4–1). For example, a small stream of water with large kinetic energy is more valuable for hydroelectric power than is a slow-moving, large system. This value is in the economics of conversion. Second, the more uncertain one is about the distribution of energy in a system, the more difficult it is to utilize that energy effectively. This uncertainty can be related to the randomness of a system or to its reliability (in the case of winds or other natural events like lightning). It may refer to microscopic uncertainty, which is more familiar to physicists.

In this chapter thermodynamic principles will be developed, following the work of Gibbs [1,2] and emphasizing the approach of Jaynes [3,4] as presented in technical form by Tribus [5].

The energy of a control volume is sometimes useful for accounting procedures. It will be used in this description when needed, but energy associated with particles rather than with space will be the rule unless otherwise stated.

The energy associated with a system of particles will be assumed to be a function of (1) the external coordinates—the volume, surface, area, length, electrical field, gravitational field, and any other constraints on the particles of the system; (2) the kinds of particles—the molecular, compositional, or

Table 4–1
Energy Density

Energy Form	Storage Medium	Energy Density
		$\times 10^6$ J/Kg
Linear kinetic	100 mph	0.001
Gravitational	100 m elevation	0.0098
Rotational kinetic	5,000 rpm	0.053
Magnetic field	2 WB/m^2 in air	0.0016 J/liter
Electric field	6.5×10^8 V/m mylar	0.006 J/liter
Chemical		
Battery	Lead–lead dioxide	0.16
Heat of reaction	Gasoline	44
Heat of recombination	$H + H \rightarrow H_2$	216
Nuclear		
Ionization	$V_i = 10$ ev	990 J/kg mole
Fission	U_{235}	83,000 J/kg
Fusion	$D + T \rightarrow He_4 + 17.6$ Mev	340,000 J/kg

other category of particles of concern (normally referred to as the number of each distinguishable particle being examined); and (3) the energy states available to each of these particles (in quantum mechanics these are energy states, but in biology they may be energy configurations or arrangements). The definition of states depends on our knowledge of energy states available to particles. In the absence of specific understanding of these, a classification must be established and an ordering delineated.

Thermodynamic Analysis

States are defined as possible energy states of a system. These states may or may not be occupied by particles. The energy states will be designated as

$$\varepsilon_1 = E(X_k, N_a, N_b, N_c, \cdots N_m).$$

where X_k are the external coordinates.

N_a, N_b, $\cdots$ are the number of particles of each kind.

i is the description of a set of discrete energy levels.

The number of possible states is often unlimited, but for purposes of clarity we will number possible states according to energy value from lowest $(i = 1)$ to highest (n), corresponding to low energy to high energy.

For a given system of N particles, the number of particles in each possible state is designated N_i. We do not know N_i in any real case, so we will designate $P_i = N_i/N$ and call P_i the probability that the ith state is occupied. Then the statement

$$\sum_{i=1}^{n} P_i = 1 \tag{4.1}$$

If we can calculate P_i, this means we would have the most probable number in this state, not the actual number. If the particles have energy, then summing (over all states) the number of particles in each state N_i times the energy of that state ε_i will give the total energy. Instead of discussing the total energy, we will multiply the state energy ε_i by the probability of its being occupied and sum over all possible states, calling this

$$\sum_{i=1}^{n} P_i \varepsilon_i = \langle \varepsilon \rangle, \tag{4.2}$$

the expected value of the energy or expected energy per particle.

To help illustrate this concept, let us refer to figure 4–1(a), which shows a box with shelves (representing states). The box contains marbles representing particles. Now let us consider shaking the box (that is, imparting a certain energy to the particles). The particles will distribute themselves on the shelves with a distribution that will depend on the weight of each particle. Lighter particles are more likely to reach a high level than are heavier ones. The new configuration and system of particles may be as shown in figure 4–1(b). The energy state is represented by the gravitational potential energy of all particles. The potential energy of each shelf is different, and the number of particles in each level may be different.

Next, the idea that particles may be different is introduced. Some may weigh more than others, or they may stick together to form new particles. We must consider that we know such things as the number of particles of each we put into a system; but after the system changes, we may not know their form, in that molecular configurations may change with chemical reactions, phase changes, and the like. If we put 1 mole of oxygen A in a container with 2 moles of hydrogen B, we know the number of molecules of each, or the

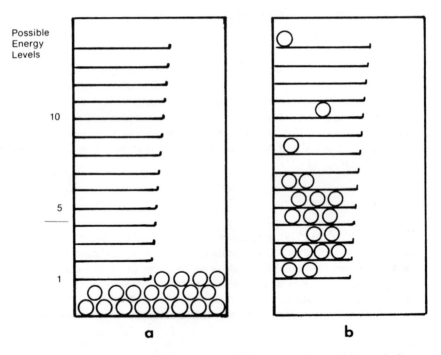

Note: The box on the left is an initial state with marbles representing particles on the bottom. Shelves represent energy levels. The box on the right shows the energy state after shaking energy is supplied.

Figure 4–1. An Illustration of the Distribution of Particles in Energy States

number of particles N_A and N_B. If we look at the energy levels and see which are occupied at a particular instant, we would see N_{A1} in level 1, N_{A2} in level 2, and so on; N_{B1} in level 1, N_{B2} in level 2, and so on; and some H_2O C with say N_{C1} in level 1 and N_{C2} in level 2, and so on.

The important idea at this point is that we do not know the exact number of molecules of each type once we introduce the notion that they may react to form other particles (in this case, for example, H_2O or H_2O_2. We can then only talk about an expected number of each particle type. Note that the number of different types of particles that may be formed cannot be determined a priori but must be found from experimental evidence.

Using the expected value in this case means

$$\langle N_A \rangle = \sum_{i=1}^{n} P_i(N_{Ai}), \qquad (4.3a)$$

$$\langle N_B \rangle = \sum_{i=1}^{n} P_i(N_{Bi}), \qquad (4.3b)$$

$$\langle N_C \rangle = \sum_{i=1}^{n} P_i(N_{Ci}). \qquad (4.3c)$$

N_{Ai} then means the number of A particles in the energy state i. Then

$$N = \langle N_A \rangle + \langle N_B \rangle + \langle N_C \rangle \cdots, \qquad (4.4)$$

where N is the total number of particles present. Of course, we do not know the number of particles of type A in state i any more than we know the total number N_i in that state.

Note that the $\langle \ \rangle$ quantities are macrostate descriptions and the subscripted i quantities are microstate descriptions. We only observe the macrostates and are primarily interested in the transactions between macrostates of systems. The microstates enter when we try to determine how macrostates change in the transactions between macrostates.

In thermodynamic problems involving chemical reactions, one question is: What is the equilibrium chemical composition as a result of the interactions of particles of different types? One is interested in determining the temperature and pressure required in a reactor to produce a certain product. Thermodynamics is formulated to determine this equilibrium condition. If one mixes chemicals based on these rules and finds that the product is not as expected, then it is argued that the equilibrium has not been reached. Time is required for completion of reactions. An understanding of chemical kinetics [6,7] is required to deal with the time dependence of reactions.

In this formulation the particles may be considered as chemical species.

Consideration of components rather than species will be introduced later to simplify calculation procedures. As an example of the differences, in the stoichiometric-reaction equation,

$$2H_2 + O_2 \rightleftharpoons 2H_2O.$$

three "species," H_2, O_2, and H_2O, are assumed to be present. The uncertainty is in part about how many particles of each of these species are present. If, additionally, other chemical compounds or ions like H_2O_2 or H^+ are under consideration, then the species list will change. The requirement of dealing with stoichiometric equations is necessary in this initial formulation. These chemical stoichiometric equations provide the additional information required. In some formulations components rather than species can be considered. *Components* refers to the amounts of the independent chemical entities, in this case hydrogen and oxygen. These amounts are fixed and independent. There is no uncertainty about the number of atoms of each. In the species case the number of each species is generally dependent on the number of each other species. This dependence is indicated by the constraints of the stochiometric equations.

The number of components can usually be determined as the number of species minus the number of independent chemical equations that relate to the species. In more general terms, components represent the species required to complete a given reaction.

From the analytical point of view, the component description would be preferred, as it would reduce the calculations required. From the practical point of view, however, it is important to determine the chemical composition in terms of useful combinations of the components, not of the components themselves. A chemist or chemical engineer is interested in the chemicals produced by a reaction, not in the number of components. We often know these components (they are what is added to the batch or mixture). The uncertainty concerns the resultant products and how fast the reaction will proceed.

In order to proceed in the design of a process, the engineer must next examine those factors or parameters under his control. From practical experience, once the initial chemical input is specified, he finds there are certain other parameters such as pressure and temperature that are useful and will influence the outcome. Thermodynamics deals with how these parameters influence the distribution of the energy available among the many possible configurations. We would like a theory that will specify the most probable distribution of energy. Given the energy $\langle E \rangle$ available to a system of N particles, the composition and state of the system will then be determined. This most probable state cannot be guaranteed; in fact, the system may never be in this state. The worth of the model is, then, in whether this prediction is close enough for practical purposes. If after calculating the most probable state of a system, one does the experiment and finds that the system is not in this state, one can draw several possible conclusions:

1. It takes a longer time to reach this most probable state, and the use of a catalyst may be required in the process.
2. There may be a flow of energy through the system that alters the state of the system. (This latter is dealt with in the next chapter, on irreversible processes.) These are particularly important processes in dealing with biological and ecological phenomena.
3. There may be several states with nearly equal probabilities, and the system may be observably fluctuating among these. A fluctuation is expected and can be calculated to help confirm this if necessary.
4. The system may truly be found in a state that is far from the most probable. As will be shown, this possibility decreases in likelihood as the number of particles increases. It represents an opportunity in rare cases, but it is usually not consistent enough for practical purposes.

This last point deserves some further comment here with respect to design and science. Any process has a certain variability. An experiment that is repeated rarely if ever produces identical results. "Good" experiments are those that are repeatable—"repeatable" only in the sense that the observables recorded are the same. A good experimentalist tries to minimize the fluctuations by proper design. Similarly, a chemical engineer designing a process attempts to produce a consistent product. Money is spent on process control to reduce variability. The same control that reduces the fluctuation in, say, the strength of an elastomer may also reduce the maximum strength of the plastic obtainable. In medicine it is most important with present techniques to use drugs that are consistent in their effects rather than to seek drugs that are specifically effective for certain individuals. The tailoring of drugs to specific individuals is not pursued because of its expense. Similarly, a drug manufacturer cannot risk introducing a drug that may prove extremely beneficial to some individuals yet fatal to others unless there is some control on its use. Thus consistency or ability to produce most probable reactions is an important characteristic of present technology.

There are exceptions, and there are people—gamblers—who capitalize on these exceptions. They may win by playing the exceptions. The problem in gambling lies in knowing—consistently—when the exceptions occur. Katz [8] has summarized this by stating that it is often not the probabilities that are unknown, but that the uncertainty of the order of the events is often the problem. In economics [9] a similar thought occurs in the concept of risk. Decision theory [10] is based on knowing the probabilities of each outcome once one knows the decisions. The problem is the uncertainty of the order of the outcomes.

In the case of equilibrium theromodynamics, if the number of particles is sufficiently large and the number of energy levels available is unlimited, then the outcome, given enough time, is satisfactorily predicted by making the guess that the most probable distribution will be realized.

Maximizing Uncertainty

The previous discussion involves making a prediction about the outcome of how particles are distributed among energy levels, given that you know the total energy and the number of each type of particle. What are the rules for prediction that will serve as a guide for this prediction? These have been outlined by Shannon [11], as follows:

1. Uncertainty about an event with possible outcomes a_i should depend upon the probabilities p_i associated with these outcomes.
2. The uncertainty about independent events taken together should be the sum of the uncertainties about each event taken separately.
3. Uncertainty should be a monotonic increasing function of the number of outcomes.

These desiderata lead one to the measure of uncertainty [3]:

$$S = k \sum_{i=1}^{n} P_i \ln P_i. \qquad (4.5)$$

This measure, termed the entropy when k is Boltzmann's constant, in thermodynamics expresses the measure of uncertainty about how the particles are distributed among the energy levels (n is the number of energy levels available). The problem then is: What is the best guess you can make about this distribution, given the information available? Tribus [5] has shown that the least biased guess is that which maximizes one's uncertainty. To make any other guess presupposes the availability of other information that has not been included. For example, if you already know that the primary reaction expected is

$$2H_2 + O_2 \rightleftharpoons 2H_2O$$

under the conditions of interest, rather than the reaction $H_2 + O_2 \rightleftharpoons H_2O_2$, then the number of different species of concern will be different. This is handled by using the methods of uncertainty maximization with different constraints.

The solution to this maximization problem has been shown by Jaynes [3], using the methods of Lagrange multipliers, to be

$$P_i = e^{-\Omega - \beta\varepsilon_i - \sum_c \alpha_c n_{ci}}.$$ (4.6)

The sum $\sum_c$ is over all expected species. Substituting this into equation 4.1 gives the Lagrange multiplier (partition function)

$$\Omega = \ln \sum_i^n e^{-\beta\varepsilon_i - \sum_c \alpha_c n_{ci}}$$ (4.7)

(α_c is the Planck potential of the species c.) Then from equations 4.2 and 4.3 there follows:

$$\langle \varepsilon \rangle = -\frac{\partial\Omega}{\partial\beta},$$ (4.8)

$$\langle n_c \rangle = -\frac{\partial\Omega}{\partial\alpha_c}.$$ (4.9)

At this point another concept must be introduced from mechanics to help relate the Ω function to measurable system properties. This concept is the relation of force and displacement to energy. In classical mechanics a change in the energy of a system is related to force and the displacement of the system from its initial state as

$$E_k = -\int F_k \, dX_k.$$ (4.10)

The force may also be defined in terms of energy as

$$F_k \equiv -\frac{\partial E_k}{\partial X_k}.$$ (4.11)

In the context of energy levels, this can be written as

$$F_{k,i} = -\frac{\partial\varepsilon_i}{\partial X_k}.$$ (4.12)

This implies that if you know how the energy levels depend on the displacements, then you can determine the forces. A simple example is the energy change of a system with volume in response to pressure forces.

Similar relations hold between surface tension and area, electrical forces and charge displacement, and so forth. The classical field theories define these forces and displacements.

Using this definition of force, with equation 4.7 and the concept of expected force,

$$\langle F_k \rangle = \sum_{i=1}^{n} P_i F_{ki} = \sum_{i=1}^{n} P_i \ \frac{\partial \varepsilon_i}{\partial X_k}, \qquad (4.13)$$

gives

$$\langle F_k \rangle = \frac{1}{\beta} \ \frac{\partial \Omega}{\partial X_k}. \qquad (4.14)$$

The expected force $\langle F_k \rangle$ is the force implied by observation of the displacement X_k of a system. Computation of Ω at this point would require knowledge of the parameter β.

The parameter β is shown in statistical mechanics to be equivalent to $1/kT$ (T is the equilibrium thermodynamic temperature) for the case of a perfect gas [3]. The important point to consider for equilibrium systems is that if two systems with different βs are allowed to exchange energy, but not particles, equilibrium will be reached when both βs are equal ("zeroth law of thermodynamics").

From equation 4.6 it is seen that if β is positive, then the larger value of β, the more probable it is that the lower energy levels will be occupied relative to the higher energy levels. If energy levels are ordered so that the lowest energy level corresponds to a lower subscript, then β is interpreted as shown in figure 4-2. If $\beta = 0$, then all energy levels are equally likely to be populated. (This corresponds to $T \to \infty$ —that is, very high temperature as required in a fusion reactor. As the temperature is lowered or β increased, the lower energy levels are more likely to be occupied. If β is negative, corresponding to negative temperatures, then it is more likely that higher-level energy states are occupied. (Using this concept, semiconductor or laser materials are sometimes referred to as *negative-temperature devices*.) This interpretation of a negative-temperature device is incorrect for two reasons. The first is that there are only a few higher-energy levels that are more highly occupied in these devices. Second, and the more important, these are non-equilibrium states. This concept is discussed further in chapter 17 with respect to biological systems, which also exhibit similar characteristics.

Differentiation of equation 4.8 with respect to α_c and equation 4.9 with respect to β shows an equivalence,

$$\frac{\partial \langle \varepsilon \rangle}{\partial \alpha_c} = \frac{\partial \langle n_c \rangle}{\partial \beta}. \qquad (4.15)$$

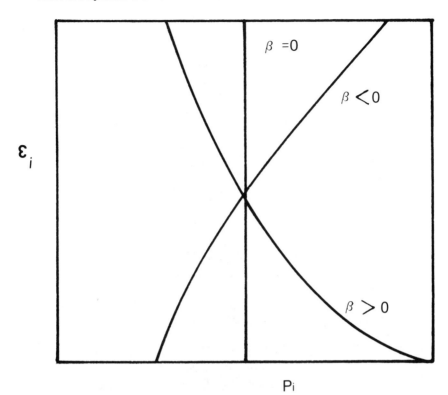

Figure 4–2. Interpretation of β in Terms of the Distribution of Particles among Possible Energy States ε_i

This indicates that changes in energy with respect to composition are related to changes in composition with β (or temperature).

Other useful conclusions may be reached by considering the variance in the energy and composition around the expected values. It can be shown [5] that the variance

$$\sigma^2 \langle \varepsilon \rangle = \frac{\partial^2 \Omega}{\partial \beta^2} ,$$
(4.16)

and

$$\sigma^2 \langle n_c \rangle = \frac{\partial^2 \Omega}{\partial \alpha_c^2} .$$
(4.17)

From equations 4.8 and 4.9, together with equations 4.16 and 4.17, one can show

$$\frac{\partial \langle \varepsilon \rangle}{\partial \beta} = \sigma^2 \langle \varepsilon \rangle,$$
(4.18)

$$\frac{\partial \langle n_c \rangle}{\partial \alpha_c} = \sigma^2 \langle n_c \rangle.$$
(4.19)

Thus, since the variance is always positive, the energy decreases as β increases and the computed number of particles of type c decreases as α_c increases. In the exchange of energy and matter between two systems, energy will flow from the system at lower β to the system at higher β. Similarly, c particles will flow from a system at low α_c to a system with a higher α_c. The first is a heat-transfer process and the second a mass-transfer process. Thus α_c plays a role in mass-transfer processes similar to that of β in heat-transfer processes.

In classical thermodynamics the concept of a perfect-gas thermomenter is used to determine the temperature of a system by allowing the thermometer to come to equilibrium with the system through energy exchange alone. In chemical-equilibrium measurement a similar concept of a semipermeable membrane separating a known substance from an unknown substance is used to measure the chemical potential. This analogous relationship can be demonstrated by consideration of entropy changes in a system.

From equations 4.5 and 4.6, using the expected-value operation, the entropy can be expressed as

$$S = k\left(\Omega + \beta \langle \varepsilon \rangle + \sum_c \alpha_c \langle n_c \rangle \right).$$
(4.20)

Differentiation gives

$$dS = k\left(d\Omega + \beta d\langle \varepsilon \rangle + \langle \varepsilon \rangle d\beta + \sum_c \alpha_c d\langle n_c \rangle + \sum_c \langle n_c \rangle d\alpha_c \right).$$

(4.21)

Now, using the fact that from equation 4.7

$$d\Omega = \frac{\partial \Omega}{\partial \beta} d\beta + \sum_k \frac{\partial \Omega}{\partial X_k} dX_k + \sum_c \frac{\partial \Omega}{\alpha_c} d\alpha_c,$$
(4.22)

and the identities from equations 4.8, 4.9, and 4.10,

$$d\Omega = -\langle \varepsilon \rangle \, d\beta + \beta_k \langle F_x \rangle \, dX_k - \sum_c \langle n_c \rangle \, d\alpha_c ,$$

gives

$$dS = k \left(\beta d\langle \varepsilon \rangle + \beta \sum_k \langle F_k \rangle \, dX_k + \sum_c \alpha_c d\langle n_c \rangle \right) . \quad (4.23)$$

If β is equated to $1/kT$, this can be written as

$$TdS = d\langle \varepsilon \rangle + \sum_k \langle F_k \rangle \, dX_k + \sum_c \langle \alpha_c \rangle k \, Td\langle n_c \rangle . \quad (4.24)$$

α_c is then observed to be equivalent to $-\mu_c/kT$, where μ_c is the familiar Gibbs chemical potential. Equation 4.24 can then be written in the usual themodynamic form:

$$TdS = d\langle \varepsilon \rangle + \sum_k \langle F_k \rangle dX_k - \sum_c \mu_c d\langle n_c \rangle . \quad (4.25)$$

Equation 4.25 is more useful for chemists dealing with constant-temperature reactions, in which case the chemical potential acts as the equilibrium measure. In systems with temperature changes occurring, the α_c or Planck-potential formulation is more appropriate for establishing equilibrium conditions. Only in constant-temperature cases is the chemical-potential interpretation useful. This is particularly important in nonequilibrium thermodynamics, where μ_c is identified with a "force" analogous to the forces associated with $\langle F_k \rangle$.

A further clarification of the role of Ω may be obtained from consideration of equation 4.20 using $\beta = 1/kT$. Then,

$$\Omega = - \frac{\langle \varepsilon \rangle - TS - \sum_c \mu_c \langle n_c \rangle}{kT} \quad (4.26)$$

is a nondimensional free-energy representation. This equation will be useful later in formulating thermodynamic-nonequilibrium functions.

Returning to equation 4.25, let us integrate this for the special case in which all intensive variables are constant. This corresponds to the addition of more particles in the same states as those already present. This integration gives

$$TS = \langle \varepsilon \rangle + \sum_k \langle F_k \rangle X_k - \sum_c \mu_c \langle n_c \rangle . \quad (4.27)$$

Combining equations 4.26 and 4.27 gives

$$\Omega = \frac{\Sigma \langle F_k \rangle X_k}{kT}.$$

For the special case in which only pressure focus ($\langle F_k \rangle = P$) and volume ($X_k = V$) changes are of concern, this reduces to

$$\Omega = \frac{PV}{kT}.$$

The probability of each energy-level occupation then becomes, from equation 4.6,

$$P_i = e^{-\frac{PV}{kT} - \frac{\varepsilon_i}{kT} + \Sigma_c \frac{\mu_c}{kT} n_{ci}} \tag{4.28}$$

This expression provides an important connection between entropy and available energy, which is discussed in detail in chapter 6.

Equation 4.26 for the special case of pressure forces and volume changes may be rearranged to

$$\Sigma \mu_c \langle n_c \rangle = \langle \varepsilon \rangle - TS + \langle F_k \rangle X_k.$$

if only the energy states change and the probabilities remain the same, then

$$\Sigma \mu_c \langle n_c \rangle = \langle \varepsilon \rangle - TS + PV.$$

The function $\Sigma \mu_c \langle n_c \rangle$ is called the *Gibbs potential* or the *Gibbs free energy* and is used extensively in chemical thermodynamics, particularly in establishing equilibrium and stability criteria. Thus

$$G \equiv \Sigma \mu_c \langle n_c \rangle = \langle \varepsilon \rangle - TS + PV. \tag{4.29}$$

In classical thermodynamics a further useful function, *enthalpy*, defined as $H \equiv \langle \varepsilon \rangle + PV$, is introduced, giving

$$G = H - TS. \tag{4.30}$$

The enthalpy is a useful function in steady-flow processes for accounting purposes. Since it combines two quantities of a different nature, it contributes to difficulties in understanding, particularly with respect to the availability of this energy for useful purposes.

In thermodynamics it has proved useful to classify energy exchanges between a system and its environment into two categories: work (W) and heat (Q). Energy transactions that result in a change in the energy of a system are stated in energy-conservation terms as

$$d\langle \varepsilon \rangle = dQ - dW. \qquad (4.31)$$

The usual differentiation is that, of these two energy exchanges, heat is an interchange of energy as a result of a temperature (or β) difference between two systems. This definition becomes difficult in nonequilibrium situations and in the case of frictional losses. Clarification of these exchanges can be aided by substitution of equation 4.31 into equation 4.25:

$$TdS = dQ + \left(\Sigma \langle F_k \rangle dX_k - dW \right) - \Sigma \mu_c d \langle n_c \rangle \qquad (4.32)$$

In this form the combination of the work term with the force-displacement term illustrates that the difference between the expected change in energy due to the force displacement and the actual work is the actual contribution to the entropy change. This difference is the *frictional loss* and contributes to an increase in uncertainty about how the energy is distributed. Eventually this uncertainty increase may be due to a heat exchange or a dissipation of energy. The important point is that this energy transfer may also end up as strain energy in a material, chemical changes, or some other energy storage. As long as we do not have the means or interest to use this energy, it is a loss. The assumption that it is a frictional loss is in part an expression of the low value of this energy.

Equation 4.32 also helps clarify the concept of *reversible work*. Reversible work is expressed mathematically as

$$dW = \Sigma \langle F_k \rangle dX_k.$$

The reversible work done by the system is equal to the expected forces times the diplacements that occur. In real situations (with friction) the actual force is less than the expected force. As a result, the actual work done is less than the expected or reversible work. Consider the simple example shown in figure 4–3 of a mass M_1 colliding with a mass M_2 on a pendulum. From mechanics one can compute the height $\langle h \rangle$ that the mass M_2 should be raised by this collision. In the actual experiment the mass M_2 only rises to a height h'. The difference is usually attributed to frictional losses. In reality this energy may be in vibrational or strain energy of the two masses. This energy loss thus depends on how closely we examine the masses. Our uncertainty has increased as to where this energy is or how to use this energy for a useful

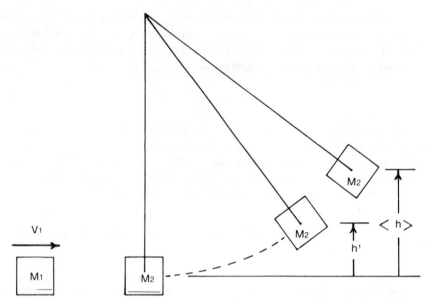

Figure 4–3. Illustration of the Actual and Expected Work Transferred in the Collision of a Mass M_1 with a Mass M_2 Suspended as a Pendulum

purpose. The loss thus depends on our objectives and our ability to utilize the changed energy forms.

A further description of the pendulum and impacting mass can be illustrated as follows. Consider first the system as the impacting object. The expected work that the mass M_1 is calculated to provide in a reversible impact is $M_2 g \langle h \rangle$. The actual work provided to M_2 is $M_2 g h'$. The difference is

$$TdS_1 = M_2 g \langle h \rangle - M_2 g h'.$$

The entropy increase in M_1 is then positive and represents uncertainty about the energy difference between that expected to have been done and that actually done.

Consider next the pendulum as the system. The actual work done on the pendulum is $M_2 g \langle h \rangle$; of this work only the energy $M_2 g h'$ is reversibly done. The difference is in some other form—vibration, rotation, or the like. Only the energy $M_2 g h'$ is available for work from the pendulum, as shown:

$$TdS_2 = M_2 g \langle h \rangle - M_2 g h'.$$

Taking into account in this formulation the sign convention that work done on the system is negative, then in both cases the entropy increase is positive!

The first law of thermodynamics, represented by equation 4.31, defines heat and work and includes the hypothesis that energy is conserved. The second law of thermodynamics hypothesizes that the entropy of an isolated system either increases or remains the same. This second law in information theory implies that the uncertainty about the distribution of energy among a set of particles either increases or remains the same if no external forces or energy exchanges occur. From equation 4.32 it is obvious that the entropy of a general system can decrease if the thermal energy (dQ) exchange is out of the system. If this decreases the system temperature, the distribution of the remaining energy is more skewed than originally. The combined work and expected-energy-change terms are either zero or positive. The chemical contribution produces a positive entropy change if the species chemical potential α_c decreases, analogous with the β change in a thermal exchange. Generalizations concerning changes in the number of species present as a result of chemical reactions are not possible with the information currently at hand.

Conclusions about the Gibbs free energy are useful at this point because of its extensive use in chemical problems. Differentiation of equation 4.29 gives

$$dG = d\langle \varepsilon \rangle - TdS - SdT + PdV + VdP. \qquad (4.33)$$

If a process proceeds at constant temperature and pressure (a good approximation for many chemical-reaction processes), this reduces to

$$dG = d\langle \varepsilon \rangle - TdS + PdV. \qquad (4.34)$$

using equation 4.31, this can be written as

$$dG = dQ + (PdV - dW) - TdS. \qquad (4.35)$$

In this form the Gibbs-free-energy change in a thermodynamic system that is adiabatically isolated $(dQ = 0)$ is

$$dG = (PdV - dW) - TdS.$$

The difference between PdV and dW is the network in process. The change in the Gibbs free energy when the entropy change is zero represents the maximum energy convertible to work in a constant temperature and pressure process.

From the perspective of the loss of capability of a system to do useful work, if a system changes from one stable state to another at constant P and T with no useful effect, then

$$dG = -TdS.$$

The change in the Gibbs free energy then represents the irreversible change in the system. An increase in entropy represents a decrease in the Gibbs free energy of a chemical system.

In chapter 6 the concept of available energy is discussed in detail. Available energy represents the potential of a system to produce thermodynamic work. The Gibbs free energy represents this potential for processes that are carried out at a constant pressure and temperature.

A chemical reactor ordinarily is not used to produce useful work or for heating purposes. Heat flux to or from a reactor is often used for process control and is an important engineering consideration. In an isolated chemical reactor, if a spontaneous chemical process requires that dS be positive or zero, then the change in the Gibbs free energy is required to be either negative or zero. Equilibrium in chemical reactions or the direction of spontaneous chemical reactions is determined by evaluation of the Gibbs free energy of the species involved in a reaction. In the reaction of the combustion of ammonia at standard pressure and temperature,

$$NH_3 + O_2 \rightleftharpoons \tfrac{1}{2}N_2O + \tfrac{3}{2}H_2O.$$

From standard tables of free energy of formation ΔG_f° [17] at 1 atm. and 25°C,

	ΔG_f° kcal/mole
NH_3	−3.97
N_2O	24.76
H_2O	−54.67
O_2	0

ΔG° of the reactants is −3.97 kcal/mole and ΔG° of the products is −65.60 kcal/mole. The Gibbs free energy of the products is lower than that of the reactants. A spontaneous reaction is then expected to produce product species H_2O and N_2O. This procedure is usually formalized to taking the difference between the Gibbs free energy of the products and expected reactants as

$$\Delta G = G_{\text{products}} - G_{\text{reactants}}.$$

If this is negative, the reaction will proceed in the direction assumed. If it is positive, the reaction will proceed in the opposite direction toward equilibrium. There is no guarantee that the reaction will proceed at an observable rate, or even at all. For example, if NH_3 and O_2 are mixed and allowed to react, there is no assurance that the reaction assumed will occur. Depending on the proportions of ammonia and oxygen present, two other reactions are also possible:

$$NH_3 + \frac{7}{4}O_2 \rightleftharpoons NO_2 + \frac{3}{2}H_2O \qquad \Delta G° = -65.59 \text{ kcal/mole},$$

or

$$NH_3 + \frac{5}{4}O_2 \rightleftharpoons NO + \frac{3}{2}H_2O \qquad \Delta G° = -57.26 \text{ kcal/mole}.$$

In reality only the last two reactions occur, and then only extremely slowly, except at high temperatures. The magnitude of the calculated Gibbs-free-energy difference for hypothesized reactions is not indicative of which reaction will actually occur.

Each reaction results in a decrease in the Gibbs free energy and, as a result, if it occurs at constant temperature and pressure, this amount of energy must be removed from the system. If no attempt is made to use this energy in a work process, it is assumed that there is a thermal-energy loss. This is termed an *exothermic reaction*. If the system is insulated, reaction kinetics generally cause the reaction to speed up as the temperature increases. Another alternative in this type of reaction is that the energy may be utilized to drive a simultaneous *endothermic* (heat-absorbing) *reaction*.

Caution must be used in generalizing this to all reactions. The change in Gibbs free energy consists of several parts: a change in G may occur because of a change in the distribution of the energy among bond energies, or because of a change in the random kinetic energies. The first may result in new available-energy levels without changing the thermal state. This would appear as a change in the TS term without affecting the thermal-energy state.

Tribus [12] has noted that, in the case of a monatomic gas, heat and work may be distinguished by the fact that the expected energy is a sum of products;

$$\langle \varepsilon \rangle = \Sigma P_i \varepsilon_i .$$

A change in the expected energy may occur because of a change in the probabilities P_i or a change in the energy levels ε_i. If only the energy levels are changed, but the probabilities remain the same, then the entropy,

$$S = -k\Sigma P_i \ln P_i ,$$

can remain constant. In a monatomic gas this is a *work* process. The other alternative—the energy levels remaining the same but the probabilities changing—is a *heat* process. In most actual energy exchanges both P_i and ε_i change, and it is difficult to separate the two effects on this basis. In chemical processes the change in bond energies usually results in important entropy changes. In many combustion processes the bond energy is released as the principal energy source.

Another useful way to look at the Gibbs-free-energy change is observed by rewriting equation 4.34 as

$$\frac{dG}{T} = \frac{d\langle \varepsilon \rangle + PdV}{T} - dS. \qquad (4.36)$$

Using the definition of enthalpy as $\langle \varepsilon \rangle + PV$, this becomes

$$\frac{dG}{T} = \frac{dH - VdP}{T} - dS.$$

For a constant-pressure chemical reaction this becomes

$$\frac{dG}{T} = \frac{dH}{T} - dS. \qquad (4.37)$$

The left-hand dG/T represents the total entropy change of the chemical system *and* its surroundings. The term dH/T represents the change in entropy of the surroundings due to an exchange of energy with the system. The term dS of course represents the change in the entropy of the system itself.

In combustion processes (with heat loss) the main source of the energy is from bond energy released, which occured as ΔH for a reaction called the *heat of combustion*. This energy is lost to the environment. It includes an expansion against the environment as well as a heat exchange.

Tabulated heats of combustion for chemical reactions are made on the basis that the reactants and the products are at the same temperature before reaction as they are at the completion of the reaction.

Transport-Process Thermodynamics

The concept of energy transfers implies that energy can be transferred from one system to another if the state of the two systems is different. *Different* implies either that the possible quantum states for each are different or that the distribution of particles over the quantum states is different. Mixing processes and heat-transfer processes are typical of the consequences of two systems interacting that have different energy distributions or allowable energy states. In the case of monatomic gases there is the possibility of transfers in which the energy states alone change, or the probabilities of energy-state occupation change. Since the expected energy of a system is

$$\langle \varepsilon \rangle = \Sigma P_i \varepsilon_i,$$

change may be represented as

$$d\langle \varepsilon \rangle = \Sigma \varepsilon_i dP_i + \Sigma P_i d\varepsilon_i.$$

From the entropy

$$S = -k\Sigma P_i \ln P_i,$$

if only the energy states change and the probabilities remain the same, then the entropy would not change. In a monatomic gas the changing of the ε_i can occur independent of P_i. These are the reversible changes associated with PdV work processes. This is a consequence of the internal energy being a function of the temperature alone. In this case

$$TdS = d\langle \varepsilon \rangle + PdV.$$

Thus

$$\Sigma P_i d\varepsilon_i = -PdV.$$

The interaction of the two systems may be considered the interaction of their probability distributions. In fact, if there is no interaction of these distributions, no transfer occurs. A simple illustration of this is the process that occurs when a discontinuity is produced in a fluid in a shock tube. If the temperature is different on the two sides of the discontinuity, then the distribution of particle velocities is different on each side. As shown in figure 4-4, if there is a Maxwellian velocity distribution on both sides, then the slower moving particles in 2 will interact with the high-speed particles in 1. This interaction causes a smearing out of the discontinuity, so that as time passes, the disturbance becomes more spread out. A similar view may be taken of the interaction of fluid layers moving with different average velocities. Momentum exchange occurs between layers because the distribution of velocities (or momentum of particles) in each layer is different. This exchange tends to bring the probability distributions into equilibrium and is opposed by the shear-force energy input.

Diffusion may similarly be considered as the interaction of different probability distributions that at equilibrium become uniform. If particles are

Shock Front

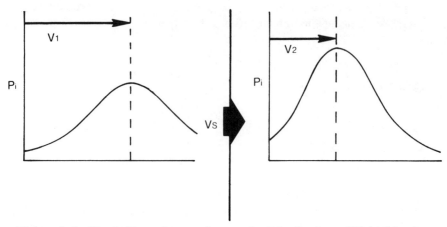

Figure 4–4. Shock Front Interaction on the Distribution of Velocities in a
Moving Fluid, Illustrating the Irreversibilities in Such a Process

supplied and removed continuously, as in precipitation, sedimentation, or
chemical reaction, then there can exist a steady-state situation with a spatial
variation in the probability distribution among energy states.

Phenomenological relations between gradients of state variables and the
energy interchange have been established. These classical relations are as
follows [133].

$$\text{Diffusion of mass} \quad j = -D \text{ grad } C_i$$

$$\text{Charge flow} \quad i = \frac{1}{\rho} \text{ grad } E$$

$$\text{Heat flow} \quad q = -k \text{ grad } T$$

These relationships are approximations applicable only when gradients
are small or the fluxes are low. In effect, they are most useful for systems
near equilibrium. In transport problems the deviation from these relationships is often accounted for by assuming the proportionality constants are
functions of the state variables. An example is the assumption that the
thermal conductivity is a function of the temperature, or that the electrical
resistance of an electrical element depends on the temperature. In transport
sciences it is recognized that convection effects introduce deviations because
of bulk transport. Typical examples are the introducttion of mass-transfer
coefficients $j = M(c_1 - c_2)$, heat-transfer coefficients $q = h(T_1 - T_2)$, and
electrochemical-transfer coefficients $i = c(E_1 - E_2)$.

In all these systems entropy creation occurs because of the increase in the uncertainty about how the energy is distributed among the particles. The available energy decreases when the separate systems with different available energies interact. Interactions lead to an increase in the random motion of molecules and appear as a temperature increase above the environmental temperature. Interaction with the environment then removes energy, and a further entropy and available-energy change occurs.

When the flux of particles occurs from one energy state to a lower energy state, energy must be given up to particles in the new region. The simplest example is the flow of particles from a tube into the environment, as shown in figure 4–5. The part of the energy in the flow of air in the tube is directed kinetic energy. This is convertible in theory to mechanical motion by an impulse turbine with an efficiency not limited by the second law.

When the stream interacts with particles in the environment, this energy is exchanged and becomes randomly distributed. We often associate this random kinetic energy with thermal energy and say that the available energy has been *dissipated*. In steady flow the rate of particles times the available energy per particle. In the situation illustrated the available energy is kinetic energy $1/2V^2$ per unit mass, and the flux is the mass rate of flow in. The rate of dissipation is $mV^2/2$.

In the situation of a charge moving in a resistor, as shown in figure 4–6, the flow of charge depends on the gradient of the potential. The energy of this

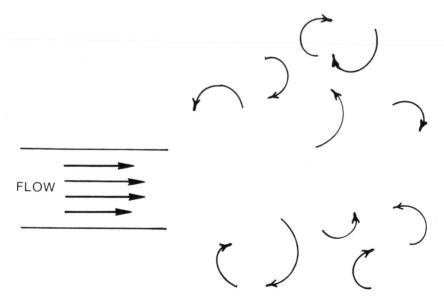

FLOW

Figure 4–5. Illustration of *Dissipation* when Directed Air Flows from a Tube into the Atmosphere

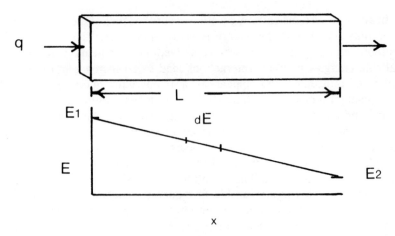

Figure 4–6. Electrical Resistor Charge Flow–Potential Relationship

charge decreases as it moves from a high potential E to a low potential $(E - dE)$. The available energy of the charge at the lower potential $(E - dE)$ is less, so that the rate of decrease in available energy is idE. In more familiar terms, this is the current times the voltage difference. At an equilibrium steady state this rate of energy loss is equal to the heat loss to the environment, and the system temperature is higher than that of the environment.

In irreversible thermodynamics [14–16] the generalized force is designated X_k and the flux J_k. The rate of available energy loss is $A = J_k X_k = T\dot{\sigma}_c$. The entropy generated by the system is then

$$\dot{\sigma}_c = \frac{J_k X_k}{T} .$$

Calculation of the power dissipated in the resistor may help clarify the role of the gradient and the force flux relationship. The flow of charge at any point along the resistor is

$$i = \frac{A}{\rho} \frac{dE}{dX}$$

where A is the cross-sectional area and ρ the resistivity. The power dissipated in any differential element dx is then idE. Since the change in the potential $dE = \text{grad } E \cdot dx$, then substituting that

$$\text{grad } E = i\rho/A,$$

then

$$\int_0^L i^2 \frac{\rho}{A} dx = i^2 \frac{\rho L}{A} = i^2 R$$

is the power dissipation in the resistor.

In the cases near equilibrium where the flux J_k is proportional to the force X_k as $J_k = L_{kk} X_k$, the power dissipation can be written as

$$T\dot{\sigma}_c = L_{kk} X_k^2 = \frac{J_k^2}{L_{kk}}$$

In the electrical case L_{kk} corresponds to the conductance or inverse of the resistance.

Similar expressions for the dissipation can be obtained for laminar flow in a pipe, where the generalized force is the pressure and the flux is the mass flow.

Then

$$\dot{m} = C_1 \Delta P$$

and

$$T\dot{\sigma}_c = \dot{m} \Delta P = \frac{\dot{m}^2}{C_1} = C_1 (\Delta P)^2$$

The dissipation rate increases at higher mass-flow rates in the form of turbulent motion, and the pressure difference required to produce a given flow is increased. At high flow rates the linear relation between the flow rate and the pressure difference is not valid. The product of the pressure difference times the mass-flow still gives the power dissipation, however, and is the power required from a pump to provide a given mass flow. A systematic attempt to incorporate this in a thermodynamic analysis is detailed in chapter 8.

Each of these dissipation processes may be considered as an energy-conversion system in which there is conversion of energy from one form to another (that is, thermal). Observing a resistor as a block box, one might conclude that the heat flux out was "caused" by the electrical-potential difference or the current flow. A coupling can be hypothesized between the two energy fluxes. This cannot generally be reversed, in that a temperature difference or heat flow into the resistor will not produce an equivalent current flow. A device can be constructed wherein this reverse process can occur but with certain second-law restrictions. Thermoelectric and thermionic devices and heat engines work on this principle. An available-energy analysis will indicate the degree to which such an inverse flow is possible.

An energy-conversion system can be considered as a coupled system in which the output flux is proportional to the associated output force and also to the input force as

$$J_1 = L_{11} X_1 + L_{12} X_2.$$

Similarly, the input flux is coupled as

$$J_2 = L_{21} X_1 + L_{22} X_2.$$

This coupling is discussed in detail in chapter 7 in relation to limits of energy-conversion processes [16].

In electrical-power elements and in chemical reactors thermal interaction with the environment must be controlled to reduce available-energy losses. Maintenance of a system in a desired state, that is, keeping electronic components from melting, keeping chemical reactants in liquid form, optimizing chemical yields, and so forth involves difficult technological energy-exchange problems. Often the difficulty in transferring this thermal energy to the environment limits the production rate in electromechanical-power or chemical-conversion systems. Similarly, frictional heating processes associated with mechanical motion or fluid motion may also limit the rate at which these processes can be carried out.

Cooling technology used to maintain a system within a suitable temperature range usually does not extract any useful effect from the available energy, which is then completely dissipated. In most cooling systems available energy is expended by pumps, blowers, fans, and the like to accomplish this heat transfer. The cooling system for an internal-combustion engine reduces the net output of the engine. A double loss is thus incurred: both the heat loss and the energy expended to produce the heat loss. In winter the thermal energy is sometimes used for heating purposes, producing a useful effect.

The development of total energy systems for providing the energy requirements of shopping centers and industrial complexes utilizes the thermal energy produced by the electrical-generation system to provide heating or, through an absorption system, cooling of the building. Available-energy analysis can assist in the evaluation of the overall effectiveness of alternative designs for these systems (see chapter 5).

References

1. Gibbs, J.W. *Collected Works,* vol. II, *Elementary Principles of Statistical Mechanics,* 1901. New Haven: Yale University Press, 1948.

2. Gibbs, J.W. *Collected Works*, vol. I, *On the Equilibrium of Heterogeneous Substances*, 1878. New Haven: Yale University Press, 1948.

3. Jaynes, E.T. "Information Theory and Statistical Mechanics." *Phys. Rev.* 108(1957):171.

4. Jaynes, E.T. "Information Theory and Statistical Mechanics." *Phys. Rev.* 106(1957):620.

5. Tribus, M. *Thermostatics and Thermodynamics*. Princeton, N.J.: Van Nostrand, 1961.

6. Glasstone, S.; Laidler, K.J.; and Eyring, H. *Theory of Rate Processes*. New York: McGraw-Hill, 1941.

7. Moelwyn-Hughes, E.A. *Physical Chemistry*. New York: Pergamon Press, 1957.

8. Katz, A. *Principles of Statistical Mechanics*. San Francisco: W.H. Freeman, 1967.

9. Raiffia, H., and Schlaifer, R. *Applied Statistical Decision Theory*. Cambridge, Mass.: Harvard Business School, 1961.

10. Tribus, M. *Rational Descriptions, Decisions and Designs*. New York: Pergamon Press, 1969.

11. Shannon, C. "A Mathematical Theory of Communication." *Bell System Tech. J.* 27(1948):623.

12. Tribus, M. "Information Theory as the Basis for Thermostatics and Thermodynamics." *J. Appl. Mech.* 128(1961):1.

13. Bird, R.E.; Stewart, W.E.; and Lightfoot, E.N. *Transport Phenomena*. New York: Wiley, 1960.

14. DeGroot, S.R., and Mazur, P. *Non-Equilibrium Thermodynamics*. Amsterdam: North-Holland, 1962.

15. Prigogine, I. *An Introduction to Thermodynamics of Irreversible Processes*. New York: Interscience, 1968.

16. Onsager, L. "Reciprocal Relations in Irreversible Processes." *Phys. Rev.* 37(1931):407.

17. Reid, R.C., and Sherwood, J.K. *Properties of Gases and Liquids,* 2nd ed. New York: McGraw-Hill, 1966.

5

Chemical Potential of Substances

We live in a world of phases, so much more astonishing than the explosion of rockets, that we cannot, unless we are Gibbs or Watt, stop every moment to ask what becomes of the salt we put in our soup or the water we boil in our teapot and we are apt to remain stupidly stolid when a bulb bursts into a tulip or a worm turns into a butterfly.
     —Henry Adams, *The Phase Rule Applied to History* (1908)

Chemical Equilibrium

The solution to many pollution problems has been to dilute the effluent by mixing it with air or water until a "safe" concentration has been reached. This solution recognizes the chemical potential of high concentrations of chemical species relative to the organic-chemical potential of biological organisms (especially people). Faced with the waste products of a process, chemical manufacturers recognize that in most cases these waste products have chemical thermodynamic potential relative to the environment. The incineration of waste products to produce heat is one way to use this potential for useful purposes.

The dilution process reduces the potential for producing useful thermodynamic work or heat. Pollution, as stated earlier, can sometimes be considered a resource out of place.

The determination of the chemical potential of a chemical compound is almost exclusively done experimentally. The role of thermodynamics is to delineate how this chemical potential changes with dilution and with the thermodynamic variables of temperature and pressure.

The science of chemical thermodynamics is based on the work of J. Willard Gibbs [1]. The chemical potential utilized in most chemical engineering is referred to as the Gibbs free energy. In this chapter the procedures for calculating the Gibbs free energy of substances are outlined. More detailed treatments are available in the chemical-thermodynamic literature [2–5].

This section summarizes some important relations in chemical thermodynamics that are useful in calculating thermodynamic functions and establishing relations between variables.

Consider a reversible reaction of hydrogen and oxygen that might be approached in a fuel cell:

101

$$H_2 + \tfrac{1}{2}O_2 \rightleftharpoons H_2O.$$

Equation 4.32 is written in the form

$$\Sigma\mu_c \, d\langle n_c\rangle = (dQ - T dS) + \left(\Sigma\langle F_k\rangle \, dX_k - dW \right). \quad (5.1)$$

If the reaction is reversible, the terms in parentheses on the right are zero (that is, $dQ/T = dS$ and $\Sigma\langle F_k\rangle \, dX_k = dW$). In this case,

$$\mu_{H_2} d\langle n_{H_2}\rangle + \mu_{O_2} d\langle n_{O_2}\rangle + \mu_{H_2O} d\langle n_{H_2O}\rangle = 0. \quad (5.2)$$

The chemical equation says that 1 mole of hydrogen and $\tfrac{1}{2}$ mole of oxygen combine to form 1 mole of water. Thus if $d\langle n_{H_2}\rangle = -1$ mole, then $d\langle n_{O_2}\rangle = -\tfrac{1}{2}$ mole, and $d\langle n_{H_2O}\rangle = +1$ mole. Then substituting into equation 5.2 gives

$$\mu_{H_2O} = \mu_{H_2} + \tfrac{1}{2}\mu_{O_2}.$$

Equivalently, in terms of Planck potentials,

$$\alpha_{H_2O} = \alpha_{H_2} + \tfrac{1}{2}\alpha_{O_2}.$$

since equilibrium requires temperature equilibrium as well.

These types of relations are important in determining chemical potentials of components in chemical mixtures. This equality relationship of chemical potentials also defines equilibrium between different phases of a single substance. An example of this is the chemical-potential equilibrium beween liquid water and water vapor that occurs in evaporation-condensation processes. This relation is a basic principle that allows the use of semipermeable membranes to measure chemical potentials of species in solution. The semipermeable membrane permits equilibrium of specific molecules to be obtained without permitting other molecules to interfere. An important application of semipermeable membranes is the separtion of salts in solution. In these cases the solution is permeable in the membrane, and the salts are not. Reverse-osmosis desalination technology is based on this principle.

The calculation of the Gibbs free energy and chemical potentials for mixtures and solutions is important in establishing the useful energy that can be obtained in a reaction. It is often necessary to calculate these properties at temperatures and pressures different from the standard state for which extensive tables are available. To illustrate how these calculations are done, let us consider the formation of N_2O_4, an oxide of nitrogen formed in a combustion chamber:

$$N_2 + 2O_2 \rightleftharpoons N_2O_4.$$

At standard temperature (25°C) and pressure (1 atm.) all these components are in gaseous form and may be treated as perfect gases. Extension to solutions is usually made by analogy, with introduction of the concepts of affinity and activity. From thermodynamic tables [6–11] the change in the Gibbs free energy of formation of each species can be found. Then the difference between the free energy of products and reactants at the standard state can be calculated:

Species	ΔG_f° kcal/mole
N_2	0
O_2	0
N_2O_4	2.2

$\Delta G_f^{\circ} = 0$ for oxygen and nitrogen because this is the reference equilibrium state for these species. Now,

$$\Delta G = G_{\text{Products}} - G_{\text{Reactants}} = 2.2 \text{ kcal/mole.}$$

This indicates that at standard conditions this mixture will not react to form the nitrogen oxide indicated unless energy is added.

Since many reactions require pressures and temperatures different from standard conditions, the next question is: At what temperature and pressure would this reaction proceed to form the nitrogen oxide indicated? To answer this question let us examine how the Gibbs free energy changes with temperature and pressure. Differentiating equation 4.30,

$$G = H - TS,$$

gives

$$dG = dH - TdS - SdT. \tag{5.3}$$

For a pure species, G is a function of the pressure and temperature, so that

$$dG = \left.\frac{\partial G}{\partial P}\right)_T dP + \left.\frac{\partial G}{\partial T}\right)_P dT. \tag{5.4}$$

We are interested in

$$\left.\frac{\partial G}{\partial P}\right)_T.$$

Now Gibbs equation 4.25 for a pure substance with the force-displacement of pressure-volume is

$$dE = TdS - PdV,$$

(5.5)

which may be written using the enthalpy function as

$$dH - VdP = TdS.$$

(5.6)

Then equation 5.3 may be written in terms of differential changes in P and T as

$$dG = VdP - SdT.$$

(5.7)

Comparing equations 5.4 and 5.7 gives

$$\left. \frac{\partial G}{\partial P} \right)_T = V$$

(5.8)

and

$$\left. \frac{\partial G}{\partial T} \right)_P = -S.$$

(5.9)

Integrating first for a constant temperature process,

$$\int dG = \int VdP.$$

If the species behaves like a perfect gas, then $V = nRT/P$. Substituting gives

$$\int dG = \int \frac{nRT}{P} dP.$$

Integrating at constant temperature gives

$$\Delta G = nRT \ln (P_2/P_1).$$

(5.10)

If the initial pressure P_1 is 1 atm. and the final pressure P_2 is expressed in atmospheres, this reduces to

$$\Delta G = nRT \ln P_2,$$

(5.11)

or, in terms of the change from the reference formation state,

$$\Delta G = \Delta G_f^{\circ} + nRT \ln P_2. \tag{5.12}$$

For the nitrogen oxide reaction,

$$\Delta G_T = 2.2 + (1)RT \ln P_2 - (2)RT \ln P_2 - (1)RT \ln P_2.$$

Then

$$\Delta G_T = 2.2 - 2RT \ln P_2 \text{ kcal/mole}$$

for each mole of product formed. Equilibrium between reactants and products is present when $\Delta G_T = 0$, which occurs when the pressure is raised to $P_2 \backsim 6$ atm. If the pressure is increased above this value, the reaction shifts toward the product side. The pressure may then be used to control the direction in which a reaction will proceed.

In a combustion process in an internal-combustion engine, the higher the compression ratio, the more nitrogen oxides are produced. Reducing the compression ratio is one way to reduce the oxides of nitrogen in the exhaust. This reduction in the oxides of nitrogen is an important factor in improving air quality.

From examination of the chemical equation, one observes that with reaction, a decrease in the total number of moles occurs. Increasing the pressure moves the reaction in the direction of a smaller number of moles, or a smaller volume. In combustion of hydrocarbons like propane, for example,

$$C_2H_6 + 7/2 \, O_2 \rightleftharpoons 2CO_2 + 3H_2O,$$

increasing the pressure will move the equilibrium to the left with less complete combustion, since there are 4.5 moles formed on the left when 5 moles on the right react.

The temperature effect on the direction of the reaction is found by using the relation

$$\left. \frac{\partial G}{\partial T} \right)_P = -S$$

and the Gibbs free energy relation $G = H - TS$ to give

$$\left. \frac{\partial G}{\partial T} \right)_P = \frac{G - H}{T}. \tag{5.13}$$

This can be rearranged to give the relation

$$\left(\frac{\partial (G/T)}{\partial T} \right)_P = -H/T^2. \qquad (5.14)$$

If the species act as perfect gases, $H = c_p T$, and

$$\left(\frac{\partial (G/T)}{\partial T} \right)_P = -c_p/T.$$

Integrating gives

$$G/T - G^0/T_0 = -c_p \ln (T/T_0).$$

The Gibbs free energy at the temperature T relative to that at the temperature T_0 is then

$$G = T(G^0/T_0 - c_p \ln (T/T_0)). \qquad (5.15)$$

In the nitrogen oxide example, the temperature at which this reaction will proceed can now be found. If ideal diatomic gases are assumed, then $c_p \sim (7/2)R$, and $R \sim 2$ cal/mole-K.
Then

$$G_{N_2O_4} = T \left[\frac{2.2 \times 10^3}{298} - 7 \ln (T/298) \right],$$

$$G_{N_2} = -T(7) \ln (T/298),$$

$$G_{O_2} = -2T(7) \ln (T/298).$$

Setting $\Delta G = 0$ and solving for the temperature gives $T = 176$ K. If the temperature is greater than this, the reaction will proceed toward the formation of the oxide. Note that for the actual temperature calculation specific-heat variations with temperature from experimental data will be required in the integration of equation 5.14.

In the combustion of hydrocarbon fuels the energy release ΔG in the reaction usually increases with pressure. This is one of the reasons for carrying out combustion of fuels at high pressures in a combustion chamber. It is not the only reason, however; increased energy density of the machine and increased efficiency are also thermodynamic factors. For many fuels, reactions proceed more rapidly at high temperatures and pressures, another advantage of compression and heating of reactants.

The determination of changes in Gibbs free energy with temperature and pressure for liquids, solids, and complex gases requires either spectroscopic data on energy levels or experimental thermodynamic data. For more details on calculation procedures, particularly numerical methods, consult reference [12].

Pressure dependence will be discussed further here because of its extension to concentration dependence. Concentration dependence is particularly important for evaluating the value of energy and mineral resources. It is also basic to the determination of the feasibility of recovery or recycle of waste products.

The concept of partial pressure is used extensively in solution chemistry and combustion analysis as well as in wastewater-conditioning work. The concept is based on the observation that in many solutions and mixtures the separate species behave thermodynamically as if they were occupying the total volume of the solution or mixture. For a mixture of n_a moles of one species and n_b of a second, which act as perfect gases in a volume V, the equation state is

$$n = n_a + n_b = PV/RT. \qquad (5.16)$$

The pressure contribution of each species to the total pressure is proportional to the number of moles of each species. This is usually designated by the partial pressure, so that

$$P_a + P_b = P, \qquad (5.17)$$

with

$$n_a + n_b = n \qquad (5.18)$$

and

$$V_a + V_b = V. \qquad (5.19)$$

Since $n = PV/RT$ and $n_a = P_aV/RT$, $n_b = P_bV/RT$. With the mole fractions defined as $x_a \equiv n_a/n$ and $x_b \equiv n_b/n$, consideration of equations 5.18 and 5.19 indicates that partial pressures are equivalent to mole fractions if the reference pressure is 1 atm.

Partial pressures have significance in solubility problems involving gases in liquid solutions. The partial pressures determine the quantity of gases that can be dissolved in a liquid, an important factor in the aeration of bacterial or microbial cultures. The design of sewage-treatment systems and biochemical reactors requires knowledge of the effect of pressure on the solubility of oxygen and nitrogen in water. Solubility differences of gases can be used for separation processes as well. For example, the solubility of oxygen in water is higher than that of hydrogen. Therefore, the concentration of hydrogen in a hydrogen-oxygen gas mixture can be increased by bubbling the gas mixture through water.

If two species are mixed at constant temperature and pressure, a loss in available energy is incurred. To examine this loss, let us return to the Gibbs function,

$$G = H - TS.$$

Differentiating gives

$$dG = dH - TdS - SdT,$$

but

$$dH = dE + d(PV).$$

If the species act as perfect gases,

$$dE = c_v dT,$$

and $d(PV) = 0$ if $PV = nRT$ and T is constant. For a constant pressure and temperature mixing, then,

$$dG = TdS. \tag{5.20}$$

From the Gibbs equation,

$$dE = TdS - PdV,$$

but

$$dE = 0.$$

Therefore,

$$TdS = PdV \quad \text{and} \quad dG = -PdV. \tag{5.21}$$

Substituting that

$$P = \frac{nRT}{V},$$

the change in the Gibbs free energy for n_a moles of A when mixed with n_b moles of B becomes

$$\int dG_A = -\int_{V_a}^{V_{TOT}} \frac{(n_a RT)}{V} \, dV.$$

(5.22)

The volume of the mixture is the sum of the volumes of each of the species before mixing $V_a + V_b = V_{TOT}$. In terms of moles, then,

$$\frac{n_a RT}{P} + \frac{n_b RT}{P} = \frac{(n_a + n_b) RT}{P},$$

so

$$\frac{V_a}{V_{TOT}} = \frac{n_a}{n_{TOT}}.$$

Integrating to find the change in the Gibbs free energy due to the mixing of A gives

$$G_A = G_A^\circ + n_a RT \ln \frac{n_a}{n_{TOT}}.$$

(5.23)

Similarly, for species B the Gibbs free energy change is

$$G_B = G_B^\circ + n_b RT \ln \frac{n_b}{n_{TOT}}$$

(5.24)

The total Gibbs free energy change on mixing is then

$$\Delta G_{TOT} = \Delta G_A + \Delta G_B = RT \left[n_a \ln \frac{n_a}{n} + n_b \ln \frac{n_b}{n} \right].$$

(5.25)

Since the mole fraction of each species is less than one, the Gibbs free energy of each species is decreased on mixing, and the opportunity to do work has been lost.

Extensions to ideal mixtures of liquids can also be made. In cases of multiple-species mixtures, the total Gibbs free energy can be expressed as

$$\Delta G = RT \Sigma_i n_i \ln \left(\frac{n_i}{\Sigma_k n_k} \right).$$

(5.26)

Substitutions of partial pressures or mole fractions may be made where these variables are more convenient.

We will be interested in chemical processes involving extraction of minerals or coupounds from solutions. Examples are given in chapters 11 and 12 for extraction of contaminants and minerals from water solutions.

Solution-chemical Gibbs free energy changes are usually expressed in mole fractions, with x_i, the concentration of solute in the solvent, as

$$G_i - G_i^\circ = \Delta G_i = RT \ln x_i \qquad (5.27)$$

The use of ammonia as a fertilizer in recent years has added to lake eutrophication because of the evaporation of ammonia to the air and the runoff. Both the ammonia gas and concentrated ammonia solutions are absorbed by nearby waters. Let us take the example of an ammonia solution in water. How much does the Gibbs free energy of a mole of NH_3 and 10 moles of water change when the ammonia goes into solution? From thermodynamic tables at 1 atm. and 25°C,

$$G_{NH_3}^\circ = -3.9 \text{ kcal/mole,}$$

$$G_{H_2O}^\circ = -54.6 \text{ kcal/mole.}$$

The Gibbs free energy change for the ammonia is

$$\Delta G_{NH_3} = 1RT \ln 1/11 = -RT \ln 11.$$

For the water it is

$$\Delta G_{H_2O} = 10RT \ln 10/11 = -10RT \ln 11/10.$$

The total change is then

$$\Delta G_{TOT} = \Delta G_{NH_3} + \Delta G_{H_2O} = -RT \ln 11 - 10RT \ln 11/10$$

$$\Delta G_{TOT} = -1.97 \text{ kcal.}$$

The Gibbs free energy change for dilute solutions is mainly due to the solute-concentration change. This approximation is often used in work with dilute solutions.

The Gibbs free energy change on mixing also represents the energy required to separate the constituents. This energy is not linearly proportional to the concentration. Much more energy proportionately is required to extract a constituent from a dilute solution. A plot of this relation is shown in figure 5–1. Note that this separation energy is independent of the particular solute and solvent in the approximation of ideal solutions.

This figure is surprising because of its symmetry and the large energy

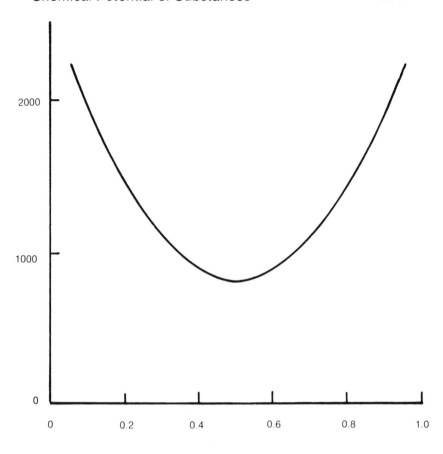

Concentration Xi

Note: Initial solute concentration is X_i.

Figure 5–1. Gibbs Free Energy Required to Separate n_i Moles of Solute from $[(1 - X_i)n_i]/X_i$ Moles of Solvent

requirements for separation of substances at high concentrations. This is counterintuitive, as one would expect the energy required to remove a substance from a solution to decrease as the solution became more concentrated. This apparent contradiction is explained by the fact that this curve represents the energy for separation into two pure components when the total number of moles of the solution is not fixed. More moles of the solution are separated at both high and low concentrations. If this energy of separation is put on a per mole basis of the solute, it will appear as in figure 5–2. This agrees with expectations. The exponential character of the energy

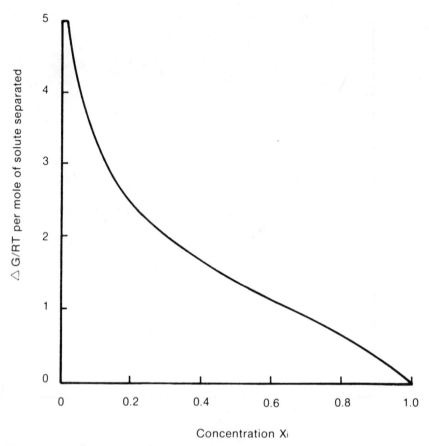

Figure 5–2. Gibbs Free Energy Required to Separate a Mole of Solute from a Solution Having an Original Solute Concentration X_i

requirement at high dilutions is most important for environmental considera-tios, where the goal of zero pollutants is advocated. Only with infinite energy expenditures is this possible. This mathematical characteristic is an important relationship included in the definition of resources and economic recovery discussed in chapters 9 and 10. Most economic analyses are linear models, assuming that twice the money and energy expended will remove twice the pollutants present. This is obviously incorrect, since the energy required to reduce concentrations of pollutants is dependent on the energy required in a significant nonlinear manner.

An associated principle demonstrated in extracting minerals from seawater is that the problem of extracting minerals from a saline solution is different from the problem of converting salt water to a low-concentration freshwater. In saline-water conversion to fresh water, neither product constituent is required to be a pure substance. The seawater remaining is a concentrated brine, and the useful water has a concentration from 10 to 500 ppm, depending on the user requirements. Absolute purity is not an economically feasible requirement. Extracting minerals of high purity, however, requires a more complete removal of the water from the solution, a much more costly process. The energy requirement for this is not infinite as suggested from figure 5–2, because, although the minerals may be extracted in nearly pure form, the water itself is not in pure form at the end of the process. In addition, the equation is only an approximation for real solutions and is less applicable at higher concentrations. The recovery of materials from mixtures is discussed in more detailed chemical-process terms in chapter 11.

The Equilibrium Constant

From equation 4.29 the Gibbs free energy G is related to the chemical potential μ as

$$G = \Sigma n_i \mu_i.$$

In mixtures of perfect gases,

$$\mu_i = \mu_i^\circ + RT \ln P_i,$$

where P_i is the partial pressure of the ith species of gas, where

$$\Sigma P_i = P \qquad \text{(Dalton's law)}.$$

Similarly, for dilute solutions,

$$\mu_i = \mu_i^\circ + RT \ln C_i,$$

where C_i is the concentration of species i in solution, and

$$\Sigma C_i = 1.$$

Note that in a molar solution

$$C_i = 1 \text{ mole/liter, and } \mu_i = \mu_i^\circ.$$

The chemical potential at standard states for liquids is taken at a concentrations of 1 mole/liter. The concentrations must then be in the same units.

Similarly, μ_i° is at 1 atm. pressure when the chemical potential is expressed as $\mu = \mu_i^\circ + RT \ln P_i$. Caution must be observed in interpreting P_i as a partial pressure. It is only a fraction when the total pressure is 1 atm.

If the mixtures are not ideal, then one can conceptualize a solution in which the behavior is as if there were different concentrations of each species present. These imaginary concentrations are called *activities* and are defined

as $a_i \equiv \gamma_i C_i$.

Solution-chemical potentials are then written

$$\mu_i = \mu_i^\circ + RT \ln \gamma_i C_i = \mu_i^\circ + RT \ln a_i.$$

In real gas mixtures similar concept is used in which the equivalent behavior is represented by *fugacities*:

$$f_i \equiv \gamma_i P_i,$$

and

$$\mu_i = \mu_i^\circ + RT \ln \gamma_i P_i = \mu_i^\circ + RT \ln f_i.$$

Care should be taken to remember that μ_i° is a function of temperature.

In chemical processing one is interested in the yield of a reaction or the moles of product formed in a reaction. To illustrate the relationship of the yield to the Gibbs free energy, and the equilibrium constant, let us consider the classical reaction of hydrogen and nitrogen, in gas form, to produce ammonia:

$$3H_2 + N_2 \rightleftharpoons 2NH_3.$$

The question is: If hydrogen and nitrogen are supplied in the proportions given, and a reaction carried out at constant pressure and temperature, how much ammonia will be present at equilibrium? To approach this systematically, let 2ϕ equal the number of moles of NH_3 present at any time. If a total of 3 moles of H_2 and 1 mole of N_2 is supplied, then the number of moles of reactant gases present at any time will be

$$n = 4 - 2\phi.$$

This follows from the conservation of mass in the chemical reaction represented by the balance equation

$$\Sigma v_i n_i = 0.$$

v_i are the stoichiometric coefficients, positive for the product species and negative for the reactants. In this example,

$$2n_{NH_3} - 3n_{H_2} - 1n_{N_2} = 0.$$

The number of moles of each species present at any time n_i is

$$n_i = n_i^\circ + v_i\phi.$$

n_i° is the initial number of moles of each species present. In this case,

$$n_{H_2}^\circ = 3, \ n_{N_2}^\circ = 1, \ n_{NH_3}^\circ = 0,$$

and

$$n_{H_2} = 3(1 - \phi), \ n_{H_2} = 1 - \phi, \ n_{NH_3} = 2\phi.$$

The free energy at any time is then

$$G = \Sigma(n_i^\circ + v_i\phi)\mu_i.$$

This free energy will change as the chemical reaction proceeds. The change can be written in terms of a change in ϕ only:

$$dG = \Sigma v_i\mu_i d\phi.$$

The reaction will reach equilibrium when $dG = 0$. Then

$$\Sigma v_i\mu_i = 0.$$

This is true only at constant pressure and temperature. If the species present behave ideally, then

$$\Delta G = \Sigma v_i\mu_i = \Sigma v_i\mu_i^\circ + RT\Sigma v_i \ln P_i,$$

or

$$\Delta G = \Delta G^T + RT\Sigma v_i \ln P_i.$$

ΔG^T is the Gibbs free energy difference of the reaction at the reaction temperature. At equilibrium $\Delta G = 0$, and then $\Delta G^T = -RT\Sigma v_i \ln P_i$. *Using* the properties of logarithms, this may be written

$$\Delta G^T = -RT \ln (P_1^{v_1} P_2^{v_2} \ldots).$$

With the convention that v_i is negative for reactants, this may be arranged with the reactant and product terms as

$$\Delta G^T = -RT \ln \frac{\overset{p}{\Pi} P_p^{v_p}}{\overset{r}{\Pi} P_p^{v_p}}$$

In the ammonia reaction this is

$$\Delta G^T = -RT \ln \frac{P_{NH_3}^2}{P_{H_2}^3 P_{N_2}^1} .$$

The term

$$k \equiv \frac{\overset{p}{\Pi} P_p^{v_p}}{\overset{r}{\Pi} P_r^{v_r}}$$

is defined as the *equilibrium constant*.
Then

$$\Delta G^T = -RT \ln k.$$

If the Gibbs free energies of the reactants and products are known at standard conditions, then the equilibrium constant representing the extent of the reaction can be found at the standards temperature. If the reaction takes place at a different temperature, the approximation given in the first section of this chapter or chemical tables or charts of equilibrium constants as shown in figure 5–3 may be utilized.

In many chemical reactions (particularly gas reactions) the enthalpy difference ΔH between products and reactants is nearly independent of the temperature of the reaction. In these reactions $\Delta H = \Delta H°$, and the Gibbs free energy difference is due to entropy changes. Using the results of appendix 5A,

$$d(G\beta) = \left. \frac{\partial(G\beta)}{\partial \beta} \right)_P d\beta + \left. \frac{\partial(G\beta)}{\partial P} \right)_\beta dP,$$

$$d(G\beta) = Hd\beta + \frac{n}{P} dP.$$

Then, integrating from standard temperature and pressure,

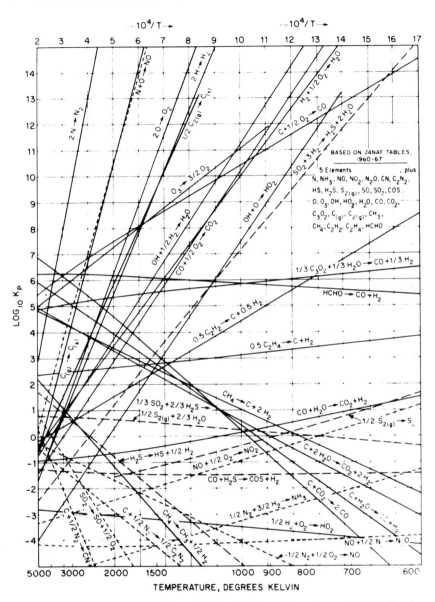

Source: From Michael Modell, Robert C. Reid, *Thermodynamics and Its Applications*, copyright 1974, p. 396. Reprinted by permission of Prentice-Hall, Inc., Englewood Cliffs. N.J.

Note: Note the linear relation between the logarithm of the equilibrium constant and the reciprocal of the absolute temperature.

Figure 5–3. Equilibrium-Constant Variation with Temperature for a Selected Set of Reactions

$$G\beta - G^\circ\beta_0 = \int_{\beta_0}^{\beta} H d\beta + n \ln \frac{P}{P_0} \ .$$

Then, using the subscript notation G_p (product), G_r (reactant),

$$G\beta)_r = G^\circ\beta_0)_r + \int_{\beta_0}^{\beta} H_r d\beta + n_r \ln \frac{P_r}{P_0} \ ,$$

$$G\beta)_p = G^\circ\beta_0)_p + \int_{\beta_0}^{\beta} H_p d\beta + n_p \ln \frac{P_r}{P_0} \ ,$$

$$\Delta(G\beta) = G\beta)_p - G\beta)_r \ ,$$

$$\Delta(G\beta) = G^\circ\beta_0)_p - G^\circ\beta_0)_r + \int_{\beta_0}^{\beta} (H_p - H_r) d\beta + n_p \ln \frac{P_p}{P_0}$$

$$- n_r \ln \frac{P_r}{P_0} \ .$$

If $H_p - H_r = \Delta H^\circ$ is independent of temperature, then

$$\Delta(G\beta) = \Delta(G^\circ\beta_0) + \Delta H^\circ(\beta - \beta_0) + \ln\left(\frac{P_p}{P_0}\right)^{n_p} + \ln\left(\frac{P_r}{P_0}\right)^{n_r} \ .$$

The equilibrium constant for reactions where ΔH is independent of temperature is then found for $\Delta(G\beta) = 0$. With β on a mole basis $\beta = 1/RT$,

$$0 = \frac{\Delta G^\circ}{RT_0} + \Delta H^\circ\left[\frac{1}{RT} - \frac{1}{RT_0}\right] + \ln\left[\frac{\left(\dfrac{P_p}{P_0}\right)^{n_p}}{\left(\dfrac{P_r}{P_0}\right)^{n_r}}\right] \ .$$

With an equilibrium constant

$$k = \left(\frac{P_r}{P_0}\right)^{n_p} \bigg/ \left(\frac{P_p}{P_0}\right)^{n_r} \ ,$$

then

$$\ln k = - \frac{\Delta G^\circ}{RT_0} - \Delta H^\circ\left[\frac{1}{RT} - \frac{1}{RT_0}\right] \ .$$

The straight-line relationship between $\ln k$ and $1/T$ for the simple chemical conversions shown in figure 5–3 indicates that the ΔH is independent of T for these reactions. For exothermic reactions, where $\Delta H°$ is negative, it is clear that increasing the temperature reduces the equilibrium constant and hence reduces the product quantity at equilibrium. Conversely, for endothermic reactions, where $\Delta H°$ is positive, increasing the temperature moves the reaction to the product side. This is generally known as LeChatelier's principle.

The yield of the ammonia reaction can be determined from equilibrium constants as follows. Partial pressures in perfect gas mixtures are first related to mole fractions as

$$\frac{P_i}{P} = \frac{n_i}{n}.$$

In this expression P is the total pressure and n is the total moles present at any time. In the example n changes from 4 moles initially to 2 moles if the reaction were complete. If there are 2ϕ moles of ammonia present, then

$$\frac{P_{NH_2}}{P} = \frac{2\phi}{4 - 2\phi},$$

$$\frac{P_{N_2}}{P} = \frac{1 - \phi}{4 - 2\phi},$$

$$\frac{P_{H_2}}{P} = \frac{3(1 - \phi)}{4 - 2\phi}.$$

The equilibrium constant is then

$$k = \left(\frac{2\phi}{4 - 2\phi}\right)^2 P^2 \bigg/ \left[\frac{3(1 - \phi)}{4 - 2\phi}\right]^3 (P^3)\left(\frac{1 - \phi}{4 - 2\phi}\right) P$$

$$= \frac{(2\phi)^2(4 - 2\phi)^2}{(3 - 3\phi)^3(1 - \phi)P^2}$$

The variation of the yield represented by ϕ as a function of temperature and pressure can be found from the equilibrium constant and the relationship $\Delta G^T = -RT \ln k$.

For the reaction to form ammonia, table 5–1 illustrates the variation in yield. The Haber process used to produce ammonia relies on a higher temperature than indicated from this table because of the low reaction rate at

Table 5–1

Yield of the $3H_2 + N_2 \rightleftharpoons 2NH_3$ Ammonia-Product Reaction as a Function of the Reaction Temperature and Pressure

T(K)	Equilibrium Constant k	ϕ		
		P = 1 atm.	P = 200 atm.	P = 500 atm.
498	0.206	0.205	0.902	0.942
798	4.57×10^{-5}	0.004	0.40	0.57
1,298	2.1×10^{-7}	0.0003	0.052	0.115

low temperatures. Equilibrium calculations must therefore be supplemented by an understanding of reaction kinetics in the design of chemical-processes technology. In the Haber process pressure of 1,000 atm. and temperature of 800K, with suitable catalysts and recycle of materials, are required for achieving economic yields.

The effect of increasing the pressure on the yield for this reaction is illustrated by reference to the equilibrium-constant equation. At a temperature of 498K the equilibrium constant is 0.206. Solving for ϕ with a pressure of 1 atm. gives $\phi = 0.205$, whereas at a pressure of 200 atm. $\phi = 0.924$. Increasing the pressure increases the equilibrium product output. This is also indicated by the fact that the reaction goes from an initial 4 moles toward 2 moles of product, which occupies a smaller volume.

In chemical-engineering terms the percentage conversion of product is 100ϕ. The use of recycle to increase the conversion is discussed in detail in chapter 10 on chemical processing.

The Gibbs free energy of a chemical system changes as the reaction proceeds. As indicated previously, the Gibbs free energy of the reactants must be higer than that of the products for the reaction to proceed. Equilibrium is reached, of course, when the Gibbs free energy reaches a minimum.

The effect of inert constituents like water in solution problems and nitrogen in air-mixing problems is discussed in chapter 11. The effect on the equilibrium composition is also assessed. These chemical-equilibrium principles are then applied to energy storage and chemical-energy transfer problems in chapter 16.

Catalytic Action of Chemicals in the Environment

Perhaps the most misunderstood aspect of pollutants is their catalytic effect in certain circumstances. A particularly striking example is the concern about the depletion of the upper-atmosphere ozone layer. This layer is produced by the absorption of ultraviolet radiation by oxygen molecules in the paired reactions:

$$O_2 + h\upsilon \rightarrow 2O,$$

$$O + O_2 \rightarrow O_3 + \text{energy}.$$

The ozone is then returned to molecular oxygen by many chemical reactions driven in part by the absorption of ultraviolet radiation. These removal reactions include chain reactions like

$$NO + O_3 \rightarrow NO_2 + O_2,$$

$$NO_2 + O \rightarrow NO + O_2.$$

The net effect of these reactions is the removal of ozone from the stratosphere. If the ozone molecules are removed in this fashion rather than through absorption of ultraviolet light, there will be an increase in ultraviolet-light penetration to the earth's surface. This nitrogen oxide–ozone removal reaction is of concern in supersonic-transport operation in the stratosphere, since nitrogen oxides are formed by combustion processes in jet engines at high pressures.

Examination of this pair of reactions shows that a molecule of NO introduced is returned through the second reaction. The result is that a single NO molecule can theoretically react successively with an infinite number of ozone molecules. Actually, up to 50,000 reaction cycles are probable before the NO molecule reacts with other molecules and is removed. The catalytic recycling action then multiplies the effect 50,000 times. A similar recycling reaction occurs as the primary removal system for chlorinated fluorocarbons in the environment. This is the problem of Freon propellant use in aerosol can dispensers. In the Freon situation, chlorine molecules appear to be the predominant recycling catalyst in the reaction pair

$$ClO + O_3 \rightarrow ClO_2 + O_2,$$

$$ClO_2 + O \rightarrow ClO + O_2.$$

The result is similar to the NO reaction system.

Recycling reactions, of course, present a basic difficulty in water-eutrophication problems. The recycling of phosphates and nitrate fertilizers in water systems accelerates biomass production. Algae are particularly instrumental in this recycling process because of their short recycle times compared with those of large water plants.

Catalytic recycling systems are also used to technical advantage in the refining of aluminum and the removal of many metals from oxides. Most proposed future systems for producing hydrogen from water either thermo-chemically or by photolysis rely on the use of recycling catalytic reactions.

Important Other Ideas

Carbon monoxide is not a reducing agent above a certain temperature, as the reaction at low temperature,

$$CO + \tfrac{1}{2}O_2 \rightarrow CO_2,$$

is reversed, as can be shown from Gibbs free energy calculations. This is important in a reducing process like pyrolysis of waste products to produce fuels, as well as in pollution-control techniques aimed at reducing CO in exhaust gases. At high temperatures the addition of oxygen will not reduce the carbon monoxide present.

The *adiabatic temperature*, or the temperature a reaction will reach if the reactor is insulated, is an important parameter in metallurgical furnace and smeltering operations. If the melting point of a metal is above this temperature, the reaction supplying heat will not be effective. Similarly, if the process requires maintenance of a temperature above a certain level to keep the material in a molten condition, then only the heating capacity of a given fuel above the melt point is important. This places certain fuels at a disadvantage in foundry use compared with electric-arc furnaces, which can operate at around 3,000°C, allowing a greater percentage of the heat to be available for high-temperature applications.

The rapid heating possible with electric-arc heating can also reduce thermal losses. If the combustion gases are not used for reheat, they represent a loss not incurred in electric heating systems.

A temperature higher than the adiabatic temperature may of course be obtained by preheating of the air and fuel. Using oxygen instead of air also increases the adiabatic temperature. This increase is due to the fact that when air is used, the nitrogen must be heated, which reduces the energy available for heating.

References

1. Gibbs, J.W. *Collected Works*, vol. I, *On the Equilibrium of Heterogeneous Substances*, 1878. New Haven: Yale University Press, 1948.
2. Guggenheim, E.A. *Thermodynamics, An Advanced Treatment for Chemists and Physicists*. Amsterdam: North-Holland, 1950.
3. Modell, M., and Reid, R.C. *Thermodynamics and Its Applications*. Englewood Cliffs, N.J.: Prentice-Hall, 1974.
4. Prigogine, I., and Defay, R. *Chemical Thermodynamics*. New York: Wiley, 1962.

5. Benson, S.W. *Thermochemical Kinetics*. New York: Wiley, 1976.
6. Rossini, F.D.; Pitzer, K.S.; Arnett, R.L.; Braun, R.M.; and Pimentel, G.C. *Selected Values of Physical and Thermodynamic Properties of Hydrocarbons and Related Compounds*. Pittsburgh, Pa.: Carnegie Press, 1953.
7. Lewis, W.K.; Radasch, A.H.; and Lewis, H.C. *Industrial Stoichiometry*, 2nd ed. New York: McGraw-Hill, 1954.
8. Reid, R.C., and Sherwood, J.K. *Properties of Gases and Liquids*. New York: McGraw-Hill, 1966.
9. Stull, D.R., and Prophet, H. *JANEF Thermochemical Tables*, 2nd ed. NSRDS-NBS, 1971.
10. "Selected Values of Chemical Thermodynamic Properties." NBS Technical Note 270-3, 270-4. Washington, D.C.: U.S. Government Printing Office, 1968–1969.
11. West, R.C., ed. *Handbook of Physics and Chemistry*. Cleveland: Chemical Rubber Company, 1979.
12. Van Zeggeren, F., and Storey, S.N. *Computational Methods of Chemical Equilibrium*. London: Cambridge University Press, 1970.

Appendix 5A:
Gibbs Free Energy in
Terms of Pressure
and β

Gibbs free energy may be expressed in terms of pressure and β. The use of β instead of the temperature allows the simplification of some calculations. Let us begin by writing equation 4.30 using the identity $\beta \equiv 1/kT$.

$$G = H - TS = H - S/k\beta.$$

Multiplying by β on both sides gives

$$G\beta = H\beta - S/k. \qquad (5A.1)$$

Take the differential of this expressing,

$$d(G\beta) = d(H\beta) - (dS)/k. \qquad (5A.2)$$

Using the differential of the enthalpy,

$$dH = dE + PdV + VdP,$$

and Gibbs equation,

$$(dS)/k\beta = dE + PdV,$$

gives, from equation 5A2,

$$d(G\beta) = Hd\beta + \beta VdP. \qquad (5A.3)$$

Then it follows that

$$\left. \frac{\partial(G\beta)}{\partial\beta} \right)_P = H,$$

and

$$(5A.4)$$

$$\left. \frac{\partial(G\beta)}{\partial P} \right)_\beta = \beta V. \qquad (5A.5)$$

For a perfect gas $\beta V = n/P$, and the last equation can be written as

$$\frac{\partial(G\beta)}{\partial P}\Bigg)_{\beta} = n/P. \tag{5A.6}$$

For a constant β (temperature) process,

$$\int d(G\beta) = \int \frac{n}{P} dP = \frac{R}{k} \int \frac{dP}{P}. \tag{5A.7}$$

For a constant pressure, process,

$$\int d(G\beta) = \int H d\beta. \tag{5A.8}$$

The enthalpy must be known as a function of β in order to do this integration. For a perfect gas the enthalpy can be written as

$$H = E + PV = \frac{c_v}{k\beta} + \frac{n}{\beta},$$

and

$$c_v + R = c_p.$$

Then equation 5A.8 can be written (for constant pressure) as

$$\int d(G\beta) = \frac{c_p}{k} \int \frac{d\beta}{\beta}. \tag{5A.9}$$

6

Available Energy

He found himself in a land where no one had ever penetrated before; where order was an accidental relation obnoxious to nature; artificial compulsion imposed on motion; against which every free energy of the universe revolted, being merely occasional, resolved itself back into anarchy at last.
—Henry Adams, *The Education of Henry Adams* (1906)

Available-Energy Functions

Any process in which the energy of a system is changed is reflected in the change of two factors, the *quantity* of the energy and its *quality*. The quantity is measured by reference to the classical quantities of internal energy, kinetic energy, and potential energy. This total energy will be defined as

$$E_T = E \quad + \quad KE \quad + \quad PE. \tag{6.1}$$
$$\text{Total} = [\text{Internal}] + [\text{Inertial}] + [\text{Storage in a field}]$$

We are interested in the minimum energy required to change the energy state of a system. This may be the energy required to produce pure water from seawater or from sewage effluent. It may be the energy required to remove sulfur dioxide from an exhaust gas. In another sense we may be interested in determining the energy value of a fuel or the value of hot sodium in a nuclear reactor. The hypothesis is made that any substance not at thermodynamic equilibrium with its environment has an energy value. The objective of the available-energy function is to provide a measure of this value.

A system to demonstrate this is shown in figure 6–1. The system representing a processing device is shown enclosed in a dotted line. The substance transformed as product is shown as the cross-sectioned area. In this sytem a Van't Hoff chamber is used to remove pure components with a semipermeable membrane and a piston configuration [1,2].

Conservation of mass will be indicated by saying that the input flow $\dot{n}_c$ either appears as product flow $\dot{n}_p$ or waste flow $\dot{n}_w$ or is stored (dn/dt). The waste stream will be indicated as the input minus that converted or stored, $\dot{n}_c - \delta\dot{n}_c$.

The dotted-line region is indicated as the thermodynamic system. Material enters from the left either as a single stream or as multiple input

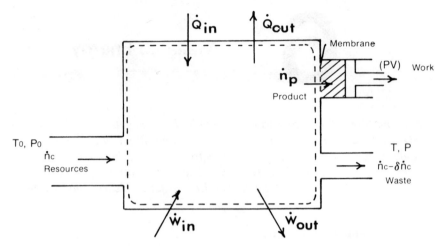

Figure 6–1. Available-Energy-System Diagram where Resources Are Introduced, Waste Is Generated, and a Product Is Produced with Separation by a Semipermeable Membrane

reactants as required. The products leave via the cross-sectioned port and the waste products leave from the port on the right.

A mass balance on the system gives

$$\frac{dn}{dt} = \Sigma \dot{n}_c - \left\{ \Sigma(\dot{n}_c - \delta \dot{n}_c) + \Sigma \dot{n}_p \right\}, \tag{6.2}$$

or

$$\frac{dn}{dt} = \Sigma \delta \dot{n}_c - \Sigma \dot{n}_p. \tag{6.3}$$

If mass is not accumulating in the system, then $dn/dt = 0$, and $\Sigma \delta \dot{n}_c = \Sigma \dot{n}_p$. In this formulation the emphasis is on the changes in the components that enter the system.

Since in any process energy for processing may be supplied at certain points and also recovered, both inputs and outputs of work and heat are considered separately. The input in this derivation will be considered at the temperature T_0 and pressure P_0 of the environment.

Allowing for time variation of the system energy, the energy conservation is written as:

{rate of change of the energy of the system} = {flow of energy in

− flow of energy out}.

In symbolic notation,

$$\frac{dE_T}{dt} = \left\{ \dot{Q}_{in} + \dot{W}_{in} + \Sigma \dot{n}_c \bar{H}_{0c} + KE_{in} + PE_{in} \right\}$$
$$\text{all input components}$$

$$- \left\{ \dot{Q}_{out} + \dot{W}_{out} + \Sigma \dot{n}_c \bar{H}_{0c} - \Sigma \delta(\dot{n}_c \bar{H}_{0c}) \right\}$$
$$\text{all output components} \tag{6.4}$$

$$+ \left(\Sigma \dot{n}_p \bar{H}_p \right) + KE_{out} + PE_{out}. \Bigg\}$$

$\bar{H}_{0c}$ represents the partial molal enthalpy of each input component at the
  the reference state, including the product component in the mixture $\bar{H}_{0p}$.

$\bar{H}_p$ represents the partial molal enthalpy of the output product.

$\dot{n}_c$ represents the mass flow of each component in moles per unit time.

Expansion of the expression representing the change in energy of the material leaving as waste shows that:

$$\Sigma \delta(\dot{n}_c \bar{H}_{0c}) = \Sigma \dot{n}_c \delta \bar{H}_{0c} + \Sigma \bar{H}_{0c} \delta \dot{n}_c. \tag{6.5}$$

However, as shown in appendix 6A,

$$\Sigma \dot{n}_c \delta \bar{H}_{0c} = 0. \tag{6.6}$$

Therefore,

$$\Sigma \delta(\dot{n}_c \bar{H}_{0c}) = \Sigma \bar{H}_{0c} \delta \dot{n}_c. \tag{6.7}$$

The energy-conservation equation then becomes

$$\frac{dE_T}{dt} = \Sigma \dot{n}_p (\bar{H}_{0p} - \bar{H}_p) + (\dot{Q}_{in} - \dot{Q}_{out}) + (\dot{W}_{in} - \dot{W}_{out})$$

$$+ (KE_{in} - KE_{out}) + (PE_{in} - PE_{out}). \tag{6.8}$$

A similar balance on the entropy of the system is required to determine how the quality of the energy is affected. The result is an equation

$$\frac{d\sigma}{dt} = \Sigma \dot{n}_p [\bar{S}_{0p} - \bar{S}_p] + \frac{\dot{Q}_{in} - \dot{Q}_{out}}{T_0} + \dot{\sigma}_c \qquad (6.9)$$

<u>Net entropy-flow</u>
rate from the
environment

In this equation $\dot{\sigma}_c$ represents the rate of entropy creation in the system by dissipative processes. Some of this entropy creation is required in order to produce a finite production rate. This includes electrical-resistive dissipation, mechanical friction, and heat- or mass-transfer entropy contributions. The term $(\dot{Q}_{in} - \dot{Q}_{out})/T_0$ represents the entropy change in the system when the entropy-creation term $\dot{\sigma}_c$ excludes the external heat-flow effect.

Multiplying equation 6.9 by T_0 and subtracting it from equation 6.8 combines the measures of quantity and quality in a useful form for thermodynamic calculations:

$$\frac{dE_T}{dt} - T_0 \frac{d\sigma}{dt} = \Sigma \dot{n}_p [[\bar{H}_{0p} - T_0 \bar{S}_{0p}] - [H_p - T_0 \bar{S}_p]]$$

$$- \dot{W}_{net} + T_0 \dot{\sigma}_c + (KE_{in} - KE_{out}) \qquad (6.10)$$

$$+ (PE_{in} - PE_{out})$$

where $\dot{W}_{net}$ is the net work produced by the system.

The terms on the left of the equation represent the rate of increase in available energy of the system itself. For the two terms on the left, the first dE_T/dt represents the rate of increase in the total energy; however, part of the energy accumulated is unavailable for work processes. This is the second law of thermodynamics in a different form, in which terms are combined.

For the terms on the right of the equation, the first term represents the net change in available energy of the product over its available energy in the environment. The term $T_0 \dot{\sigma}_c$ represents the dissipation of available energy in the process and is always positive except in a reversible process, when it is zero.

The unsteady-equation form is important for determining the energy required to bring a system up to operating conditions or to close it down for repairs or adjustments. Care must be taken to include in the system all the mass that is initially specified, since in this derivation dn/dt is set equal to zero. In most analyses, however, it is assumed that a system is designed for steady-state operation, and this is neglected. The equation in this form is useful for demonstrating one advantage of continuous-flow versus batch processing in the chemical industry. In batch processing the system must be brought to operating conditions from some initial state, and this energy requirement may be large compared with its operating requirements. In these starting situations and in shutdown operations, the system usually will be operating off the optimum, and extra losses will be incurred. In any shutdown of operations the total accumulated energy in a system is usually lost because of the expense of providing energy-storage or conversion equipment.

Let us now examine systems operating in a steady state in which the KE and PE changes are negligible. The terms on the left are then zero, and the equation can be written in terms of the net work, $\dot{W}_{net}$:

$$\dot{W}_{net} = \Sigma \dot{n}_p \{[\bar{H}_{0p} - T_0 \bar{S}_{0p}] - [\bar{H}_p - T_0 \bar{S}_p]\} - T_0 \dot{\sigma}_c \quad (6.11)$$

If the process is reversible, then $\dot{\sigma}_c = 0$, and the $\dot{W}_{net}$ then represents the maximum useful work that can be obtained from the process. This means that if one has the component called a product in this example and devises a method to convert its energy to useful work, then the maximum work that is obtainable in a steady-flow process can be calculated by setting $\sigma_c = 0$. Alternatively, when $\sigma_c = 0$, $\dot{W}_{net}$ represents the minimum available energy required to produce the product from material at the datum state.

In this form the equation is equivalent to the availability in steady-flow equation of Keenan [3,28] and Darrieus [16]. Extension to chemical problems can be made by identifying the reference energy form $\bar{H}_{op} - T_0\bar{S}_{op}$ as the specific Gibbs free energy or the chemical potential of the product μ_{op}. Equation 6.11 may then be written as

$$\dot{W}_{net} = -\Sigma \dot{n}_p \{(\bar{H}_p - T_0 \bar{S}_p) - \mu_{op}\}. \quad (6.12)$$

Integrating with respect to time gives

$$W_{net} = -\Sigma(n_p \bar{H}_p - n_p T_0 \bar{S}_p - n_p \mu_{op}),$$

or with $H_p \equiv \Sigma n_p \bar{H}_p$, and $S_p \equiv \Sigma n_p \bar{S}_p$,

$$W_{net} = -(H_p + T_0 S_p - n_p \mu_{op}). \quad (6.13)$$

The term on the right is the *exergy* [4] or *essergy* [5] function:

$$B \equiv H_p - T_0 S_p - \Sigma n_p \mu_{op}. \quad (6.14)$$

This function differs from the Gibbs free energy in that the temperature used in the calculation is that of the reference state of the product components. Physically, the difference is that the Gibbs free energy change represents only the maximum work produced or minimum work required for a process carried out at constant temperature and pressure.

It should be noted that in the energy equation the enthalpy was used rather than the internal energy. This means that the energy requirement for the flow work is included in the analysis.

Consider a simple situation where only heat exchange dQ is used to produce a work output dW_{net} with no composition change. Then $dW_{net} = -dB$, or

$$dW_{net} = -dH + T_0\, dS - T_0\, d\sigma_c. \qquad (6.15)$$

If the process is reversible, $d\sigma_c = 0$ and $dS = dQ/T$, and

$$dW_{net} = -dH + T_0\, dQ/T. \qquad (6.16)$$

Let us look at this as a two-stage process (figure 6–2) in which the material with H_0 and S_0 enters a heat exchanger, where it leaves at enthalpy $H_0 + dH$ and entropy $S_o + dS$. The material then flows through a heat machine, giving a work output and returning the fluid to its original energy state H_0. If the process is reversible for the heat-exchange part,

$$dS_{0-1} = dQ/T,$$

and

$$dH_{0-1} = dQ.$$

Once the fluid has reached this point, let us find the maximum work possible from the engine. The enthalpy is decreased to H_o, so that

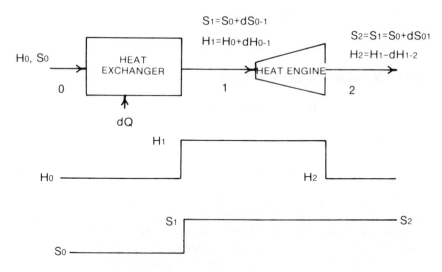

Figure 6–2. Simplified Heat Exchanger and Heat Engine System as a Serial Process

$$dH_{0-1} = -dH_{1-2}.$$

Then

$$dW_{net} = -(dH - T_0 dS) = (H_1 - H_0) - T_0(S_1 - S_0).$$

Now

$$S_1 - S_0 = \frac{dQ}{T}$$

and

$$H_1 - H_0 = dQ.$$

Then

$$dW_{net} = \frac{T - T_0}{T} dQ. \qquad (6.17)$$

This is the Carnot principle, indicating the maximum possible work from a system at a temperature T exchanging heat with the surroundings at temperature T_0.

Closed-System Available Energy

If the system under consideration is a closed system, (no mass enters or leaves the sytem), then equations 6.4, 6.9, and 6.10 reduce to:

$$\frac{dE}{dt} = \dot{Q}_{in} - \dot{Q}_{out} + (\dot{W}_{in} - \dot{W}_{out}), \qquad (6.18)$$

$$\frac{d\sigma}{dt} = \frac{\dot{Q}_{in} - \dot{Q}_{out}}{T_0} + \dot{\sigma}_c, \qquad (6.19)$$

$$\frac{dE}{dt} - T_0 \frac{d\sigma}{dt} = -\dot{W}_{net} - T_0 \dot{\sigma}_c. \qquad (6.20)$$

If the only work in a process is expansion work, then

$$\dot{W}_{net} = -\frac{d}{dt} \int (P - P_0) dV. \qquad (6.21)$$

Integration of equation 6.20 with respect to time gives, using equation 6.21,

$$dE - T_0 d\sigma + P_0 dV = PdV - T_0 d\sigma_c.$$

Then rearranged,

$$dW = PdV = -(dE - T_0 d\sigma + P_0 dV) - T_0 d\sigma_c.$$

Whe work obtainable from the expansion of the system to the environmental state is

$$W = \int PdV = (E - T_0 \sigma + P_0 V) - (E_0 - T_0 \sigma_0 + P_0 V_0) - T_0 \sigma_c.$$
$$(6.22)$$

The maximum work occurs when there is no internal dissipation such as turbulent motion, friction, or the like. In that case $\sigma_c = 0$, and

$$W_{max} = (E - T_0 \sigma + P_0 V) - (E_0 - T_0 \sigma_0 + P_0 V_0). \quad (6.23)$$

Let us define this as the *closed-system available energy* or *availability:*

$$A \equiv (E - T_0 \sigma + P_0 V) - (E_0 - T_0 \sigma_0 + P_0 V_0). \quad (6.24)$$

This available energy is equal to the available energy in steady flow minus the expansion work required on the environment. One must supply this work as a minimum to reach operating conditions for a system used to process materials or energy. This available energy is lost when an operating system is turned off and returns to the environmental state. This loss is incurred when a system is shut down for repairs or in batch processing where intermittent start-up and shutdown are required.

Differentiation between the closed-system available energy and the steady-flow available energy must be maintained. In the general case, E, σ, and V refer to the values of the parameters of the system through which the mass flows and heat is exchanged. H, S, and n refer to the values of parameters associated with the material flow through the system. Much confusion in thermodynamics is attached to failure to distinguish carefully between these two different entities. Although it is true that at some instant of time, the particles that are within the system correspond to the same particles as those which are flowing through, the distinction must always be preserved.

The idea of available energy as a measure of the utility of systems has been extended by Evans and Tribus [9,10,19]. They have clarified the relationship between entropy and essergy. To examine available energy in the essergy form, we must first note that that the concept that the available energy of a system represents the portion of the energy of a system that can be converted to thermodynamic work is misleading. Systems that have no energy relative to the environment may have available energy. The simplest example is compressed air at the temperature of the environment, which is discussed later in this chapter.

The essergy extension can be illustrated by reexamination of the entropy-maximization process. From equation 4.5,

$$S = -k\Sigma p_i \ln p_i.$$

The maximization with constraints on the energy and mass components leads to the probability of energy levels being occupied as

$$P_i = e^{-\frac{pV}{kT} - \frac{\varepsilon_i}{kT} + \sum_c \frac{\mu_c}{kT} n_{ci}}. \tag{4.27}$$

The logarithm of p_i is then

$$-\ln p_i = \frac{\varepsilon_i + PV}{kT} - \sum_c \frac{\mu_c}{kT} n_{ci}. \tag{6.25}$$

If the logarithm of the probability p_{i0} is taken at the datum state of temperature T_0 and P_0 then

$$-\ln p_{i0} = \frac{\varepsilon_i + P_0 V}{kT_0} - \sum \frac{\mu_{c0} n_{ci}}{kT}. \tag{6.26}$$

Let us then examine the function

$$A_i \equiv kT_0 \ln \frac{p_i}{p_0}, \tag{6.27}$$

which represents, in energy units, the difference between the distribution of particles in a condition T, P and in a condition T_0, P_0. Now find the expected value of A_i:

$$\langle A \rangle = \Sigma p_i A_i = kT_0 \Sigma p_i \ln p_i - kT_0 \Sigma p_i \ln p_{i0}. \tag{6.28}$$

The first term is $-T_0 S$, so the expression can be written as

$$\langle A \rangle = -T_0 S - kT_0 \Sigma p_i \ln p_{i0}. \tag{6.29}$$

Substituting from equation 6.26 gives

$$\langle A \rangle = \langle E \rangle + P_0 V - T_0 S - \Sigma \mu_{c0} n_c. \tag{6.30}$$

Using equation 4.29 for the Gibbs free energy then gives

$$\langle A \rangle = \langle E \rangle + P_0 V - T_0 S - (E_0 + P_0 V_0 - T_0 S_0),$$

which is equation 6.24 representing the available energy of a closed system.

Equation 6.28 indicates that the available energy can be interpreted as the expected availability in an energy distribution compared with that in an energy distribution at the datum state. Equation 6.27 further indicates that the availability is the difference in the energy distribution. In chapter 17 we will return to this interpretation when we examine the difference between the energy distribution among states in biological systems and the thermal-equilibrium states of the environment.

In continuum mechanics a convection derivative indicating the rate of change of a set of particles is often introduced. This ensures the conservation of mass and helps identify the forces that alter the momentum of particles. If this derivative is desired, the the steady-state assumption is not required. Following the same particles, $dn/dt = 0$, and no through flow of particles occurs, so $\dot{n}_p = 0$. The derivative d/dt then represents the rate of change with time of the system of the same particles. In spatial coordinates this is written

$$\frac{d}{dt} = \frac{\partial}{\partial t} + v_i \cdot \frac{\partial}{\partial x_i},$$

or

$$\frac{d}{dt} = \left. \frac{\partial}{\partial t} \right)_{x_i \text{ fixed}} + \bar{v} \cdot \text{grad}.$$

Note that v_i and $\bar{v}$ are the vector velocities of particles.

Other special applications and extensions to chemical systems have been made by Gaggioli [6–8] and others [11,23,24,29]. The extensive literature in German is indicative of the important role that availability or exergy plays in the European chemical industry [4,12–15,17].

Available-Energy Loss in Thermal Mixing

Whenever streams of materials at different temperatures are mixed or are allowed to come to thermal equilibrium, a loss in available energy is incurred.

Typical situations in which this loss is detrimental include heat-exchanger leakages, mixed-stream thermal recovery systems, and thermal-radiation exchange.

In order to illustrate the available-energy loss in thermal mixing, let us consider the situation sketched in figure 6–3. Two streams at different temperatures T_1 and T_2 are thermally mixed to a final common temperature T_3. T_1 is arbitrarily taken as greater than T_2.

The change in available energy can be written as

$$\Delta B = M_3(h_3 - T_0 S_3) - [M_2(h_2 - T_0 S_2) + M_1(h_i - T_0 S_1)].$$

If the system is insulated, then energy is conserved, and

$$M_3 h_3 = M_2 h_2 + M_1 h_1.$$

Mass is also conserved, so that

$$M_3 = M_2 + M_1.$$

From these two conservation conditions the available-energy change becomes

$$\Delta B = T_0 [M_2(S_3 - S_2) + M(S_3 - S_1)].$$

Let us express this in integral form to eliminate the entropy

$$\Delta B = \int dB = -T_0 \left\{ M_2 \int_2^3 dS + M_1 \int_1^3 dS \right\}.$$

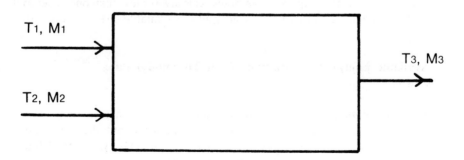

T1, M1

T2, M2

T3, M3

Figure 6–3. Simple Mixing Diagram

Now

$$dS = \frac{dh}{T} = \frac{c_p \, dT}{T}$$

for c_p a constant. Then

$$\Delta B = -T_0 c_p \left\{ M_2 \ln \frac{T_3}{T_2} + M_1 \ln \frac{T_3}{T_1} \right\}.$$

Using the energy-conservation relation gives

$$T_3 = \frac{T_1 M_1 + T_2 M_2}{M_1 + M_2}.$$

Let $x_1 = M_1/M_3$ and $x_2 = M_2/M_3$; then from mass conservation,

$$x_1 + x_2 = 1.$$

Then we can write

$$\Delta B = -T_0 c_p M_3 \ln \left[\frac{T_1 x_1 + T_2 x_2}{T_1^{x_1} T_2^{x_2}} \right]$$

Let $T_1 = T_2 + z$, where z is positive. Then

$$\Delta B = -T_0 c_p M_3 \ln \left[\frac{(1 + z/T_2)^{x_1}}{(1 + z x_1/T_2)} \right].$$

For $0 < x_1 < 1$, this expression in brackets is always less than one. Then ΔB is always negative in thermal mixing, and the available energy decreases.

Available-Energy Computations from Thermodynamic Charts

The computation of available-energy gains or losses in a thermodynamic system can be facilitated by certain observations about the states on a conventional enthalpy-entropy diagram [27,28,30,31]. Although detailed computations are best carried out with computer calculations, for design work, charts and diagrams can be useful in estimating expected changes in performance. Figure 6–4 shows a representative enthalpy (H) - entropy (S)

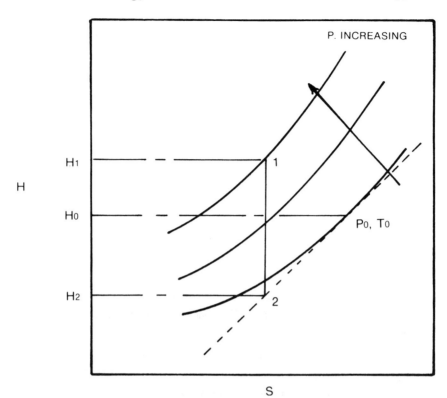

Figure 6–4. Schematic Enthalpy-Entropy Diagram for Illustration of the Graphical Determination of Available Energy

diagram with constant pressure and temperature lines superimposed. The datum state is shown at the intersection of the pressure P_0 and temperature T_0 of the environment.

It is desired to calculate the available energy of a system at state 1. This could be calculated from computation of B as

$$B = (H_1 - T_0 S_1) - (H_0 - T_0 S_0),$$

by extracting values of H_1, S_1, H_0, and S_0 from the diagram. A simpler method is possible by noting that if the Gibbs equation is written in the form

$$dH = TdS + Vdp,$$

and the identity

$$dH = \left. \frac{\partial H}{\partial S} \right)_p dS + \left. \frac{\partial H}{\partial P} \right)_s dp$$

is used, then

$$\left. \frac{\partial H}{\partial S} \right)_p = T.$$

This partial derivative represents the slope of constant-pressure lines on the $H - S$ diagram. This explains the divergence of constant-pressure lines as the temperature is increased.

A line on the $H - S$ diagram through the T_0, P_0 tangent to the constant-pressure P_0 line can be represented by the equation

$$H = C + \left. \frac{\partial H}{\partial S} \right)_p S,$$

$$H = C + T_0 S,$$

where C is a constant, the enthalpy at $S = 0$. This line is shown as the dashed line on the diagram.

Along this line $C = H - T_0 S$. If the slope is drawn from the point T_0, P_0, then

$$C = H_0 - T_0 S_0,$$

which is the reference available-energy state.

The available energy of any point along this line is the same as that of the datum state, which is taken as zero. The available energy of any point is then represented by the difference in its value of $H - T_0 S$ from that anywhere on this line. The simplest computation is to select a point on this line with the same entropy. The available energy of (1) is then equivalent to the difference in enthalpy from point (1) to this C line at point (2), shown on the diagram as $B = H_1 - H_2$.

By constructing the C line through the datum point, any change in available energy in a process can be readily estimated by measurement of the change in the vertical distance from process points to this line. The tangent-line construction is simplified by observing that the slope

$$\left. \frac{\partial H}{\partial S} \right)_p$$

increases with the temperature. It is also readily seen that if this is the case, an increase in the datum temperature T_0 will decrease the available energy, since the slope C of the line will then increase, giving a smaller distance from a process point to the reference line. Similarly, a decrease in the datum temperature will increase the available energy.

In Carnot limited systems, the lower the datum temperature, the more energy is available energy in a work process in the equation

$$dW_{max} = dQ\left(\frac{T - T_0}{T}\right) .$$

The result is consistent if one considers the heating of a material at constant pressure as occurs in a boiler. Let us consider figure 6–5 for two cases in which the same amount of heat is transferred at constant pressure. If the datum is at T_{01} and the working fluid is heated at constant pressure, then $Q = \Delta H$, and the system ends at temperature T_{02}. The change in available energy is then ΔB_1. If the datum state was initially at T_{02} and the same energy was added, then the final state would be at T_{03}, and the change in available energy would be ΔB_2. The available-energy change ΔB_2 for the higher datum temperature T_{02} is less than ΔB_1 for the lower datum temperature T_{01}. The useful work that can be obtained after heat is added is then less for the system initially at a higher datum temperature. The change in the datum temperature, as it occurs seasonally or diurnally, can affect the available energy of a system differently depending on the state of the system.

Another observation regarding the available-energy increase of a substance when heated is also apparent from this datum-line consideration. Heating a substance at constant pressure does not increase its available energy substantially compared with heating it at constant volume, which increases its pressure as well as its temperature. Heating a substance in an unconstrained environment is very ineffective in increasing its available energy.

Figure 6–6 illustrates some classical flow processes [5,18,22,32] in which the enthalpy and available-energy change differences are indicated. The differences show how the use of available-energy function rather than enthalpy is the better means for improving system performance.

$a - b$ The process $a \rightarrow b$ represents a throttling process in which the enthalpy remains constant. The available energy decreases, however. There is thus a loss associated with the throttle process, as illustrated by ΔB_{ab}.

$c - d$ The process $c \rightarrow d$ represents a perfect nozzle where $\Delta H_{cd} = \Delta B_{cd}$, a conversion from thermal to kinetic energy without losses.

$c - d'$ The process $c \rightarrow d'$ represents an actual nozzle process. Since the constant-pressure lines diverge with increased S, there is a greater loss in available energy $\Delta B_{cd'}$ than in enthalpy loss $\Delta H_{cd'}$ as

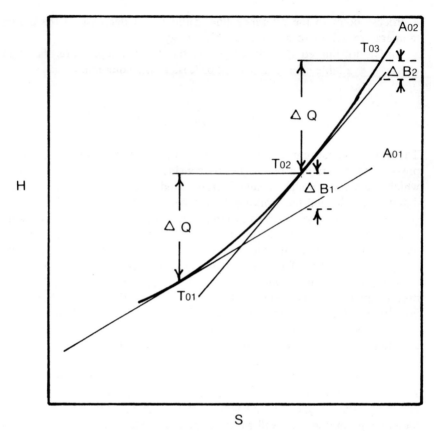

Figure 6–5. Enthalpy-Entropy Diagram Illustrating the Change in the Available Energy of a System to Which Equal Energy Is Added as Heat at Constant Pressure at Different Datum States

$$\Delta B_{cd'} = \Delta H_{cd'} + \Delta M.$$

A simple experiment often performed in thermodynamics laboratories is the measurement of power delivered by a small air motor or turbine that is supplied by compressed air. The student is asked to determine the efficiency of this device by measuring the air-flow rate and the input-to-output change in state of the air.

The student invariably finds, to his consternation, that the power output is greater than the power extracted from the air as determined by the mass rate of flow times the enthalpy change. Confidence in the steady-flow energy equation is immediately shaken and only partially restored by resorting to explanations about entropy changes, heat exchange, and so forth.

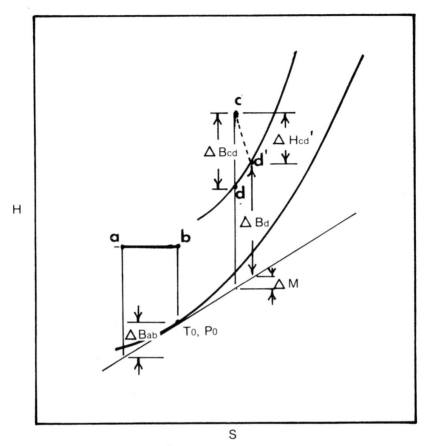

Figure 6–6. Illustration of Available-Energy Changes in Classical Thermodynamic Flow Processes

A question in air-motor performance is whether the heat flow from the environment to the air motor is beneficial. If beneficial, it would indicate that air motors that operate below environmental temperature should have fins attached to absorb a maximum amount of heat from the environment. If detrimental to performance, the air motor should be insulated from the environment.

Let us consider an air turbine system shown in figure 6–7. The energy source is assumed to be compressed gas at the environmental temperature T_0 but a higher pressure $P_1 > P_0$. The available energy at state (1) may be computed as

$$B_1 = H_1 - T_0 S_1$$

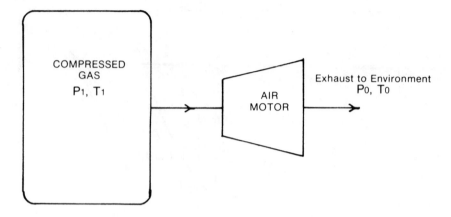

Figure 6–7. Air-Motor Flow Schematic

relative to the environment

$$B_0 = H_0 - T_0 S_0.$$

The maximum work obtainable from gas at conditions (1) is then

$$W_{max} = B_1 - B_0 = (H_1 - H_0) - T_0[S_1 + S_0].$$

Since $H_1 = H_0$,

$$W_{max} = T_0(S_0 - S_1).$$

This is represented graphically by the vertical distance shown on the $H - S$ diagram (figure 6–8).

From this diagram it is clear that the maximum work ΔB_s achievable from an isentropic process with the exit at P_0 is less than $T_0(S_0 - S_1)$. If the process is isentropic, then for a perfect gas the maximum work output is

$$\Delta B_s = \int dB = \int dH_s = \int_{T_s}^{T_0} c_p \, dT = c_p(T_0 - T_s),$$

but

$$T_s = T_0 \left(\frac{P_1}{P_0} \right)^{\gamma - 1/-\gamma} \quad \therefore \ \Delta B_s = c_p T_0 \left(1 - \left(\frac{P_1}{P_0} \right)^{\gamma - 1/\gamma} \right).$$

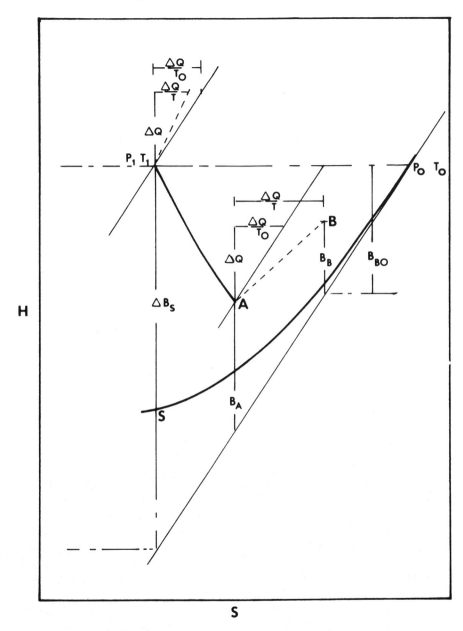

Figure 6–8. Enthalpy-Entropy Diagram Illustrating Air-Motor Performance
with and without Heat Addition from the Environment

The other obvious process is one at constant temperature. This would be represented by the horizontal line and could in theory be achieved by heat flow in from the environment. Then

$$\Delta B_T = \int dB_T = \int T ds = RT_0 \ln \frac{P_1}{P_0}$$

For $\gamma > 1$, a physical requirement,

$$\Delta B_{T=\text{const}} > \Delta B_{s=\text{const}}$$

Now $dH = V dp + T dS$. For the constant temperature case:

$$dW = -dH + dQ$$

$$dW = -V dp - (T dS - dQ)$$

$$dW = -\frac{RT_0}{p} dp - (T dS - dQ)$$

$$dW = RT_0 \ln \frac{P_1}{P_0} - (T dS - dQ)$$

If reversible, $T dS = dQ$. If not reversible, $T dS > dQ$, and the work output will be reduced. The question posed initially can now be addressed by reference to figure 6–8.

In order to pick up heat from the atmosphere, the temperature is reduced by expansion. Let us say this is to point A. Then at this lowered temperature the available energy is B_A. If heat ΔQ is then picked up reversibly, then the entropy increase is $\Delta Q/T$. This entropy increase is greater than $\Delta Q/T_0$ since T is less than T_0. The available energy B_B at B is then less than it was at A before the heat is picked up.

The heat addition thus reduces the available energy. There is therefore a loss that occurs if heat is absorbed from the atmosphere. Hence there is nothing to be gained by improving the conduction of the air motor to the environment. It should be noted that there is an advantage to heating the air motor if the heating increases the temperature above the environment temperature, as shown by the upward paths from (P_1, T_1).

In the situation with the air motor, the available energy of the initial state or maximum work is calculated by assuming a constant-temperature process. This does not mean that adding heat from the environment will improve the efficiency.

As we have shown, in the air-motor case, as in the usual turbine situation, the closer to an isentropic process, the better the air-motor performance.

The air-motor example demonstrates an important principle in that the energy of the compressed gas is the same as that of the environment. The potential to do work is attributable to its *available energy*, not its *energy*. The available energy is in this case due to its entropy being lower than in the compressed state.

In a closed system the available energy $A = E - T_0 S + P_0 V$ may be represented in graphical form for design purposes by plotting A versus T/T_0 and P/P_0. The minimum value of A will occur at $T/T_0 = 1$ and $P/P_0 = 1$, since the system will then be at equilibrium with its environment.

In two dimensions this may be represented as shown in figure 6–9. Lines of constant available energy can then be constructed and the available energy

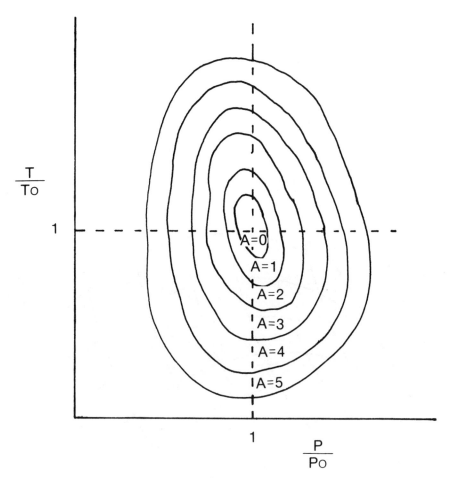

Figure 6–9. Closed-System Availability Contour-Plot Illustration

in a given process traced as on a topological map. Paths of maximum or minimum rate of dissipation of available energy can then be identified for optimization purposes.

An even more useful design plot is the three-dimensional plot of A versus T/T_0 and P/P_0 shown in figure 6–10. For a perfect gas the change in available energy with temperature and pressure can be evaluated analytically. Since

$$dA = dE - T_0 dS + P_0 dV,$$

and

$$TdS = dE + PdV.$$

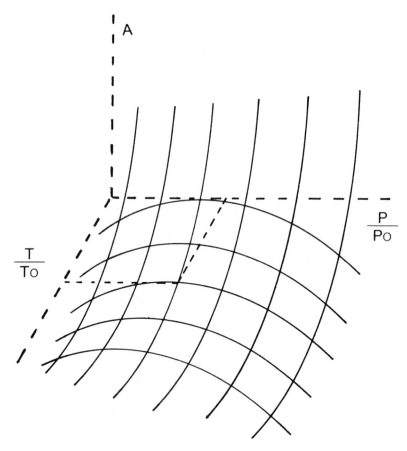

Figure 6–10. Closed-System Availability Diagram

with $dE = c_v dT$ and $PdV = RdT - VdP$. Using these to eliminate V and S in the available-energy change gives

$$dA = \left[c_v \left(1 - \frac{T_0}{T} \right) + \left(\frac{P_0}{P} - \frac{T_0}{T} \right) R \right] dT$$

$$+ \frac{R}{P} \left(T_0 - \frac{P_0}{P} T \right) dP.$$

A constant-pressure heating process is represented in figure 6–9 by a vertical line, and a constant-temperature expansion process is represented by the horizontal lines.

The formulation of the change in available energy with temperature and pressure can be used to examine the use of gas for energy storage. It is generally though that compressed gas provides a useful energy-storage system. Similarly, energy storage in heated gas is considered a reasonable storage mechanism. A question that is seldom asked is whether storage in the form of a low-pressure system or a low-temperature system might be more effective. To examine this, let us consider a differential change in available energy as

$$dA = \left. \frac{\partial A}{\partial T} \right)_p dT + \left. \frac{\partial A}{\partial P} \right)_T dP.$$

Comparing this with the previous equation gives

$$\left. \frac{\partial A}{\partial T} \right) = c_v \left(1 - \frac{T_0}{T} \right) + \left(\frac{P_0}{P} - \frac{T_0}{T} \right) R,$$

and

$$\left. \frac{\partial A}{\partial P} \right)_T = \frac{R}{P} (T_0) \left(1 - \frac{P_0}{P} \frac{T_0}{T_0} \right).$$

If the storage is at temperature $T = T_0$ in compressed gas, then

$$\left. \frac{\partial A}{\partial P} \right)_T = \frac{R}{P} (T_0) \left(1 - \frac{P_0}{P} \right).$$

This is shown plotted in figure 6–11.

For a given pressure increment, a greater change in A occurs for a

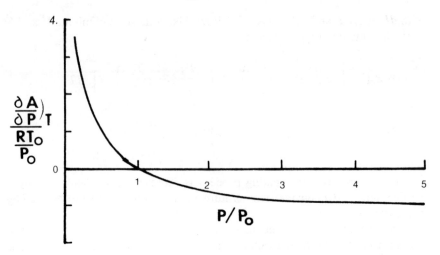

Figure 6–11. Rate of Availability Change with Pressure in Gas Energy-Storage Systems as a Function of Pressure

negative ΔP than for an equivalent positive pressure ΔP. Similarly, in processes for lowering the pressure of a system, the available energy approaches $-\infty$ as $P/P_0 \to 0$. This illustrates the large energy required to create a vacuum.

In the case of storage at low temperatures, consider pressure $P = P_0$. Then

$$\frac{\partial A}{\partial T}\bigg)_p = c_v\left(1 - \frac{T_0}{T}\right) + R\left(1 - \frac{T_0}{T}\right) = c_p\left(1 - \frac{T_0}{T}\right)$$

This is a maximum at $T/T_0 \to 0$. Change at low temperatures as shown in figure 6–12.

Low-temperature storage gives greater available-energy differences for a given increment ΔT. The cooling of a gas below the environmental temperature requires more available energy than heating it through the same ΔT.

The importance of energy conservation has resulted in a wide range of available-energy applications [20,21,25,26,31]. The use of the available-energy concept in the design of new industrial processes will require new ways to visualize and predict their performance. This section has attempted to illustrate a few of the simple graphical techniques. Computer-aided design with new graphical methods should allow further development in this area.

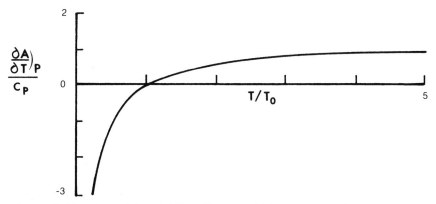

Figure 6–12. Rate of Availability Change with Temperature in Gas Energy-Storage Systems as a Function of Temperature

References

1. Evans, R.B. "A Proof that Essergy is the Only Consistent Measure of Potential Work." Ph.D. diss., Dartmouth College, Hanover, N.H., 1969.
2. Obert, E. *Internal Combustion Engines.* Scranton, Pa.: International Textbook Company, 1950.
3. Keenan, J.H. *Thermodynamics.* New York: Wiley, 1941.
4. Rant, Z. "Fxergie, ein neues wort fur, technische Arbeitfahigkeit." *Forsh. Ing-Wes.* 22(1956):32.
5. Tribus, M. *Thermostatics and Thermodynamics.* Princeton, N.J.: Van Nostrand, 1961.
6. Gaggioli, R.A. "The Unrestricted Engineering Bernoulli Equation." *Chem. Eng. Sci.* 16(1961):167.
7. Gaggioli, R.A. "The Concept of Available Energy." *Chem. Eng. Sci.* 16(1961):87.
8. Gaggioli, R.A. "The Concepts of Thermodynamic Friction, Thermal Available Energy, Chemical Available Energy and Thermal Energy." *Chem. Eng. Sci.* 17(1962):623.
9. Evans, R.B. "A Contribution to the Theory of Thermo-Economics." Report no. 62-36, University of California Department of Engineering, 1962.
10. Tribus, M., and Evans, R.B. "Thermo-Economics of Saline Water Conversion." *Ind. Eng. Chem., Process Des. Develop.* 4(1965):195.

11. Riekert, L. "The Efficiency of Energy Utilization in Chemical Processes." *Chem. Eng. Sci.* 29(1974):1613.
12. Szargut, J. and Styrylska, T. "Angenaherte Bestimmung der exergie von Brennstoffen." *Brennstoff-Warme-Kraft* 16, no. 12(1964):125.
13. Baehr, H.D. "Zur Definition Exergeti scher Wirkungsgrade-Eine Systematische Untersuchung". *Brennstoff-Warme-Kraft* 20, no. 5(1968): 197.
14. Grassmann, P. "Die Exergie und das Flussbild der technisch nutzbaren Leistung." *Allgen Waermetch* 9, no. 4(1959):79.
15. Bosnjakovii, F. "Practical Importance of Exergy." *Brennstoff-Waerme-Kraft* 13, no. 11(1961):481.
16. Darrieus, G. "A Rational Definition of Steam Turbine Efficiencies." *Engr.* 130(September 1930)283.
17. Szargut, J. "Grenzen fuer die Anwendungsmoeglichkeiten des Exergie begriffs." *Brennstoff-Waerme-Kraft* 19, no. 6(1967):309.
18. Obert, E., and Gaggioli, R. *Thermodynamics.* New York: McGraw-Hill, 1961.
19. Evans, R.B. "A New Approach for Deciding upon Constraints in the Maximum Entropy Formalism." In *The Maximum Entropy Formalism*, ed. R.D. Levine and M. Tribus, p. 169. Cambridge, Mass.: MIT Press, 1979.
20. Broydyanskii, V.M. "A Single Criterion for Assessing the Performance of Power Plants." *Teploenergetika* 14, no. 3(1967):71.
21. Gyftopoulos, E.P.; Lazaridis, L.J.; and Widmer, T.F. "Potential Fuel Effectiveness in Industry." Report to the Energy Project, Ford Foundation. Cambridge, Mass.: Ballinger, 1975.
22. Reistad, G.M. "Availability: Concepts and Applications." Ph.D. diss., University of Wisconsin, 1970.
23. Riekert, L. "Conversion, Conservation and Loss of Energy in Chemical Processes." In *Future Energy Production Systems*, vol. II, ed. J. Denton and N. Afgan, p. 675. New York: Academic Press, 1976.
24. Riekert, L. "The Conversion of Energy in Chemical Rections." *Energy Conv.* 15, no. 3–4(1976):81.
25. Berg, C.A. "A Technical Basis for Energy Conservation." *Mech. Engr.* 96, no. 5(1974):30.
26. Clarke, J., and Horlock, J.H. "Availability and Propulsion." *J. Mech. Engr. Sci. 17*, no. 4(1975):223.
27. Keenan, J.H. "A Steam Chart for Second-Law Analysis." *Mech. Engr.* 54(1932):195.
28. Keenan, J.H. "Availability and Irreversibility in Thermodynamics." *Brit. J. Appl Phys.* 2(1951):181.
29. Hammitt, A.G. "Chemical Energy Engines." *Adv. Energy Conv.* 7(1968):191.

30. Haywood, R.W. "A Critical Review of the Theorems of Thermo-dynamic Availability with Concise Formulations." part 1. *J. Mech. Engr. Sci.* 16, no 3(1974):160; part 2. *J. Mech. Engr. Sci.* 16, no. 4(1974):258.

31. Bruges, E.A. *Available Energy and the Second Law Analysis.* New York: Academic Press, 1959.

32. Clausius, R. *The Mechanical Theory of Heat,* ed. T.A. Hirst, English trans. London: Blackie and Sons, 1961.

33. Hussein, M.; Wood, R.J.; Ocallagh, P.W.; and Probert, S.D. "Efficiencies of Exergy Transductions." *Appl. Energy* 6(1980):371.

Appendix 6A:
A Note on Partial
Quantity Variation
with Composition

Consider a volume of material that has a composition of several different species. It is sometimes important to determine how the thermodynamic state variables change with composition changes. To handle these situations, the concept of partial quantities is introduced.

Suppose we have a material of i different species. Let us denote the number of moles of the ith species as n_i. If the material has a state with a thermodynamic property, say, B, then this property, if extensive, depends on the partial properties of each species B_i. This B may represent quantities like enthalpy, volume, Gibbs free energy, and so forth.

$$B \equiv \sum_i B_i n_i \qquad \text{summed over all species } i. \qquad (6A.1)$$

A change in B may be represented by a differential, which by the chain rule of differentation gives

$$dB = \sum_i B_i dn_i + \sum_i n_i dB_i. \qquad (6A.2)$$

Now the thermodynamic variable B is a function of $i + 2$ variables, since the composition i, given any two thermodynamic variables, defines the state of the system. For example, if we know the temperature and entropy of a material as well as its composition, then its enthalpy, pressure, specific volume, and so forth are known.

The usual quantities of state used in thermodynamic problems are the temperature and pressure because of their ease of measurement. If this is the case, then B can be represented as a function of P, T, and n_i; that is,

$$B = B(P, T, n_i). \qquad (6A.3)$$

The differential of B may then be represented by

$$dB = \frac{\partial B}{\partial P}\bigg)_{T, n_i} dP + \frac{\partial B}{\partial T}\bigg)_{P, n_i} dT + \sum_i \frac{\partial B}{\partial n_i}\bigg)_{T, P, n_j} dn_i. \qquad (6A.4)$$

Equating equations 6A.2 and 6A.4 gives

155

$$\Sigma B_i dn_i + \Sigma n_i dB_i = \frac{\partial B}{\partial P}\bigg)_{T,n_i} dP \; + \frac{\partial B}{\partial T}\bigg)_{P,\,n_i} dT + \Sigma \frac{\partial B}{\partial n_i}\bigg)_{T,\,P,n_j} dn_i \, . \quad (6A.5)$$

Now is should be recognized that

$$\frac{\partial B}{\partial n_i}\bigg)_{T,\,P,\,n_j}$$

is an expression that tells how B changes as the quantity of species i changes. This is exactly B_i. Therefore,

$$B_i \equiv \frac{\partial B}{\partial n_i}\bigg)_{T,\,P,\,n_j} \, .$$

Equation 6A.5 then becomes

$$\Sigma n_i dB_i = \frac{\partial B}{\partial P}\bigg)_{T,n_i} dP + \frac{\partial B}{\partial T}\bigg)_{P,\,n_i} dT$$

If the change in the material occurs at constant pressure and temperature, then

$$\Sigma_i n_i dB_i = 0. \quad (6A.6)$$

This is an important case, when the pressure and temperature are controlled or, as in many chemical situations, the reaction is carried out at atmospheric pressure and ambient temperature.

For these cases the variation or change in the variable B is represented by

$$dB = \Sigma B_i dn_i. \quad (6A.7)$$

In this situation the change in B for each component need not be determined. In particular, the change in B with composition will remain constant.

Relations 6A.6 and 6A.7 are often considered as Gibbs-Duhem relationships, but as just show they are a result of the definitions of partial quantities and mathematical properties of total differentials of functions of several variables. As an example, if the enthalpy of a system is

$$H = \Sigma H_i n_i.$$

The differentials are

$$dH = \Sigma H_i \, dn_i + \Sigma n_i \, dH_i$$

$$dH = \frac{\partial H}{\partial T}\bigg)_{P, n_i} dT + \frac{\partial H}{\partial P}\bigg)_{T, n_i} dP + \Sigma \frac{\partial H}{\partial n_i}\bigg)_{T, P, n_j} dn_i . \quad (6A.8)$$

If the temperature T and pressure P remain constant in a process, or are the same at the beginning and end of the process,

$$dH = \Sigma H_i \, dn_i . \qquad (6A.9)$$

Then from equation 6A.8,

$$\Sigma n_i \, dH_i = 0.$$

Appendix 6B:
Datum and
Reference-State
Discussion

Future economic and energy accounting must concern itself with the implications of environmental-state variability. The alteration of the environment by processing systems is a vital question to be resolved in terms of its effect on the processing systems as well as on personal health and welfare. The question of assimilative capacity revolves around the alteration of the environment.

The available energy of a substance relative to its environment is of special concern in ecological problems. The detrimental potential of a waste flow must be related to an acceptable condition. An acceptable condition is generally prescribed by air- and water-quality standards. Tables 6B–1 and 6B–2 are typical environmental standards set by U.S. regulatory agencies.

An environmental-state definition requires the specification of T_0, P_0, S_0, μ_0 and the concentration of each substance in the environment.

Let us first examine the temperature datum T_0. In a power plant two temperatures are immediate candidates. The first is the temperature of the atmosphere at the plant site. The second is the temperature of the water body used for condenser cooling. The water body is usually used for cooling not simply because it is colder than the air, but also because the cost of the heat exchanger for water transfer is less than the cost of an air-interface heat exchanger.

For technical reasons it is sometimes useful to use the temperature of a river or deep well as the environmental temperature. This introduces a simplification in comparative technology assessment but is not useful for basic decisions in which environmental effects are involved. Cold river water is a resource that is being utilized and exploited in cooling processes. It has available energy relative to the air environment that is depleted by the cooling process. At present this available energy is assumed as free, requiring for its use only the cost of the heat exchangers and their operation. In the past heating of the water was slight because the withdrawal and use flow was slight compared with the main stream flow. This is no longer the case, and the mainstream available-energy change is now often significant.

In many sewer lines the average temperature in winter is in the range of $50°–70°F$. This is a very attractive thermal source for heat-pump use, but widespread operation of heat pumps using the sewer line would adversely effect sewage-treatment processes. The reaction rates of biological and chemical systems in a sewage-treatment plant are temperature dependent. If the inlet-flow temperature were reduced, the size of the treatment reactors

Table 6B–1
National Ambient-Air-Quality Standards

	Primary	Secondary
Suspended particles ($\mu g/m^3$)		
Annual geometric mean	75	
Maximum, 24 hr[a]	260	150
Sulfur oxides ($\mu g/m^3$)		
Annual arithmetic mean	80	
Maximum, 24 hr[a]	365	
Maximum, 3 hr[a]		1,300
Carbon monoxide (mg/m^3)		
Maximum, 8 hr[a]	10	10
Maximum, 1 hr[a]	40	40
Photochemical oxidants ($\mu g/m^3$)		
Maximum, 1 hr[a]	160	160
Nitrogen oxides ($\mu g/m^3$)		
Annual arithmetic mean	100	100
Hydrocarbons ($\mu g/m^3$)		
Maximum, 3 hr[a]	160	160
Lead ($\mu g/m^3$)		
Maximum calendar-quarter average	1.5	1.5

Source: U.S. Environmental Protection Agency, Washington, D.C. (1979).
[a]Not to be exceeded more than once a year per site.

Table 6B–2
Chemical Standards for Drinking Water, U.S. Public Health Service

Substance	Maximum Concentration (mg/liter)	
	Recommended	Permissible
Alkyl benzene sulfonate	0.5	
Arsenic	0.01	0.05
Barium		1.0
Cadmium		0.01
Carbon chloroform extract	0.2	
Chloride	250	
Chromium		0.05
Copper	1.0	
Cyanide	0.01	0.2
Iron	0.3	
Lead		0.05
Manganese	0.05	
Nitrate	45.0	
Phenols	0.001	
Selenium		0.01
Silver		0.05
Sulfate	250	
Total dissolved solids	500	
Zinc	5.0	

Source: Drinking Water Standards," Publication no. 956. U.S. Department of Health, Education and Welfare, Public Health Service (1962).

would need to be increased to allow more complete reaction. The alternative of heating the sewage would, of course, lead to a net increase in energy use.

As a further note, this is a case in which it makes sense for an individual to use a local resource (warm sewage flow) to his own advantage, but use by the whole community would be unacceptable. If the whole community installed sewage-flow heat pumps, the result would be detrimental. Not only would additional energy be required at treatment plants, but freezing of sewer lines would likely become a significant problem.

The environmental temperature (T_0) selection can be taken as the temperature of the air at the point of withdrawal or discharge when the utilization process is not operating. Any alteration of the environment can then be assessed relative to the changes induced by the process. Judgment about whether the discharge is deleterious or beneficial can then be made on a more consistent scale. A power plant that warms the water in a lake, reducing the ice cover in winter and hence improving the oxygenation, may be beneficial. In the summer, however, the same heating can result in reduced oxygen content in the water, which may be harmful. The availability changes associated with a power plant can then be assessed on a better basis in relation to other environmental factors. Heating per se is not to be taken as bad; it is the environmental effect of this heating that must be evaluated.

The use of available energy in a technical accounting system requires the establishment of a state at which the available energy is zero. Keenan[1] in his consideration of available energy referred to this as the *dead state*. This dead state is only a concept. There is no actual state at which the available energy of an earth environment is zero. Evans[2] in his work carefully points out that the available energy of a substance is related to the difference in the distribution of energy from that when it is in equilibrium with the environment.

The available energy of a specific quantity of a substance relative to any fixed environment can be calculated with little difficulty. One can calculate the available energy of a lump of coal relative to an environment where its components are in equilibrium. The available energy of all the coal in the earth's crust however, cannot be calculated since if the components were chemically reacted in the atmosphere, the resultant environment would not be predictable. The effect on the atmosphere of the CO_2 released by the coal's combustion, for example, could not be determined.

For most engineering computations only a relative available-energy change is important, and the selection of an environmental state can be arbitrarily made. There are several situations where an absolute available energy would be useful. These occur where effectiveness measures are made to compare technology and in the communication of the value that available energy is assigned in economic-costing analysis. Several researchers have attacked this problem of establishing a *reference* state that would represent an environment where the availability of all substances would be fixed [3–9].

They hypothesize separate equilibrium environments for the atmosphere, the ocean, and the earth's crust. This reference is established based on the criteria that the most-devalued form of each compound found in the environment be used and that the reference state approximate the natural environment. A shortcoming of this approach is that the resultant reference state does not guarantee that an arbitrary substance in any state will have a positive available energy.

A difficulty arises in part because the environment is not in an equilibrium state. There is a continual flux of energy through any space attributable either to solar-energy flux, convection, diffusion, or conduction processes. A true environmental temperature would involve an attempt to evaluate the average temperature of the solar system, universe, or galaxy. In principle, this would probably be close to $4K$ if it were at equilibrium (which it is not). An average earth temperature can be argued, but in practice the selection of 298.15K, the temperature at which standard chemical-thermodynamic properties have been tabulated, is much more convenient. In the Antarctic this would not prove useful. In any event, as a general rule it is probably satisfactory to use $T_0 = 300$ K as the reference temperature.

When environmental changes are crucial, as in situations where the local climate is changed by a process, or the performance of a system is strongly effected by the local environment then a local environment state needs to be used.

A datum or reference state is needed for computations. An environmental state is needed for environmental evaluations. It is attractive to utilize standard-table chemical-thermodynamic properties as the datum state. This assigns zero enthalpy to pure components at a standard pressure and temperature. This has been done by Sussman [9], and it is especially convenient for practical purposes.

The definition of the chemical reference state is further complicated by the fact that the atmosphere (or air at any location) is not in chemical equilibrium. Table 6B–3 indicates a typical chemical composition of the near-ground atmosphere. This is neither an equilibrium distribution of chemical species nor in equilibrium with the soil interface. The nonequilibrium is maintained by solar-energy flux processes and biological activity.

For substances that exist as stable compounds in the environment the available energy is evaluated simply by straightforward computations. The enthalpy and entropy can be calculated from standard tables using procedures outlined in chapter 5. The difference in chemical potential can also be calculated utilizing the difference in the concentration between the specified state and the reference state. For compounds that are not stable in the environment, the computation of the chemical potential requires the assumption of a chemical reaction to stable compounds in the environment. Specific important environmental examples include carbon monoxide and sulfur dioxide. Neither of these compounds exists in a stable acceptable environ-

Table 6B–3
Typical Air Composition Near the Earth's Surface

Component	Parts per Million (PPM)
Nitrogen	780×10^3
Oxygen	210×10^3
Water	20×10^3
Argon	9.3×10^3
Carbon dioxide	320
Neon	18
Helium	5.2
Krypton	1
Xenon	8×10^{-2}
Nitrous oxide	25×10^{-2}
Methane	1.5
Hydrogen	0.5
Nitrogen dioxide	1×10^{-3}
Ozone	2×10^{-2}
Sulfur dioxide	2×10^{-4}
Carbon monoxide	0.1
Ammonia	1×10^{-2}

Source: National Academy of Sciences, "Atmospheric Chemistry: Problems and Scope," Committee on Atmospheric Sciences, National'Research Council, Washington, D.C. 1975).

ment. Compounds that are formed by the reaction of these compounds with components of the environment, however, do exist. Carbon monoxide is reacted with oxygen to carbon dioxide, and sulfur dioxide is either reacted with water to form sulfuric acid (an unacceptable environmental component) or with calcium carbonate to form an acceptable gypsum compound. In the absence of an environmental component that will react to give an acceptable environmental compound, a component must be supplied.

The computation of the chemical potential of CO_2 at environmental conditions would be found using standard tables as

$$\mu_{CO_2,0} = H_{CO_2}(T_0) - T_0 S_{CO_2}(T_0,P_0) + RT_0 \ln X_{CO_2,0}.$$

For CO however, which is not present in environmental conditions, the chemical potential at the environmental state is found by computing the chemical potential from the equilibrium reaction with oxygen to form CO_2. The chemical potential equilibrium equation for the reaction

$$CO + 1/2\ O_2 \rightleftharpoons CO_2$$

is

$$\mu_{CO} = \mu_{CO_2} - 1/2\ \mu_{O_2}.$$

The chemical potential for CO is then found from the chemical potentials of

CO_2 and O_2 at the environmental conditions of temperature, pressure and concentration.

For purely anthropomorphic reasons it is the deviation from a life-supporting, healthful environment that is of importance. It is not an equilibrium atmosphere that is needed for life support, but a nonequilibrium environment that can be exploited by life processes. A Precambrian atmosphere is probably closer to an equilibrium situation but is probably not human-life sustaining because of its high carbon dioxide, hydrogen, methane, and ozone content.

We would like, then, to use as a reference state for the atmosphere a standard that is life supporting. One could take that of the typical near-ground atmosphere, as in table 6B–3. For the earth, let us take a homogeneous earth-crust standard, as in table 1–2, when dealing with mineral resources. For seawater exploitation or utilization processes, let us take the standard seawater composition, as in table 1–4.

A question that arises from having standard compositions for air, water, and earth is how to assess impacts relative to each other. If a mineral is extracted from the earth's crust, it has an available energy relative to the crust composition. In the processing it may have wastes discharged to both the air and the ocean. Part of the processing function can be associated with accelerating the movement toward equilibrium between, say, the earth and the ocean. This results in a loss of available energy. The available accounting has several parts:

1. the loss of available energy of the material in the process of mixing and reaction with the water;
2. the loss of available energy of the earth because of the extraction of high-available-energy minerals;
3. the increase in available energy of the atmosphere because of increased concentrations of air constituents;
4. processing-available-energy losses.

These are all relative changes in available energy and as such present no computational difficulty. The difficulty lies in evaluating the importance of water relative to air pollution. Available-energy change represents the available energy required to return the environment to its original condition.

Note that a datum pressure is also required for determination of available energy. The choice of 1 atm. is usually made without any discussion. The choice is, of course, as difficult as that of temperature or concentration, if one considers the availability in the stratosphere. One atmosphere pressure is selected because of (1) the advantage in calculation from tabulated properties (2) the small effect atmospheric-pressure variations have on the available energy, and (3) the practical use of availability at the earth surface in most applications.

Environmental conditions may, of course, change over the lifetime of a power plant or chemical-processing plant. The corrosive properties of the water in a river may change, altering the preprocessing requirements to reduce scaling boiler-feed water or cooling uses. In a chemical plant the input-water purity may need to be improved through chemical treatment. Thermal pollution from upstream power stations, chemical use, or sewage-treatment use may change the stream temperature. The location of the inlet and discharge pipes for a desalination plant may need to be relocated if a buildup of salinity occurs in the plant vicinity. This involves the idea of locating the reference point spatially. The accounting must allow for changes in the reference state if one is to use it for environmental-change purposes. Any discharge with available energy relative to the datum could in theory be utilized by another process if the discharge has value, or the cost to recover is less than the transportation and processing cost.

In a similar way the energy complex idea for utilization of waste heat for heating greenhouses or space heating involves utilization of energy that would otherwise alter the environment or reference state. Those environmentalists who argue for no environmental change are arguing for no discharge except at the reference state, or for discharge at the assimilation rate. The power company that continues to utilize the reference as the river, when in fact it is the atmosphere, misses the point if cooling towers are used to reduce the thermal pollution. Cooling-pond or cooling-tower construction costs are an integral part of the process cost, not an extra cost.

Human economic activity may sometimes be classified as activity that takes advantage of the improbable distributions of the earth's resources. Ultimately, the utilization of resources means the homogenization of these improbably distributed resources. In this homogenized form they are not useless but can be transformed to useful forms through the sun-earth exchange processes. These natural exchange processes produce concentration of elements. Examples include solar-evaporation processes and natural accumulation of manganese nodules at ocean depth. Biological processes that concentrate iodine, mercury, and so forth are also included.

The biochemical cycles driven by solar energy maintain the nonequilibrium state of the environment. The atmosphere is not in thermodynamic equilibrium with either the hydrosphere or the earth's crust. At equilibrium there would be little free oxygen in the atmosphere. Nitrogen in the atmosphere would mostly be reacted with water to nitrates or nitric acid [3–5]. Water vapor in the presence of carbon dioxide leads to precipitation with a pH of 5.6 rather than a neutral pH of 7. Acid precipitation [10–12], which is altered by technological processes that release nitrogen oxides and sulfur oxides, further changes environmental conditions.

Biochemical assimilation processes are particularly sensitive to small quantities of toxic substances that accumulate in surface layers. Air-water interfaces are crucial areas of bacterial assimilation. These bacterial processes are responsible for the breakdown of organic substances, which is a basic respiration function of the environment.

In summary, an environmental state is a dynamic nonequilibrium state that is altered by technological processes. For technological accounting purposes this reference state is usually assumed to be a static condition. The reality of these differences must be recognized more fully in the future.

References

1. Keenan, J.H. *Thermodynamics*. New York: Wiley, 1941.
2. Evans, R.B. "The Proof That Essergy Is the Only Consistent Measure of Potential Work." Ph.D. dissertation, Dartmouth College, 1970.
3. Ahrendts, J. "Reference States." *Energy, Int. J.* 5 (1980): 667.
4. Szargut, J. "Grenzen für die Anwendungsmoeglichkeiten des Exergiebegriffs." *Brennstoff-Waerme-Kraft* 19, no. 6 (1967): 309.
5. Riekert, L. "The Conversion of Energy in Chemical Reactions." *Energy Conversion* 15, no. 3–4 (1976): 81.
6. Riestad, G. "Availability and Irreversibility in Thermodynamics." Ph.D. dissertation, University of Wisconsin, 1970.
7. Wepfer, W.J. and Gaggioli, R.A. "Reference Datums for Available Energy." In *Thermodynamics: Second Law Analysis*, ed. R.A. Gaggioli. Washington, D.C.: American Chemical Society, 1980.
8. Rodriguez, L. "Calculation of Available-Energy Quantities." In *Thermodynamics: Second Law Analysis*, ed. R.A. Gaggioli. Washington, D.C.: American Chemical Society, 1980.
9. Sussman, M.V. "Steady Flow Availability and the Standard Chemical Availability." *Energy, Int. J.* 5 (1980): 793.
10. Likens, G.E.; Bormann, F.H.; and Johnson, N.M. "Acid Rain." *Environment* 14 (1972): 33.
11. National Academy of Sciences, "Atmospheric Chemistry: Problems and Scope." Committee on Atmospheric Sciences National Research Council, Washington, D.C., 1975.
12. Cowling, E.B. "Acid Precipitation and Its Effects on Terrestrial and Aquatic Ecosystems." In *Aerosols: Anthropogenic and Natural, Sources and Transport*, ed. T.J. Kneip and P.J. Lioy. *Ann. NY Acad. Sci.* 338 (1980): 540.

7 Energy Conversion

At the rate of progress since 1800, every American who lived into the year 2000 would know how to control unlimited power. He would think in complexities unimaginable to an earlier mind.

–Henry Adams, "The Law of Acceleration,"
The Education of Henry Adams (1906)

Energy Conversion and Thermodynamics

The science of thermodynamics is basic to the optimization of energy-conversion systems. Energy conversion includes the transformation of energy in one form to energy in a desired form. Table 7–1 indicates some of the possible conversions and the process technology utilized. I will not discuss the details of the conversion processes, which are available in specialized texts [1–3]. I will, however, attempt to outline a general way to handle the thermodynamics of these energy conversions.

Thermodynamics as a science was originally founded by Carnot [4] to deal with the conversion of thermal energy (fire) to mechanical energy. The thermal energy in most converters is an intermediate energy form. The initial form is chemical energy of fuels, and the final form is electrical or mechanical energy.

Most thermodynamics texts deal primarily with the limitations of conversion from thermal to mechanical energy. The second law of thermodynamics is formulated to include limits on the efficiency of this conversion.

We extend the available-energy equation to deal with the efficiency limits of both thermal and direct conversion systems. The assumption is made that in the future energy conversion will be based on systems that do not employ the intermediate step of conversion to thermal energy.

The important measure of performance will be the available energy delivered compared with the available energy supplied. Even with direct energy conversion there will be losses associated with dissipative physical processes. The dissipation appears as the entropy-generation (or creation) term.

This entropy-generation term in the entropy-balance equation and the available-energy equation provides the basis for evaluation of the "losses" in a system. In the formulation of these equations, a generation of entropy was assumed within the system; hence, entropy was not conserved. In the available-energy formulation, the term $T_0 \dot{\sigma}$ could be equivalently associated

Table 7-1
Energy-Conversion Matrix (5, 6, 7)

Convert	To							
	Gravitational	Kinetic	Thermal	Chemical	Nuclear	Electrical	Electromagnetic	Pressure
Gravitational		Falling objects						
Kinetic	Ballistics		Friction (brakes) Braking	Dissociation by radiolysis		Electrical generator MHD	Accelerating charge (cyclotron) Phosphor	Atmospherics Stagnation pressure
Thermal	Thermal stratification	Thermal expansion (turbines)		Phase change (boiling) Dissociation		Thermoelectricity Thermionics Thermomagnetism Ferroelectricity	Thermal radiation	Piston expansion process
Chemical	Chemical stratification Osmosis	Muscle Explosions	Combustion			Battery Fuel cell	Chemilum- inescence (fireflies)	Explosion Shock tube
Nuclear		Radioactive emission	Fission Fusion	Radiation, Biology Ionization		Direct collection devices Nuclear battery	Gamma reactions Co^{60} source, A-bomb	Atomic bomb
Electrical		Electric motors Electrostriction (sonar transmitter)	Resistance heating	Electrolysis (aluminum production) Battery charging			Electromagnetic radiation Electrolum- inescense	Piezostriction Loudspeaker
Electromagnetic	Atmospherics	Mass driver Solar wind drive	Thermal absorption	Photosynthesis (plants) Photochemistry (photographic film)	ν neutron reactions $(Be^9 + \nu \rightarrow Be^8 + \eta)$	Photoelectricity Radio antenna Solar cell		
Pressure		Artillery	Injection molding	Osmosis	Fusion	Piezoelectric		

as the dissipation of available energy. In both cases the concept is introduced that something appears or disappears within a system without crossing the system boundaries. It is not associated with mass flux through the system or with energy flow, although it is clearly dependent on these fluxes or flows.

Examination of the available energy (equation 6.10) gives a better feeling for the physical representation of this dissipation term than the entropy-balance equation:

$$\frac{dE}{dt} - T_0 \frac{d\sigma}{dt} = \dot{m} \left[(H_1 - T_1 S_1) - (H_2 - T_0 S_2)\right] - T_0 \dot{\sigma}_c - \dot{W}_{net}.$$

Let us reduce this to simple conditions and then build up to greater complexities.

First, consider a system with thermodynamic parameters that do not change with time, and from which material leaves with the same properties with which it entered. This gives

$$T_0 \dot{\sigma}_c = \dot{W}_{net}.$$

The meaning is that the work required to move the material through the system has been lost or dissipated. This is associated with pumping losses or transmission or transport losses. The transportation of coal from the mine to the power plant requires work that does not change the available energy of the coal in the process. Similarly, the pumping work for oil or gas in a pipline to move it from the well to the consumer does not ordinarily appear as greater available energy at the delivery point.

These have traditionally been lumped into a frictional-loss term. Examination of equation 6.10, which reduces to

$$\frac{d\sigma}{dt} = \frac{\dot{Q}_{in} - \dot{Q}_{out}}{T_0} + \dot{\sigma}_c,$$

indicates that in this situation

$$T_0 \dot{\sigma}_c = \dot{Q}_{out} - \dot{Q}_{in}$$

If this $T_0 \dot{\sigma}_c$ is energy dissipated by work applied to the system, then in order to maintain the system without changes, this must appear as a heat loss from the system. The energy equation also indicates that, if work is added, it must be lost as heat to maintain the state without change. This equality led Joule to the equivalence of heat and work in his cannon-boring experiments. It also led to the early association of friction with heat.

In a fluid case the stirring of a batch in a chemical reactor may result in motion within the reactor. This motion may be required to provide mixing or separation of light from heavy constituents through centrifugal action. The work may then appear as kinetic energy within the reactor. If this kinetic energy is allowed to decay through turbulent or laminar exchange processes to random molecular motion, it will appear as thermal energy. Heat must then be removed from the reactor to maintain steady-state conditions. Utilization of part of this kinetic energy by conversion to work could be accomplished through ingenuity, if the economics of the recovery system were favorable. In the same way, kinetic energy required to maintain a process flow rate may be partially recovered by utilization of an exit generation system. Thermal energy lost could be partially converted to available energy using a thermoelectric converter or simply using the system as the high-temperature reservoir in a conventional engine. In all these cases the economics of a recovery system dictate whether all the work is dissipated as heat.

Second, reference to the difference between expected work and actual work produced by a process (chapter 4) indicates that the entropy change in a system is

$$d\sigma = \frac{dQ}{T} + \Sigma \frac{(\langle F_k \rangle - F_k)}{T} dX_k .$$

Put into rate form, this can be written as

$$\frac{d\sigma}{dt} = \frac{1}{T} \dot{Q} + \frac{1}{T} \Sigma (\langle F_k \rangle - F_k) \frac{dX_k}{dt} . \qquad (7.1)$$

Comparison of this equation with equation 6.19 (the closed-system case),

$$\frac{d\sigma}{dt} = \frac{\dot{Q}}{T_0} + \dot{\sigma}_c ,$$

indicates that the generation term is

$$\dot{\sigma}_c = \frac{\dot{Q}}{T} - \frac{\dot{Q}}{T_0} + \frac{1}{T} \left[\Sigma (\langle F_k \rangle - F_k) \frac{dX_k}{dt} \right] , \qquad (7.2)$$

and

$$T_0 \frac{d\sigma}{dt} = \frac{T_0}{T} \left[\dot{Q} + \Sigma (\langle F_k \rangle - F_k) \frac{dX_k}{dt} \right] . \qquad (7.3)$$

In the available-energy expression for a closed system, the term

$$\frac{dE}{dt} - T_0 \frac{d\sigma}{dt} = \dot{Q} - \Sigma F_k \frac{dX_k}{dt} - \frac{T_0}{T}$$

$$\times \left[\dot{Q} + \Sigma(\langle F_k \rangle - F_k) \frac{dX_k}{dt} \right]. \qquad (7.4)$$

The net work produced by the system has been equated to

$$\dot{W}_{net} = \Sigma F_k \frac{dX_k}{dt}.$$

This expression indicates that the change in the available energy of a system can be categorized by an actual work effect and a dissipation loss. If the process is reversible, then

$$F_k \frac{dX_k}{dt} = \langle F_k \rangle \frac{dX_k}{dt}. \qquad (7.5)$$

The actual work is equal to the expected work, and the available-energy change is the expected work minus the irreversible energy added as heat:

$$\frac{dE}{dt} - T_0 \frac{d\sigma}{dt} = -\frac{T_0}{T} \Sigma \langle F_k \rangle \frac{dX_k}{dt} - T_0 \frac{\dot{Q}}{T} + \dot{Q}. \qquad (7.6)$$

Returning to equation 7.2, this definition of entropy creation is clearly the difference between the entropy change of the system $\dot{Q}/T$ and the entropy change in the environment $\dot{Q}/T_0$. This is then the entropy generated by the process. The term

$$\frac{1}{T} \left[\Sigma(\langle F_k \rangle - F_k) \frac{dX_k}{dt} \right]$$

represents the difference between the expected (reversible) work and the actual work produced by the system. It is associated with the system, hence the factor $1/T$ instead of $1/T_0$. This is often assumed a heat loss or frictional loss. In reality it may appear as turbulent motion or, in collision processes, as acoustic waves or vibrational motion that could be work if utilized.

The form of energy dissipation,

$$\frac{T_0}{T} \Sigma(\langle F_k \rangle - F_k) \frac{dX_k}{dt}, \qquad (7.7)$$

indicates the rate nature of this loss. The faster the rate, the greater the dissipation. Further, the form of force F_k times rate of displacement dX_k/dt indicates the form of the dissipation function. In irreversible thermodynamics this product is taken as the force times the flux. The force is the difference between the force required without dissipation (the expected force) and the actual force produced.

In the flow of fluid through a pipe into an elevated reservoir, the actual pressure required for a given flow is higher than the expected pressure. The work done on the fluid to raise it to an elevation is hence greater than that calculated from a force-displacement relation in a conservative field. The expected work is

$$W_k = \int \rho g \, dX,$$

where g is the gravitational force per unit mass and dX is the displacement. The expected force $\langle F_k \rangle$ is ρg. The expected rate of doing work is

$$W = \rho g \frac{dX}{dt} = \rho g V.$$

Then ρg is the force, and V is the flux.

It is observed experimentally that, if this pumping process is carried out slowly, the actual work required approaches the expected work. It is also observed that the faster the flow, the greater this difference becomes. In laminar slow flow this difference is a frictional viscous force proportional to the velocity. At higher flow rates the flow becomes turbulent, and the viscous force increases more rapidly than the flow rate. More dissipation then occurs than in laminar flow.

Let us assume a laminar flow with the force difference

$$\tau = \mu V.$$

The proportionality constant μ is conventionally the viscosity. The dissipation can then be written as the force difference times the flux:

$$T\dot{\sigma}_c = \tau V = (\mu V)(V) = \mu V^2. \tag{7.8}$$

In general terms this is usually written with the flux proportional to the force, as

$$T\dot{\sigma}_c = J_k X_k, \tag{7.9}$$

with

$$J_k = kX_k, \tag{7.10}$$

or in terms of the flux, as

$$T\dot\sigma_c = \frac{1}{k} J_k^2. \tag{7.11}$$

Similarly, it may be written in terms of the force, as

$$T\dot\sigma_c = kX_k^2 \tag{7.12}$$

Let us examine the charging of a capacitor with the systems boundary shown in figure 7–1. The entropy generation is

$$\dot\sigma_c = \frac{\dot Q}{T} - \frac{\dot Q}{T_0} + \frac{1}{T}\left[\Sigma(\langle F_k\rangle - F_k)\frac{dX_k}{dt}\right].$$

The last term on the right represents the difference between the work required to charge the capacitor ideally,

$$\langle F_k\rangle \frac{dX_k}{dt},$$

and that required actually,

$$F_k \frac{dX_k}{dt},$$

multiplied by the factor $1/T$. This difference is proportional to the product of the current I and the voltage V, the energy dissipated in the circuit resistance.

Without any heat flow the temperature of the system would continue to increase. If the charge buildup on the capacitor is removed continuously, then with heat exchange a steady state can be reached. A heat flow $\dot Q$ equivalent to IV is required to establish equilibrium. The first and last terms in the entropy-generation equation cancel; then

$$\dot\sigma_c = -\frac{IV}{T} - \left(-\frac{IV}{T_0}\right) + \frac{1}{T}(IV) = \frac{IV}{T_0}. \tag{7.13}$$

In the case of the charging of a capacitor, a potential difference from one end of the wire to the other is required to produce a flux. The current flow I is proportional to the potential difference V as

$$I = \frac{1}{R}V.$$

I is the rate of flux of charge. The energy is dissipated as heat to the

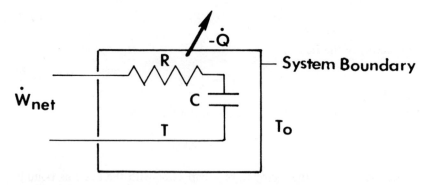

Figure 7–1. Charging of a Capacitor as a Thermodynamic Process

surroundings when steady state is reached. The rate of energy loss through heating is IV. The force $X_k = V$, and the flux $J_k = I$. Then the entropy generation is

$$\dot{\sigma}_c = \frac{IV}{T} = \frac{I^2 R}{T} = \frac{J_k X_k}{T} = \frac{kX_k^2}{T}.$$

The energy balance for the system is $\dot{Q} = -IV$. The entropy generation caused by the heat exchange to the environment is

$$\dot{\sigma}_c = \frac{IV}{T_0}.$$

From equation 6.19 the entropy change of the system is zero, as the term $\dot{Q}/T_0$ cancels with the term $\dot{\sigma}_c$. The association of the entropy generation with the system in this case is due to a process within the system but relies on the exchange of energy with the environment. The entropy creation is a function of both the system *and* its environment.

A further clarification is possible if we consider the change in the available energy of a system, such as an ingot of hot metal in a steel mill. If this metal is allowed to cool, the available energy change is represented as

$$\frac{d}{dt}(E - T_0\sigma) = -T_0\dot{\sigma}_c. \tag{7.14}$$

The entropy generation for the heat loss is

$$-T_0\dot{\sigma}_c = -T_0\left(\frac{\dot{Q}}{T} - \frac{\dot{Q}}{T_0}\right) = \left(1 - \frac{T_0}{T}\right)\dot{Q}. \tag{7.15}$$

The rate of heat transfer from the system to the environment is

$$\dot{Q} = m\frac{d(c_p T)}{dt}.$$

Then

$$\frac{d(E - T_0\sigma)}{dt} = -mc_p T_0\left(\frac{1}{T} - \frac{1}{T_0}\right)\frac{dT}{dt}. \qquad (7.16)$$

Integrating over time for a small change in temperature,

$$\Delta(E - T_0\sigma) = -mc_p\left(\frac{T_0}{T} - 1\right)\Delta T = \left(1 - \frac{T_0}{T}\right)mc_p\,\Delta T.$$

$$(7.17)$$

The available energy lost is the heat lost $(mc_p\,\Delta T)$, times the Carnot factor $[1 - (T_0/T)]$. The available energy lost is thus less than the total energy lost. This available energy could in principle have been recovered with a reversible heat engine. The difference between the initial energy relative to the environment and the final energy is Q, of which $(T_0/T)Q$ is unavailable for conversion to work. Geothermal energy is typical of this situation. The available energy of hot rock in the ground is only a small fraction of the thermal energy present. The entropy decrease in a system on cooling is Q/T, and the entropy increase of the environment is Q/T_0. The difference is the entropy created,

$$\sigma_c = \frac{Q}{T} - \frac{Q}{T_0}.$$

Returning to equation 7.6, the rate of entropy change of a system is

$$\frac{d\sigma}{dt} = \frac{\dot{Q}}{T} + \frac{1}{T}\Sigma(\langle F_k\rangle - F_k)\frac{dX_k}{dt}.$$

In irreversible-thermodynamics theory, the second term on the right is the entropy-creation term for the system designated θ.
   In force-flux notation

$$T\theta = \Sigma(\langle F_k\rangle - F_k)\frac{dX_k}{dt} = \Sigma J_k X_k. \qquad (7.18)$$

In this last formulation the environment state does not appear. This is misleading and should be noted in working with this formulation.

The energy stored in physical elements is represented in the internal-energy E value. Storage may be in many forms besides internal energy associated with molecular storage. Examples include energy storage in batteries, springs, masses, capacitors, and so forth (see chapter 16). These physical elements can store energy in a form that allows it either to be recovered in the same form or to be converted to other forms.

The rate of energy flow into and out of a system is generally expressible as the product of an intensive and an extensive variable. The intensive variable is a property of the field applied to the system. Examples are electric potential, pressure, gravitational field, magnetic field, and chemical potential. Temperature is excluded for the present since it represents a potential that is inherently related to a dissipative process. These intensive variables are assumed to express potential forces on a system. Extensive variables are associated with flow of particles into or out of a system. Examples include mass flow and charge flow. The product of the effort or force and the flow is the rate of flow of energy (power) across the system boundary.

In consideration of flow between a system and its environment, the intensive properties of both are involved. This is not a general property of physical systems. For example, the rate of flux of particles from a radioactive source may not depend on the external field. The type of energy exchange may depend on the energy of the particles and availability of energy transition states in the receiver. Examples are photovoltaic and photosynthetic processes [8–10].

The impedance to flow of energy from one system to another is one way to characterize the dissipation that occurs in a process. If the potentials or forces are nearly in balance, then the flow is slow and the dissipation is reduced. These are highly efficient processes and correspond to the idealization of processes in which the work done is

$$\int \langle F_k \rangle \, dX_k .$$

In this case $\langle F_k \rangle$ represents the expected potential difference between two systems; dX_k represents the displacement from the initial state of the system. The reversible work done by one system is equal to the work done on the other system. In the charging of a capacitor the reversible work required to move a charge q through an electric voltage difference V is

$$\int_{q_{i1}}^{q_{f1}} V \, dq .$$

q_{i1} and q_{f1} represent the initial and final charge states of system 1 shown in figure 7–2. If the process is reversible, then the work by system 2 is equivalent to that done on system 1.

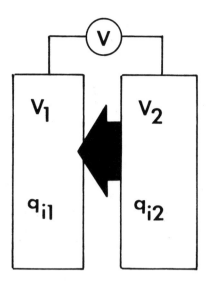

Figure 7-2. Capacitor Charging Represented by the Motion of Charge from One Electrical Potential to Another

$$\int_{q_{i2}}^{q_{f2}} V dq = \int_{q_{i1}}^{q_{f1}} V dq.$$

In spatial coordinates we attribute this flow of energy to the existence of a force that is proportional to the gradient of a scalar potential ϕ. Then the vector force $V = \text{grad } \phi$. To move a charge from one position in space to another requires energy equivalent to

$$\int F \cdot dX = \int V dq = \int q \text{ grad } \phi \cdot dX,$$

where q represents the quantity of charge. Now grad $\phi \cdot dX = d\phi$; therefore,

$$\int F dX = \int_{\phi_1}^{\phi_2} q d\phi.$$

In this case the movement of charge from a potential ϕ_2 to ϕ_1 requires work

$$q(\phi_1 - \phi_2)$$

and is independent of the path of motion.

Any time the force can be represented by the gradient of a scalar potential, the work required to displace a charge in that field will be independent of the path. These are conservative force fields.

These reversible-process systems are, of course, static and lossless systems, and the forces are only differentially unbalanced.

Atwood's Machine

The classical illustration of the effect of unbalance of forces is Atwood's machine. Let us examine the illustration (figure 7–3) for the purpose of analyzing the relationship between power transfer and efficiency.

Free-body force diagrams of each weight as a system indicate that the expected force $\langle F \rangle$ supplied to the mass m_2 is related to the momentum change as

$$\langle F \rangle - W_2 = \frac{d(m_2 V)}{dt}.$$

A similar force balance on m_1 gives

$$W_1 - \langle F \rangle = \frac{d(m_1 V)}{dt}.$$

Adding these to eliminate $\langle F \rangle$ gives

$$W_1 - W_2 = \frac{d[(m_1 + m_2)V]}{dt}.$$

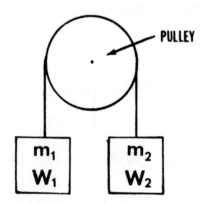

Figure 7–3. Atwood's Machine as an Example of Relationships between Efficiency and Power Transfer

For constant masses m_1 and m_2, the expected force is then found as

$$\langle F \rangle = W_2 + m_2 \frac{(W_1 - W_2)}{(m_1 + m_2)}.$$

The expected force is therefore a constant. The power delivered to m_2 by the expected force is

$$P = \langle F \rangle V = \left[W_2 + m_2 \frac{(W_1 - W_2)}{(m_1 + m_2)} \right] V.$$

The rate at which the energy E of mass m_2 is increasing is

$$\frac{dE}{dt} = \frac{d}{dt}(1/2 \ m_2 \ V^2) + \frac{d}{dt}(m_2 g h).$$

(Note: h is the height of the mass in the gravitational field g.) Simplifying gives (since $m_2 g = W_2$ and $dh/dt = V$).

$$\frac{dE}{dt} = \frac{d(m_2 V)}{dt} + W_2 V.$$

Now

$$\frac{d(m_2 V)}{dt} = \langle F \rangle - W_2.$$

The rate of energy gained by m_2 is then

$$\frac{dE}{dt} = \langle F \rangle V.$$

The efficiency, defined as the rate of energy gain divided by the rate of energy supply or power, is

$$\eta = \frac{dE/dt}{P} = \frac{\langle F \rangle V}{\langle F \rangle V} = 1.$$

This is certainly expected, since there are no losses in the system.

Let us now introduce a frictional or dissipation effect into the situation. As a simple example, let us assume a frictional force that is linearly proportional to the velocity of the weight W_2 as $F_f = kV$. With friction it is observed that the actual force acting on the weight W_2 is less by the amount of this frictional force as

$$F_{act} = \langle F \rangle - F_f = \langle F \rangle - kV.$$

The force balance on W_2 then gives

$$F_{act} = m_2 \frac{dV}{dt} + W_2,$$

or

$$\langle F \rangle = m_2 \frac{dV}{dt} + kV + W_2.$$

Solving this equation for the velocity as a function of time gives (assuming $V = 0$ at time $t = 0$)

$$V = \frac{\langle F \rangle - W_2}{k} (1 - e^{-kt/m_2}). \qquad (7.19)$$

Writing F_{act} in terms of $\langle F \rangle$ then gives

$$F_{act} - W_2 = (\langle F \rangle - W_2)e^{-kt/m_2} \qquad (7.20)$$

Let

$$F^1_{act} = F_{act} - W_2 \text{ and } \langle F \rangle^1 = \langle F \rangle - W_2$$

for later discussion on impedance. At time $t = 0$, $F_{act} = \langle F \rangle$, the expected force. As time proceeds, F_{act} decreases exponentially.

The entropy generated can now be found as

$$\dot{\sigma}_c = \frac{1}{T_0} (\langle F \rangle - F_{act})V,$$

$$\dot{\sigma}_c = \frac{1}{T_0} (\langle F \rangle - W_2)(1 - e^{-kt/m_2})V = \frac{k}{T_0} V^2.$$

The rate of energy increase of W_2 is again

$$\frac{dE}{dt} = \frac{d(m_2 V)}{dt} V + W_2 V$$

$$= (F_{act} - W_2)V + W_2 V = F_{act} \langle V \rangle ,$$

$$\frac{dE}{dt} = (\langle F \rangle - kV)V.$$

The efficiency is then the rate of energy increase divided by the expected power delivered,

$$\eta = \frac{dE/dt}{P} = \frac{(\langle F \rangle - kV)V}{\langle F \rangle V} = 1 - V/\left(\frac{\langle F \rangle}{k} \right) .$$

A plot of efficiency as a function of velocity indicates this relationship in figure 7–4.

Note that the efficiency is 1 at the start of motion, when $F_{act} = \langle F \rangle$, but reduces to zero when the weight reaches a terminal velocity of $\langle F \rangle /k$. At this velocity the expected power is all dissipated.

Returning to the rate of energy increase of the weight

$$\frac{dE}{dt} = (\langle F \rangle - kV)V,$$

it is useful to ask at what velocity the rate of energy increase will reach a maximum (the maximum-power-delivery point).

Since $\langle F \rangle$ is constant, it is easily shown that the energy-delivery rate occurs when the velocity is

$$V = \frac{\langle F \rangle}{2k}$$

or when

$$F_{act} = \frac{\langle F \rangle}{2} .$$

This is a characteristic of many linear dissipative processes. For example,

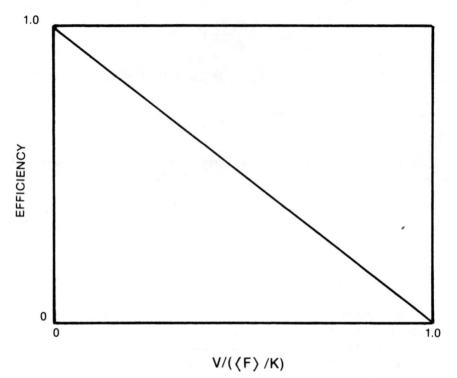

1.0

EFFICIENCY

0

0   1.0

$$V/(\langle F \rangle /K)$$

Figure 7–4. Efficiency as a Function of the Velocity of a Weight in a Simple Atwood's Machine with Linear Dissipative Losses

the maximum power delivery from a battery through a load resistance occurs when the voltage is one-half the open-circuit (or expected) voltage.

The maximum rate of energy increase is then

$$\left.\frac{dE}{dt}\right)_{max} = \frac{1}{4}\frac{\langle F \rangle^2}{k}$$

The energy-delivery rate as a function of velocity can then be described as

$$\frac{dE/dt}{(\langle F \rangle^2/k)} = \left(1 - \frac{V}{\langle F \rangle /k}\right)\left(\frac{V}{\langle F \rangle /k}\right).$$

This relation is shown in figure 7–5.

The relation between power delivery and efficiency can then be found as

$$P = \frac{\langle F \rangle^2}{k}(1 - \eta)\eta,$$

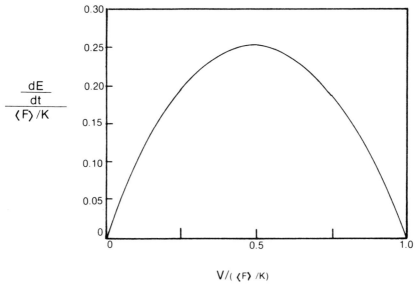

$$\frac{dE}{dt} \bigg/ \langle F \rangle / K$$

V/(⟨F⟩ /K)

Figure 7–5. Rate of Energy Delivery (Power Transfer) as a Function of the Velocity for a Linearly Dissipative System

or

$$\frac{P}{\left(\dfrac{dE}{dt}\right)_{max}} = (1 - \eta)\eta.$$

This relation is shown in figure 7–6. Maximum power delivery occurs at an efficiency of 1/2. It is also worth noting the practicality of utilizing this relation to determine if both the efficiency and power delivery can be increased in a system. If the system is inefficient (to the left of the peak-power-delivery point), then both power and efficiency can be increased. On the other hand, if the system is at a point to the right, a trade-off must be made between power delivery and efficiency.

  If maximum power flux is required, a dissipative loss must be incurred. Efficiency of transfer in these systems can be obtained by reducing the power flux. It is this trade-off between rate and efficiency that must be made in many process designs. An objective of higher efficiency may be attained by slowing up the process rate.

  Figures 7–7 and 7–8 illustrate the actual performance of an American-type multiblade wind machine compared with the ideal power system described. For such a machine the power-efficiency characteristics will also

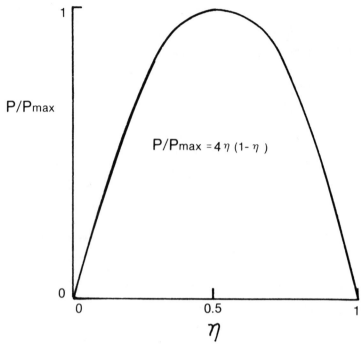

Figure 7–6. Power Flow as a Function of Efficiency for the Atwood's Machine with Linear Dissipation

apply. In the design of a wind machine, the important factor is operation at the maximum power point. At this point the efficiency will be close to 50 percent, as noted in the ideal power system.

Impedance in Energy Transfer

In electrical-circuit theory and some mechanical systems, the concept of impedance is useful for design purposes. The usual definition is that the impedance R is the parameter relating the force to the flux. In the weight problem discussed, this is

$$F^1_{act} = F_{act} - W_2 = RV.$$

Using equations 7.19 and 7.20, this becomes

$$F^1_{act} = \frac{k}{e^{kt/m_2} - 1} V,$$

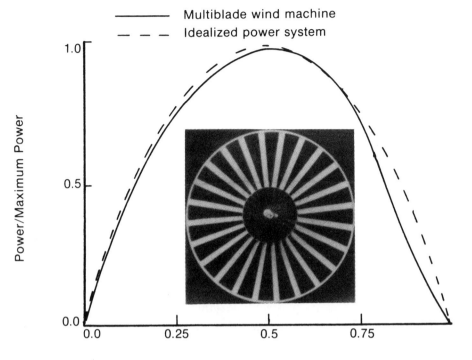

Figure 7–7. Power versus Tip Speed for a Multiblade-type American Wind Machine Compared with an Idealized Linear-Dissipation Power System

with

$$R = \frac{k}{e^{kt/m_2} - 1}.$$

In this case the impedance is time dependent, being infinite at $t = 0$ and declining to zero as $t \to \infty$. The power flow is

$$F^{\,1}_{\mathrm{act}} V = R V^2.$$

This is similar to the electrical case where the power is RI^2 when R is the electrical impedance and I is the current.

Figure 7–9 indicates how the power flux received varies with the

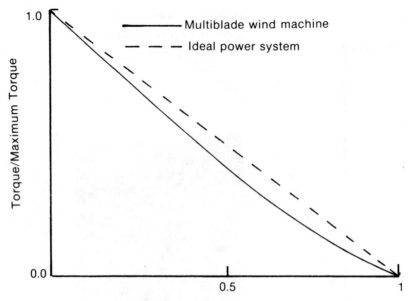

Figure 7–8. Torque versus Tip Speed for a Multiblade-type American Wind Machine Compared with an Idealized Power System with Linear Dissipation Losses

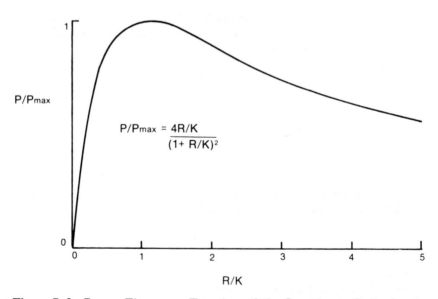

Figure 7–9. Power Flux as a Function of the Impedance Ratio for the Atwood's Machine System

impedance for the weight system. This characteristic curve indicates that the maximum power transfer occurs when the impedance R is equal to the dissipating impedance K. In circuit language, the dissipating impedance is matched to the input impedance at maximum power transfer.

An interesting geometrical sketch indicating relationships between force and flux is shown in figure 7–10. In this sketch the ratio of the actual force F_{act} on the mass m_2 to the total potential force $\langle F \rangle$ is plotted as a function of the actual flux V divided by the maximum flux V_{max} attained by the system. If there were no dissipation, the power supplied would be represented by the total area enclosed. With dissipation, the maximum power supplied is only one-quarter of this maximum, as indicated by the shaded area. As this figure indicates, the diagonal falling to the right shows how the actual force supplied to m_2 varies with the velocity. The diagonal line rising to the right indicates how the dissipating force increases with velocity. The maximum power supplied to the mass is indicated by the intersection, since lines of constant

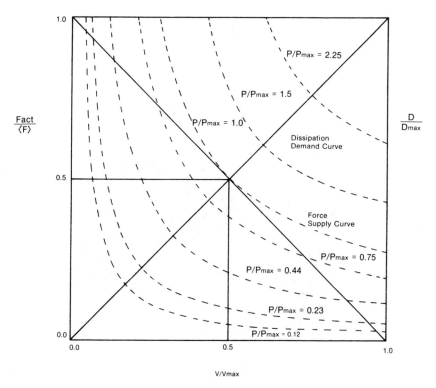

Note: Note force matching at maximum-power point.

Figure 7–10. Atwood's Machine Power-Supply Diagram, Illustrating How the Actual Force and the Dissipative Force Vary with Flux or Velocity

power are rectangular hyperbolas (dotted lines) with the product $F_{act}V$ increasing as shown. This sketch indicates the performance characteristics of the system. The ratio of dissipated power to useful power is easily determined, as is the operating point required for maximum power transfer.

These types of diagram are useful as a design tool for optimization of power transfer. The performance characteristics in practical situations are usually not straight lines as a result of nonlinearities in the force-flux relationships. Actual diagrams may appear as shown in figure 7–11. In this figure the nodes represent possible operating points for different supply and demand conditions. If the objective is maximum power transfer, the point with the intersection at the highest value of power indicated by the isopower hyperbolic curves should be chosen. The designer's job is to control the operating point by proper matching of the power-supply system to the load.

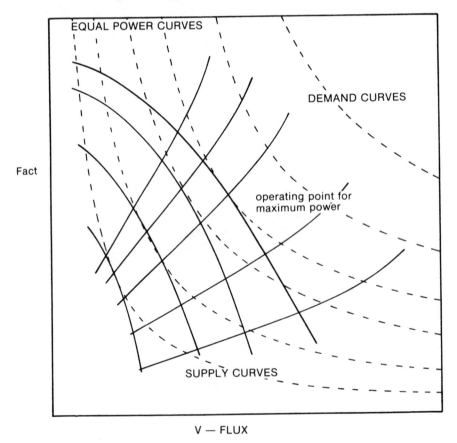

Figure 7–11. Actual Force-Flux Relations in Supply-Demand Situations Indicating Operating Points at Nodes

In many situations the demand curves change with time, and matching requires adjustment of the supply or changing the gear ratio in a mechanical system. In wind-powered energy-conversion systems the supply characteristics change with the wind, and load-impedance matching is required.

Since efficiency and economy are not considered at this point, obtaining sufficient information for final decisions requires further analysis, as shown chapter 9.

An analogous situation in impedance matching for power transfer occurs in a simple inductor circuit, as shown in figure 7–12. The mathematical analog is found by observing that the equation for this circuit is

$$V - Ri = L \frac{di}{dt} ,$$

compared with

$$\langle F \rangle - kV = m \frac{dV}{dt} .$$

All the same conditions discussed in the force-mass system hold. Linear-systems theory is discussed in many texts [1,2,5–7], in which the analogies of

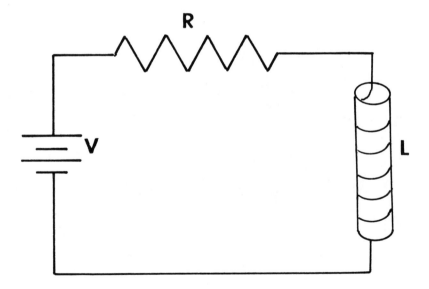

Figure 7–12. Inductor-Series Circuit Analogous to the Force-Mass System Described

Table 7-2
Generalized Force-Displacement Energy Forms and Force-Flux Power Forms

	Mechanical Linear	Mechanical Rotational	Electrical	Magnetic	Surface	Chemical
Expected force $\langle F_k \rangle$ $\quad P = \dfrac{\langle F \rangle}{A}$	Force Pressure Stress	Torque $\quad\tau$	Voltage $\quad V$	Magnetic field $\quad H$	Surface tension $\quad\sigma$	Potential $\quad\mu_c$
Displacement dX_k $\quad dV$	Distance Volume	Angle $\quad d\theta$	Charge $\quad dq$	Magnetization $\quad dM$	Surface area $\quad dA$	Extent of reaction $\quad dN_c$
Flux J_k	Linear velocity $\quad v$	Angular velocity $\quad\omega$	Current $\quad i$	Magnetic flux $\quad \dfrac{dM}{dt}$	$\dfrac{dA}{dt}$	Reaction $\quad \dfrac{dN_c}{dt}$
Dissipation force $K\dot{X}_k$	Viscous loss $\mu\dfrac{dX}{dt}$ Viscous loss $\lambda\dfrac{dV}{dt}$	Viscous loss $\quad G\omega$	Ri		$S\dfrac{dA}{dt}$	
Static energy storage	Linear spring $\dfrac{KX^2}{2}$ pressure vessel	Torsion spring $\dfrac{N\theta^2}{2}$	Capacitor $\dfrac{CV^2}{2}$			
Potential	Gravitational $\quad mg$		Electric field $\quad E$			
Kinetic force	$\dfrac{d(mv)}{dt}$	$\dfrac{d(I\omega)}{dt}$	$L\dfrac{di}{dt}$			
Kinetic energy	$\tfrac{1}{2}mv^2$	$\tfrac{1}{2}I\omega^2$				

mechanical, electrical, fluid, hydraulic, and magnetic systems are examined. A table representing these is presented for reference (table 7–2).

References

1. Angrist, S. *Direct Energy Conversion*. Boston: Allyn and Bacon, 1976.
2. Osterle, J.F. "A Unified Treatment of the Thermodynamics of Steady State Energy Conversion." *Appl. Sci. Res, sec. A*, 12(1964):425.
3. Sutton, G.W. *Direct Energy Conversion*. New York: McGraw-Hill, 1966.
4. Carnot, S. *Reflections on the Motive Power of Heat*, 1824. New York: reprinted by ASME, 1943.
5. Messerle, H.K. *Energy Conversion Statics*. New York: Academic Press, 1969.
6. Paynter, H.M. *Analysis and Design of Engineering Systems*. Cambridge, Mass.: MIT Press, 1960.
7. Macfarlane, A.G.J. *Engineering Systems Analysis*. Reading, Mass.: Addison-Wesley, 1964.
8. Tribus, M. *Thermostatics and Thermodynamics*. Princeton, N.J.: Van Nostrand, 1961.
9. Denbigh, K.G. *The Thermodynamics of the Steady State*. London: Methuen, 1951.
10. Veinik, A.I. *Thermodynamics*. NASA TT F-148, 1964.

8 Available Energy and Irreversible Thermodynamics

All that we win is a battle lost in advance with the irreversible phenomena in the background of nature.
—Henry Adams, *The Education of Henry Adams* (1906)

Steady-Flow Available Energy

Most industrial processes are designed as steady-flow systems that operate continuously in order to utilize capital investment more effectively. This section will explore the effectiveness of processes in terms of the available-energy formulation and irreversible thermodynamics.

We will begin with the general available-energy equation derived in chapter 6:

$$\frac{dE}{dt} - T_0 \frac{d\sigma}{dt} = \dot{n}_p\{(h_{0p} - T_0 S_{0p}) - (h_p - T_0 S_p)\}$$

$$- \dot{Q} - \Sigma F_k \frac{dX_k}{dt} + \dot{Q}\left(1 - \frac{T_0}{T}\right)$$

$$- \frac{T_0}{T}\left(\Sigma(\langle F_k \rangle - F_k)\right)\frac{dX_k}{dt}$$

$$+ \dot{n}_p(KE_{in} - KE_{out}) + \dot{n}_p(PE_{in} - PE_{out}). \quad (8.1)$$

Let us examine first the steady-flow cases in which no change in the kinetic energy or potential energy of the through flow occurs. The available-energy equation then reduces to

$$\dot{n}_p(h_p - h_{0p}) - \dot{n}_p(T_0 S_p - T_0 S_{0p}) = -\dot{Q} - F_k \frac{dX_k}{dt}$$

$$- \dot{Q}\left(1 - \frac{T_0}{T}\right)$$

$$- \frac{T_0}{T}\left(\Sigma(\langle F_k \rangle - F_k)\frac{dX_k}{dt}\right)$$

$$(8.2)$$

Note that $\dot{Q}$ is positive if heat flows into the system, and $F_k(dX_k/dt)$ is positive if work is done by the system.

The available energy associated with the entropy change of the through flow may be represented by several useful expressions:

$$\dot{n}_p T_0(S_p - S_{0p}) = -\dot{Q} + T_0\dot{\sigma}_c$$

$$= -\dot{Q} - \left(\frac{T_0}{T} - 1\right)\dot{Q} + \frac{T_0}{T}\left(\Sigma\langle F_k\rangle - F_k\right)$$

$$\times \frac{dX_k}{dt}. \tag{8.3}$$

Here the entropy generation $\dot{\sigma}_c$ is introduced in the first form. The second form collects terms including the Carnot factor and the heat flow, and the expected and actual forces.

Dividing this expression by the environmental temperature indicates the sources of entropy change in the through flow as

$$\dot{n}_p(S_p - S_{0p}) = \frac{\dot{Q}}{T} \qquad\qquad - \qquad\qquad \frac{\dot{Q}}{T_0}$$

$$\underbrace{\begin{array}{l}\text{Entropy increase of} \\ \text{the system due to} \\ \text{heat flow into the} \\ \text{system}\end{array}}\qquad \underbrace{\begin{array}{l}\text{Entropy decrease} \\ \text{of the environment} \\ \text{due to heat flow} \\ \text{into the system}\end{array}}$$

$$\left| \begin{array}{l}\text{Entropy change in the process due to heat} \\ \text{transfer}\end{array}\right.$$

$$+ \frac{\dot{Q}}{T_0} + \frac{1}{T}\left(\Sigma(\langle F_k\rangle - F_k)\frac{dX_k}{dt}\right)$$

$$\underbrace{\qquad\qquad\qquad}_{\begin{array}{c}\text{Entropy change due} \\ \text{to dissipative} \\ \text{processes}\end{array}}$$

Entropy generation is not usually associated with the heat flow into a system due to the temperature difference of the system from the environment. In irreversible thermodynamics, entropy generation is assigned to the system where dissipative processes occur. When internal temperature gradients and heat flows occur, a dissipation is effected. This dissipation is ultimately an energy loss to the environment and usually is considered a heat flow to the environment. In irreversible thermodynamics [1–3] the entropy generation is taken as

$$\dot{\sigma}_c = \frac{1}{T}\left(\Sigma(\langle F_k\rangle - F_k)\right)\frac{dX_k}{dt} + \frac{\dot{Q}}{T}\left(1 - \frac{T}{T_0}\right). \quad (8.4)$$

The last term may be interpreted as the difference between the expected entropy change associated with $\dot{Q}/T$ and the actual entropy change (T/T_0) $(\dot{Q}/T)$. If the last term is put into the force-flux format, it is convenient to assume that $T_0 = T + \Delta T$ is greater than T if $\dot{Q}$ is positive into the system. Then, in the limit of small ΔT.

$$\dot{Q}\left(\frac{1}{T} - \frac{1}{T + \Delta T}\right) \simeq \dot{Q}\,\frac{\Delta T}{T^2}.$$

Then

$$T\dot{\sigma}_c = \underset{\text{Flux}}{\dot{Q}}\,\underset{\text{Force}}{\frac{\Delta T}{T}}\quad\text{or locally}\quad \dot{Q}\,\frac{\operatorname{grad} T}{T}.$$

Fourier's experimental law indicates that $\dot{Q} = k\operatorname{grad} T$. Then

$$\dot{\sigma}_c = k\left(\frac{\operatorname{grad} T}{T}\right)^2.$$

The entropy generated by heat flow is thus always positive, independent of the direction of heat flow, whether into or out of the system. The entropy change of the material flowing through, of course, depends on the heat-flow direction but is modified unidirectionally by the entropy generation.

Returning to the available-energy function,

$$\dot{n}_p(h_p - h_{op}) - \dot{n}_p(T_0 S_p - T_0 S_{0p}) = \left(-\dot{Q} - F_k\,\frac{dX_k}{dt}\right)$$

$$+ (\dot{Q} - T_0\dot{\sigma}_c). \quad (8.5)$$

In this form the enthalpy change is associated with the heat flux and the work, and the entropy change with the heat flux and the entropy generation. In most practical cases, particularly in high-energy-flux cases, the σ_c term cannot be determined theoretically and must be calculated from measurement of the state of the entering and leaving material. If the actual work can be measured as well, then the heat exchange and the entropy generation can be calculated using the equivalence of the bracketed terms on each side of the equation. Measurement of the heat flux is usually a difficult problem in any conversion system.

In design problems associated with energy conversion, the objective is often the maximization of the work term $F_k(dX_k/dt)$. If the source of energy is the available energy of the input material, then the effectiveness (see chapter 13) may be written as

$$E = \frac{-F_k(dX_k/dt)}{\dot{n}_p[(h_p - T_0 S_p) - (h_{0p} - T_0 S_{0p})]}$$

$$= 1 + \frac{T_0 \dot{\sigma}_c}{\dot{n}_p[(h_p - T_0 S_p) - (h_{0p} - T_0 S_{0p})]} . \qquad (8.6)$$

Note that the heat flux need not be determined directly to determine the effectiveness for an energy-conversion device of this type.

Chemical Systems

In chemical systems it is convenient to use the variables of the chemical potential or the Gibbs free energy. With multiple-component flows the available-energy equation 8.2 can then be written as

$$\sum \dot{n}_p[(h_p - T_0 S_p) - (h_{0p} - T_0 S_{0p})] = -\sum F_k \frac{dX_k}{dt} - T_0 \dot{\sigma}_c. \qquad (8.7)$$

When the chemical potential is $\mu_p \equiv h_p - TS_p$ for each component, then

$$\sum \dot{n}_p[(\mu_p - \mu_{0p}) + (T - T_0)S_p] = -\sum F_k \frac{dX_k}{dt} - T_0 \dot{\sigma}_c.$$

Now adding and subtracting the term $\sum \langle F_k \rangle (dX_k/dt)$ from the right side and using equation 8.4 gives

$$\Sigma \dot{n}_p [(\mu_p - \mu_{0p}) + (T - T_0)S_p] = -\Sigma \langle F_k \rangle \frac{dX_k}{dt}$$

$$+ \left[\Sigma (\langle F_k \rangle - F_k) \frac{dX_k}{dt} \right]$$

$$\times \left[1 - \frac{T_0}{T} \right]$$

$$+ \dot{Q} \left[1 - \frac{T_0}{T} \right]. \qquad (8.8)$$

The right side represents the reversible work plus the dissipation energy multiplied by the Carnot factor. The left side represents the Gibbs free energy change plus a factor to account for the exit flow leaving with a different temperature than it enters.

The entropy-creation term is

$$T_0 \dot{\sigma}_c = \Sigma \dot{n}_p [(\mu_{0p} - \mu_p) - (T - T_0)S_p] - \Sigma F_k \frac{dX_k}{dt}. \qquad (8.9)$$

If the chemical potentials of all input flows are equivalent to the chemical potentials of the output flows, then

$$T_0 \dot{\sigma}_c = \dot{n}_p [h_{0p} - h_p] - F_k \frac{dX_k}{dt}, \qquad (8.10)$$

indicating that T_0 times the entropy creation is equivalent to the difference between the loss in enthalpy of the material flow and the actual work done by the system.

In a chemical reaction within a system at constant temperature and pressure,

$$\dot{\sigma}_c = -\frac{1}{T} \frac{dG}{dt}.$$

If $T = T_0$, then

$$\frac{dG}{dt} = \Sigma \dot{n}_p (\mu_p - \mu_{0p}).$$

The rate of entropy generation is then equal to the flux of particles $\dot{n}_p$ time the chemical-potential difference, and a force-flux relationship can be considered. If the flux of particles n_p is proportional to the chemical potential between two states—as occurs in osmotic systems, for example—then a dissipation proportional to the product of the force and the flux is expected. This product is then utilizable in irreversible thermodynamic formulations.

If the chemical potential varies spatially, it is observed that a flux of particles occurs until equilibrium is established. If a system is maintained with spatial chemical-potential gradients by the flux of particles and an external force, then for a differential element shown in figure 8–1 the mass balance gives

$$\dot{n}_p = -\mathscr{D}\,\frac{\mu_p + \dfrac{\partial \mu_p}{\partial X}\,dX - \mu_p}{\partial X} = -\mathscr{D}\,\frac{\partial \mu_p}{\partial X} \qquad (8.11)$$

or, generally,

$$\dot{n}_p = -\mathscr{D}\ \text{grad}\ \mu_p.$$

The negative sign is to indicate that the flow occurs from the higher to the lower potential.

In an elemental section of width dX the entropy generation is

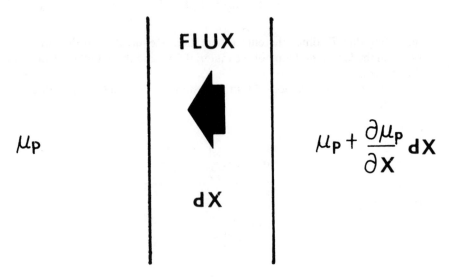

Figure 8–1.   Flux of Particles in a Chemical-Potential Field

$$\dot{\sigma}_c \, dX = - \frac{\dot{n}_p}{T_0} \frac{d\mu_p}{dX} dX = - \frac{\dot{n}_p}{T_0} \, \mathrm{grad} \, \mu_p \, dX,$$

or

$$T_0 \dot{\sigma}_c = -\dot{n}_p \, \mathrm{grad} \, \mu_p.$$

With $\dot{n}_p = -\mathscr{D} \, \mathrm{grad} \, \mu_p$, the entropy generation can be expressed as

$$T_0 \dot{\sigma}_c = \mathscr{D} \, \dot{n}_p^2$$

and is always positive because of the quadratic form.

If a chemical reaction occurs in a section of material, then no gradient may exist, but the entropy increase occurs because of the chemical reaction. If a steady state is maintained, then the products must be removed as fast as the reactants are supplied. The entropy generation at constant temperature and pressure is

$$T_0 \dot{\sigma}_c = \dot{n}_p (\mu_{0p} - \mu_p) \qquad (8.12)$$

De Donder [4] introduced the coordinates of degree of advancement and affinity to describe chemical-system changes with time. Using his notation, the degree of advancement of a reaction is

$$\xi = \frac{\dot{n}_p - \dot{n}_p^0}{\nu_p} \qquad (8.13)$$

$\dot{n}_p^0$ is the number of moles of species at the initial time, and $\dot{n}_p$ is the number at a time t later. ν_p is the stoichiometric coefficient of the p species in the reaction. The rate of change is

$$\dot{\xi} = \frac{\dot{n}_p}{\nu_p}.$$

Note that ξ has the advantage of being equal for each species taking part in a reaction.

Substituting for $\dot{n}_p$, then

$$T_0 \dot{\sigma}_c = - \dot{\xi} \nu_p (\mu_p - \mu_{0p}). \qquad (8.14)$$

The difference in chemical potential is often termed the chemical potential for the reaction, so that

$$\mu_p - \mu_{0p} = \Sigma \mu_p,$$

which is conventionally taken as the chemical potential of the products minus that of the reactants. Then

$$T_0 \dot{\sigma}_c = - \xi \Sigma v_p \mu_p. \qquad (8.15)$$

The sum $- \Sigma v_p \mu_p$ is called the *affinity* by de Donder and denoted as A, so that

$$T_0 \dot{\sigma}_c = \xi A.$$

If the rate of advancement is proportional to the affinity, then

$$\xi = kA$$

gives a flux ξ proportional to force A form again. Note again that it is assumed that the temperature is constant at the environmental temperature T_0 in the computation of $T_0 \dot{\sigma}_c$, the loss of available energy.

A similar force-flux relation for electrical-energy flow can be established. In this case the chemical-potential difference of electrons or ions varies in space. With a gradient occurring, the flux of these charged particles can be written so that, at constant temperature and other potentials,

$$T_0 \dot{\sigma}_c \, dX = \dot{n}_p (\mu_p - \mu_{0p}) = \dot{n}_p \frac{d\mu_p}{dX} \, dX,$$

or, in spatial terms,

$$T_0 \dot{\sigma}_c = \dot{n}_p \, \mathrm{grad} \, \mu_p. \qquad (8.16)$$

For ionic species the potential is usually written as

$$\tilde{\mu}_p = \mu_p + Z_p \mathscr{F} \, \psi, \qquad (8.17)$$

where μ_p is the chemical potential of neutral particles; Z_p is the ionic valence; $\mathscr{F}$ is Faraday's constant (96,500 coulombs/mole); and ψ is the electrical potential. If only the electrical-field effect is of concern, and only the flux of ionic species is involved [7,11], then

$$T_0 \dot{\sigma}_c = \dot{n}_p \, \mathrm{grad} \, (Z_p \mathscr{F} \, \psi). \qquad (8.18)$$

Experimentally it has been observed that for electrons in metals and ions in solutions with low current densities [6], the current flow is

$$i = Z_p \,\mathscr{F}\, \dot{n}_p = -\gamma \, \text{grad} \, \phi \quad \text{or} \quad i = -C \, \text{grad} \,(Z_p \,\mathscr{F}\, \psi).$$

The force-flux relations are again established as linear for systems close to equilibrium. For high current densities in liquids, convective currents are produced, and an overall empirical relation is usually established similar to convective heat-transfer coefficients.

In quantum-effect electronic devices, nonlinear behavior is inherent; but the dissipation function can still be determined experimentally as a force-flux relationship. The linear proportionality of the force to the flux, however, is not applicable.

It must be noted again that the dissipation in these systems leads to the production of thermal gradients, since otherwise a heat transfer out of the system is not possible. The heat transfer is then coupled to the particle flow. A reciprocal interaction is, of course, expected, in which the flow of thermal energy affects particle flow.

Coupled-Flow Energy Systems

Coupling of energy flows will be examined following a format introduced by Odum and Pinkerton [5] to establish relationships between power flux, effectiveness, and efficiency in energy-conversion systems.

For illustration of the energy-conversion process, let us examine a centrifugal compressor driven by an electric motor. Observed in input-output terms shown in figure 8–2, the input available-energy flux is $i\Delta V$, and change in the available energy of the air is found from equations 6.11 and 6.14 as:

$$\dot{m}[(h - T_0 S) - (h_0 - T_0 S_0)] + \dot{m}\left[\frac{V^2}{2g} - \frac{V_0^2}{2g}\right] = \dot{m}(\Delta B + \Delta KE). \tag{8.19}$$

The available energy loss is then

$$T_0 \dot{\sigma}_c = i\Delta V - \dot{m}(\Delta B + \Delta KE), \tag{8.20}$$

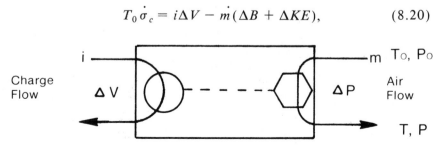

Figure 8–2. Centrifugal-Compressor Schematic Diagram

which is the difference between the available-energy input of $i\Delta V$ and the available energy gained by the air.

Consider first that air is a perfect gas and that the compressor is cooled so that the air temperature does not change from input to output (an isothermal compression). Then $T = T_0$, and

$$dh = h - h_0 = c_p(T - T_0) = 0.$$

The equation state for a perfect gas gives

$$dh = TdS + vdp,$$

so that

$$TdS = -vdp.$$

At constant temperature, with $pv = RT$ (a perfect gas),

$$S - S_0 = \int dS = -\int \frac{v}{T}\, dp = -\int \frac{R}{p}dp = -R \ln \frac{P}{P_0}.$$

If $P = P_0 + \Delta P$,

$$-R \ln \frac{P}{P_0} = -R \ln\left(1 + \frac{\Delta P}{P}\right).$$

For ΔP small compared with P_0, this is approximately

$$S - S_0 = -R\frac{\Delta P}{P_0}.$$

Then

$$\dot{m} T_0 R \frac{\Delta P}{P_0} = \dot{v}\Delta P,$$

where $\dot{v}$ is the volume rate of flow. If the kinetic-energy change is small, then the available-energy loss can be written as

$$T_0 \dot{\sigma}_c = -i\Delta V - \dot{v}\Delta P. \tag{8.21}$$

The terms on the right are then in the form of force-flux. The input power is coupled to the output power flow, and the relationship represents the performance characteristics of a machine.

The current flow is related to the voltage supplied, ΔV, and the required pressure difference, ΔP. In Odum's formulation this may be represented as

$$i = -(n + cf^2)\,\Delta V + cf\,\Delta P, \qquad (8.22)$$

and the volume flow rate as

$$\dot{v} = cf\,\Delta V - c\Delta P. \qquad (8.23)$$

These relations can be examined for physical interpretation by considering the compressor outlet pressure being raised until there is no air flow. Then $\dot{v} = 0$, giving

$$f = \frac{\Delta P}{\Delta V}$$

and

$$i = -n\Delta V.$$

Under these conditions there is no available-energy output, so that

$$T_0 \dot{\sigma}_c = -i\Delta V = \frac{c^2}{n} = n(\Delta V)^2 \qquad (8.24)$$

In this formulation f may be likened to a force amplification. In the case of a lever system, it represents the ratio of the force out to the force supplied. It is important for efficient power transfer to match the load to the supply force. In a geared system it represents the gear ratio. By changing the value of f, greater power can be supplied to a load. The ability to vary f to match load conditions is an important function of a well-designed system.

In equation 8.22 n represents a loss coefficient, or $1/n$ represents an equivalent resistance that dissipates energy. f and n are characteristics of a given machine, which can be determined by this simple experiment.

The factor c can be determined by disconnecting the power source so that $\Delta V = 0$ and measuring the flow through the machine when a back pressure is applied, since

$$\dot{v} = -c\Delta P.$$

c is then a back-flow conductance. Note that Onsager's reciprocal relationship has been assumed in this illustration, since the coefficient cf is taken as equal in both the current- and volume-flow equations.

The useful power output

$$\dot{W} = \dot{v}\Delta P$$

can be written using the volume-flow-rate dependence of ΔP and ΔV as

$$\dot{W} = cf\Delta V\Delta P - c(\Delta P)^2. \tag{8.25}$$

An interesting factor is $r = \Delta P/f\Delta V$, which represents an equivalent force ratio or output force divided by the input force. When $r = 1$, there is no useful output, and the driving force equals the back force. Energy is lost, however, since the input energy goes to frictional losses. A relationship between the power flow and the force ratio is found by eliminating ΔP from the power-flow equation using the r relation.

$$\dot{W} = cf^2(\Delta V)^2 r(1 - r) \tag{8.26}$$

Maximum power output occurs when $r = 1/2$.

When the compressor is working against a zero pressure differential, then the maximum volume rate of flow occurs. $\dot{v}_{max}$ then corresponds to

$$\dot{v}_{max} = cf\Delta V.$$

The volume rate of flow at any loading may be expressed as

$$\dot{v} = \dot{v}_{max} - c\Delta P,$$

or

$$\frac{\dot{v}}{\dot{v}_{max}} = 1 - \frac{c\Delta P}{cf\Delta V} = 1 - r.$$

A plot in figure 8–3 of this relationship is equivalent to that for the weight illustration shown in figure 7–10.

The maximum power flow occurs when the impedance of the driver is equal to the impedance of the load. In this case,

$$\frac{1}{n} = \frac{1}{cf^2/2}. \tag{8.27}$$

Equivalently, the power dissipation in the load is equal to the power dissipation in the source, in this case the motor-winding losses.

The power output can be written as

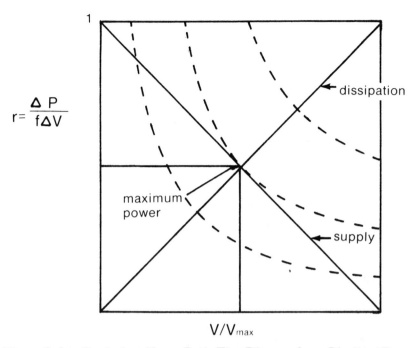

$$r = \frac{\Delta P}{f \Delta V}$$

V/V$_{max}$

Figure 8–3. Equivalent-Force-Ratio Flux Diagram for an Idealized Energy Converter

$$\dot{W} = cf^2 (\Delta V)^2 r \, \frac{\dot{v}}{v_{max}} \, . \tag{8.28}$$

At maximum power output, $\dot{v} = \dot{v}_{max}/2$ and $r = 1/2$, so that with a fixed line voltage ΔV the maximum-power-delivery point could be found by unloading the compressor and measuring the volume flow rate, then adjusting the flow conditions until one-half this maximum flow rate is reached.

The efficiency or effectiveness in this case is

$$\eta = \frac{\dot{v} \Delta P}{i \Delta V} \, .$$

In terms of r this is

$$\eta = \frac{r(1 - r)}{n/cf^2 + (1 - r)} \, .$$

At maximum power flow the maximum efficiency is 1/2. The value of 1/2 may be found by taking $r = 1/2$ and $cf^2/n = \infty$.

If the input impedance is infinite, as occurs if the loss n is zero, then the efficiency is equal to the load ratio r, or

$$\eta = - \frac{\dot{v}}{\dot{v}_{max}}.$$

This indicates that the maximum possible efficiency occurs at the minimum volume rate of flow.

An engineer's responsibility is often to deliver a certain power flow, and the optimization consists of designing a system with a high input impedance $1/n$ and a high cf^2 factor, so that the product n/cf^2 is as small as possible. In this illustration, c large indicates a low viscous loss in the compressor; f large

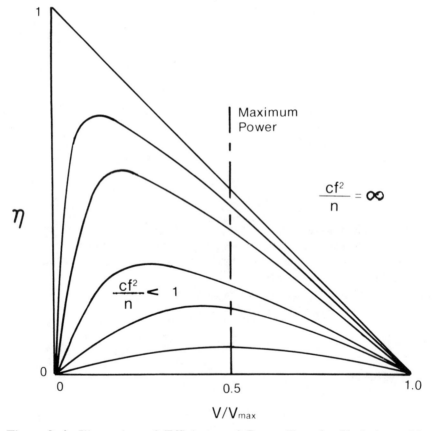

Figure 8–4. Illustration of Efficiency of Power-Transfer Variation with Flux Ratio for Different Impedence Factors (cf^2/n)

indicates a machine that can produce a high pressure at no flow. This in turn indicates an efficient rotor performance. High input impedance $1/n$ implies a large motor with large windings. The use of a large motor operated at optimum power output is then more efficient that that of a small motor that is overloaded.

Figure 8–4 illustrates the relation of the efficiency to the output flow rate as a function of the cf^2/n factor. Note that for very inefficient devices (cf^2/n small), the maximum-power-output point and the maximum-efficiency point can nearly coincide.

A classical example illustrating the interactions of electrical and fluid flows is the electrokinetic effect. In this process an electrolyte is placed in a container with a suitable porous filter or ion-exchange membrane, as shown in figure 8–5. Power flow between the electrodes produces an electrolyte flow and a pressure difference across the membrane. Alternatively, the flow of electrolyte through the membrane can produce a power flow in the electrical circuit. In both cases the efficiency is low because of the physical-chemical properties of the membrane and electrolyte. The analysis of the electric-motor compressor system is directly applicable to this experiment.

Direct energy-conversion systems without an intermediate thermal-energy form can be analyzed similarly in the limits of the linear force-flux phenomenological relations (see chapter 12). In these cases the assumption of isothermal operation requires the removal of thermal energy. This removal of energy is assumed to have been a completely dissipative process. The entropy generation is then representative of the available-energy loss of the system.

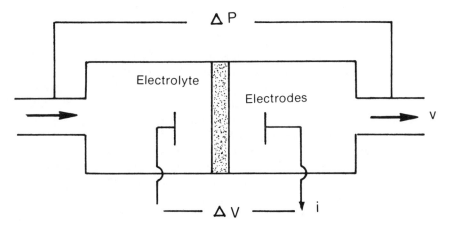

Figure 8–5. Electrokinetic-Experiment Schematic, Showing Conversion of Energy from a Flowing Electrolyte to Current Flow by Charge Separation at a Porous Plug

Irreversible Thermodynamics with Temperature Change

Thermal-energy-conversion systems require further analysis because of the need to distinguish what energy is available for a work process. Before attempting this problem, we will examine the criteria for selection of force-flux relations.

The determination of force-flux relationships is a basic problem of physical science [6,8–10]. The concept of causal relationships is utilized to establish the connections of observable phenomena to theoretical structures. Energy and entropy, as well as power, are theoretical concepts; and the observation of forces and fluxes is a test of these energy concepts. Onsager's reciprocal relationships are based on theoretical microscopic-reversibility concepts and on experimental observations. The selection of forces and fluxes in particular forms allows the use of reciprocal relationships. This greatly assists in the solving of particular problems in a manner similar to the selection of a particular coordinate system in a mechanics problem.

Meixner and Reik [7] have carefully pointed out the theoretical requirements for determining the force-flux relationships in order for Onsager's relationships to be applicable: (1) the product of the force and flux must represent the temperature times the entropy production; and (2) the summation of the products of forces and fluxes must be invariant with respect to the transformation of coordinates. To these might be added the Curie-Prigogine [3,8] requirement that the forces and fluxes be of the same tensor rank. This means that one force cannot be a scalar like temperature (rank 1), and the other a vector like mechanical force (rank 2). Similarly, the gradient of the temperature is of first order, and the gradient of a mechanical force vector is of second order. Their coupling is not expected to have a form to which the Onsager relations would be applicable.

A particular difficulty with the selection of forces and fluxes is the inclusion of the state variable temperature as a force. Temperature enters into dissipation processes in two different ways. It is recognized that in any power transfer there is a dissipation. In the context of this approach, there is a dissipation or loss of available energy. This loss is generally attributed to frictional effects and heat transfer. In effect, it is important for practical purposes to lump this as only a loss, without debating how this energy is communicated to the environment. As noted previously in a collision example, the energy may be in the form of vibrational energy, which could conceivably be retrieved. It is a question of the economics of its recovery that categorizes it as a loss.

If we take the view that this dissipation represents a thermal-energy transfer to the environment, then a temperature gradient is required that can couple with the other forces. In the compressor example the isothermal compression requires intercooling in the compressor. The coupling may be negligible, but it simplifies the problem analytically to assume a constant-temperature process.

In standard formulations the assumption is made that the dissipation depends on the temperature at which the process takes place (that is, $T\sigma$ rather than $T_0\sigma$ in the available-energy formulation).

In early energy-dissipation work by Rayleigh [10], the dissipation function was taken as the product of the physical forces and fluxes in the form

$$\phi = \Sigma M_i N_i.$$

The product has the dimensions of power, and the fluxes are related to the forces as

$$M_i = \Sigma C_{ij} N_j,$$

so that

$$\phi = \sum_i \sum_j C_{ij} N_j N_i.$$

These are particularly attractive as coordinates, since a direct analogy with Gibbs equation-of-equilibrium thermodynamics is possible in the form

$$TdS = d\langle E \rangle + \sum_c \mu_c d\langle n_c \rangle + \Sigma \langle F_k \rangle dX_k. \qquad (8.29)$$

In the introductory thermodynamics section, this was written using the Lagrange multipliers α and β, where $\alpha = -\mu/kT$ and $\beta = 1/kT$ as

$$dS = k\beta d\langle E \rangle + \Sigma k\alpha_c d\langle n_c \rangle + \Sigma k\beta \langle F_k \rangle dX_k.$$
$$(4.23)$$

If the first formulation is used, the potentials and fluxes of mass are μ_c and dn_c/dt, respectively. In the second, the force is $k\alpha_c = -\mu_c/T$, and the flux is dn_c/dt. The difference is, of course, that in the first formulation a change in composition is independent of temperature, whereas in the second formulation composition change is linked with the temperature. In the equilibrium formulation this linkage was directly indicated by the relationship of the change in internal energy with α_c to the change in composition with β as

$$\frac{\partial E}{\partial \alpha_c} = \frac{\partial n_c}{\partial \beta}$$

(equation 4.15). The linkage is basic to an understanding of chemical processes.

In the formulation of Onsager's theory [1], coupling is inherent; and he assumes the second formulation for defining the forces and fluxes, so that

$$T\dot{\sigma}_c = \Sigma J_i X_i. \tag{8.30}$$

For these processes the Rayleigh function must be reformulated so that the force-flux relation gives

$$J_i = \Sigma L_{ij} X_j \tag{8.31}$$

with

$$T\dot{\sigma}_c = \sum_{i,j} L_{ij} X_j X_i. \tag{8.32}$$

These are related to Rayleigh's ϕ function as

$$T\dot{\sigma}_c = \phi = (\Sigma C_{ij} N_j N_i) = \Sigma L_i X_j X_i. \tag{8.33}$$

Then the coefficients and the fluxes in the two systems must be altered to meet the conditions of the force-flux linearity. It has been demonstrated experimentally that Onsager's reciprocal relations apply to the formulation in terms of L_{ij}, with

$$L_{ij} = L_{ji}. \tag{8.34}$$

If only isothermal processes are considered, then the equivalence is obvious. In the case of mechanical-electrical systems such as the air-compressor example, whether the formulation is in Onsager or Rayleigh form is irrelevant. In the steady-flow case, it is the difference between the actual and expected forces that indicates the dissipation. This difference may be considered as a driving force for the dissipation.

A heat-flux formulation attempts to replace the heat-flux proportionality to the gradient of the temperature by some other gradient function of the temperature. An alternative is to treat heat-flow terms as entropy flows so that entropy generation is proportional to entropy flux.

In steady-state conditions the system state remains constant, and the entropy creation is equal to the flow of entropy into the system associated with heat transfer and mass flow. Let us call this J_s; then, from equation 7.7,

$$\dot{\sigma}_c = J_s = \frac{\dot{Q}}{T_0} + \Sigma \, \dot{n}_p \, (s_p - s_{0p}).$$

In the absence of external work, an energy balance, equation 8.1, gives

$$\dot{n}_p \, \frac{(h_p - h_{0p})}{T} = -\frac{\dot{Q}}{T},$$

since

$$\mu_p = h_p - Ts_p,$$

and

$$\mu_{0p} = h_{0p} - T_0 s_{0p},$$

$$\dot{n}_p (s_p - s_{0p}) = \dot{n}_p \left[\left(\frac{h_p - \mu_p}{T} \right) - \left(\frac{h_{0p} - \mu_{0p}}{T_0} \right) \right]$$

$$= \Sigma \, \dot{n}_p \left(\frac{\mu_{0p}}{T} - \frac{\mu_p}{T_0} \right) + \left(\frac{\dot{Q}}{T} - \frac{\dot{Q}}{T_0} \right). \qquad (8.35)$$

Then

$$J_s = \frac{\dot{Q}}{T} - \Sigma \frac{\mu_p \dot{n}_p}{T}. \qquad (8.36)$$

For small temperature differences a change in entropy creation may be written as a product of forces and fluxes; with $Q = J_q$, and using the Gibbs-Duhem relationship,

$$d\dot{\sigma} = dJ_s = J_q d\left(\frac{1}{T} \right) - J_i d\left(\frac{\mu_i}{T} \right) + J_{chem} d\alpha_c.$$

Note that both diffusion J_i and chemical-reaction J_{chem} fluxes are included.

Now

$$d\left(\frac{1}{T} \right) = -\frac{1}{T^2} dT,$$

and

$$d\left(\frac{\mu}{T}\right) = \frac{1}{T}d\mu_i - \frac{\mu_i}{T^2}dT.$$

Then the entropy generation may be written in spacial coordinates as

$$d\dot{\sigma} = -\frac{J_q}{T}\,\text{grad}\,T - \frac{J_i}{T}\,\text{grad}\left(\frac{\mu_i}{T}\right) + \frac{J_{\text{chem}}}{T}\,d\alpha_c.$$

For Onsager's relations to hold,

$$\frac{J_k}{T} = L_{kj}\,\text{grad}\,\mu_j,$$

and an entropy flux J_s is introduced. Of course, such a flux is an invention and has no physical representation.

For our purposes it is sufficient to examine the function

$$\dot{\sigma} = J_q\,\text{grad}\,\frac{1}{T} - J_i\,\text{grad}\,\frac{\mu_i}{T} + J_{\text{chem}}d\left(\frac{\alpha_c}{T}\right).$$

With a spatial system of interactions,

$$J_q = L_{qq}\,\text{grad}\left(\frac{1}{T}\right) + L_{qi}\,\text{grad}\left(\frac{\mu_i}{T}\right),$$

$$J_i = L_{iq}\,\text{grad}\left(\frac{1}{T}\right) + L_{ii}\,\text{grad}\left(\frac{\mu_i}{T}\right).$$

with $L_{iq} = L_{qi}$, and the diffusive mass flow and heat flow are coupled.

The available-energy dissipation may be written in terms of spacial variation of the thermodynamic parameters of β and α:

$$T_0\dot{\sigma} = kT_0J_q\,\text{grad}\,\beta + kT_0J_i\,\text{grad}\,\alpha_i + kT_0J_{\text{chem}}d\alpha_c$$

        (temperature   (chemical   (chemical
        gradient)   gradient)   reaction)

$$(8.37)$$

In the design of heat exchangers, the objective is often the exchange of thermal energy. These exchangers are highly irreversible systems, since in order to provide high heat flux, the temperature gradients must be high.

In other power-conversion systems the objective is the transfer of available energy from one medium to another. In these systems heat transfer is required, but the evaluation of the system must be made on an available-energy basis, as discussed in chapter 9. Note should also be made of recent work by Silver [12] that extends the formulation of irreversible thermodynamics to systems that are not necessarily close to thermodynamic equilibrium.

References

1. Onsager, L. "Reciprocal Relations in Irreversible Processes," I and II. *Phys. Rev.* 37(1931):405; 38(1931):2265.
2. DeGroot, S.R., and Mazur, P. *Non-Equilibrium Thermodynamics.* Amsterdam: North-Holland, 1962.
3. Prigogine, I. *An Introduction to Thermodynamics of Irreversible Processes.* London: Interscience, 1968.
4. de Donder, Th., and Van Rysselberghe, P. *Thermodynamic Theory of Affinity.* Palo Alto, Calif.: Stanford University, 1936.
5. Odum, H.T., and Pinkerton, R.C. "Time's Speed Regulator: The Optimum Efficiency for Maximum Power Output in Physical and Biological Systems." *Amer. Scientist* 43, no. 2(1955):331.
6. Katchalsky, A., and Curran, P.F. *Nonequilibrium Thermodynamics in Biophysics.* Cambridge, Mass.: Harvard University Press, 1965.
7. Meixner, J., and Reik, H.G. "Thermodynamik der Irreversiblen Prozesse." In *Encyclopedia of Physics,* vol. III/2, ed. S. Flugge, p. 413. Berlin: Springer-Verlag, 1959.
8. Denbigh, K.G. *The Thermodynamics of the Steady State.* London: Methuen, 1951.
9. Zubarev, D.N. *Nonequilibrium Thermodynamics,* trans. from Russian. New York: Plenum Publishing Co., Consultants Bureau, 1974.
10. Strutt, J.W. (Rayleigh). "Some General Theorems Relating to Vibrations." *Proc. Math Soc., London* 4(1873):357.
11. Miller, D.G. "Thermodynamics of Irreversible Processes." *Chem. Rev.* 160(1960):15.
12. Silver, R.S. "Collinearity and Disequilibrium in Irreversible Thermodynamics of the Steady State." *J. Phys. A; Math. Gen.* 13(1980):3253.

9

Available-Energy-
Accounting Economics

If the economists in the marketplace were to determine their estimates of shortage by looking further and further into the future, these estimates would come closer and closer to the estimates made by their colleagues, the thermodynamicists.
—R. Stephen Berry, *Bulletin of the Atomic Scientists* (1972)

Available-Energy Accounting

This chapter attempts to develop a systematic accounting method for available energy that will be applicable for environmental and economic decisions. A true available-energy accounting has never been applied in an industrial or public-decision problem. Several attempts to include available-energy factors in these areas have been made, notably the work of Tribus and Evans [1–3] in the saline-water-conversion context. The discipline in which this is considered is known as *thermoeconomics*. In thermoeconomics, thermodynamic elements (available energy) and cost factors (dollars) are combined to assist in the optimization of technical processing systems.

In these applications the two basic elements of cost are the *capital cost* and the *operating cost*. The method is useful primarily when operating cost is principally the cost of the available energy required to produce the product. The capital cost is then related to the size of the production plant required. The cost equation then looks like

$$C_T^* = C_c^*(A) + C_v^*(\dot{b}) + C_m^*. \tag{9.1}$$

The capital cost C_c^* is taken as proportional to the size A of the production region, either the surface area, as in a heat exchanger, or the volume, as in a chemical reactor. The variable cost C_v^* is a function of the available-energy flow rate $\dot{b}$ required. An additional cost C_m^* is included to account for costs not directly associated with the production process. These cost items include labor, insurance, royalties, publicity, and so forth. The cost per unit product is usually of interest and is found by dividing by the product rate p:

$$C_p^* = \frac{C_T^*}{\dot{p}} = \frac{C_c^* A}{\dot{p}} + \frac{C_v^* \dot{b}}{\dot{p}} + \frac{C_m^*}{\dot{p}} \tag{9.2}$$

215

A relation is then established between the area A and the available energy required for processing. A simple relation as shown in the available-energy-irreversibility sector is that the available energy dissipated is (see chapter 7):

$$\dot{b} = T_0 \dot{\sigma}_c = J_K X_K$$

where J_K is the flux and X_K the associated force. With a linear force-flux relation $J_K = ACX_K$, where A is the area and C the conductance, the available energy can be written as

$$\dot{b} = \frac{J_K^2}{AC}.$$

Inversely, the area can be written as

$$A = \frac{J_K^2}{\dot{b}C}.$$

The cost equation in terms of available energy is then

$$C_p^* = \frac{C_c^*}{\dot{p}} \frac{J_K^2}{C\dot{b}} + \frac{C_v^* \dot{b}}{\dot{p}} + \frac{C_m^*}{\dot{p}}. \tag{9.3}$$

For a fixed flux J_k one is left with an optimization of cost per unit product with respect to the available energy required. In a linear cost equation, this gives the usual Kelvin Law relationship with minimum cost per unit product when

$$\dot{b}_{min} = \sqrt{\frac{C_c^*}{C_v^*} \frac{J_K^2}{C}} \tag{9.4}$$

Alternatively, the area (size) factor may be included to give:

$$C_p^* = \frac{C_c^*}{\dot{p}} A + \frac{C_v^*}{\dot{p}} \frac{J_K^2}{AC} + \frac{C_m^*}{\dot{p}}. \tag{9.5}$$

Minimization of cost per unit product with respect to the area gives:

$$A_{min} = \sqrt{\frac{C_v^*}{C_c^* C}} J_K. \tag{9.6}$$

A simple example in which this optimization is made is illustrated by the classical electrical wire problem later in this chapter. A similar but linear case occurs in the minimization of cost of piping in fluid-flow applications in a chemical plant or sewage-treatment facility.

The principal difficulty in these analyses is the establishment of the relationship between the available energy required and the physical size of the processing elements. The trade-off is between efficiency and capital cost. In the piping problem a small (low-capital-cost) pipe can be used to provide a certain flow rate if one is willing to expend the extra pumping energy (high operating cost). In air-conditioning units the trade-off is between large-area heat-exchanger units, which are efficient, thus having high initial cost, and small heat exchangers, which are inefficient but have low initial cost.

This is, of course, the old marketing problem, which leads to production of least-initial-cost air conditioners that are inefficient. These sell well when the customer is unaware of the operating-cost penalty he pays, or when the customer is capital limited and purchases are based on lowest initial cost. However, the labeling of air-conditioning units with a Btu-cooling Watt-input rating can help the consumer choose only where the ratings are different for the same price. Unless the customers are sophisticated enough to determine the expected use and energy costs, they cannot make an economic decision. They can however, make a minimum-operating- or energy-cost choice if they are enegy-conservation oriented. The result is that, in the face of uncertainty about the operating cost, the lowest-initial-price alternative is often selected.

In the air-conditioner-selection process the unknowns are fairly well defined, and with a little effort an optimum economic choice can be made. Required, of course, are the expected lifetime, the cost of electricity, and the cooling load for the particular home in which it is to be installed. The latter can be approximated with a knowledge of weather data and the size and insulation characteristics of the house. All this information could easily be included in an optimization computer program, which could be supplied at a terminal in a store, with appropriate entries provided by the customer and with localized climate and machine data supplied by the manufacturer or a consulting engineer. This illustrates that the process of arriving at an economic decision is fairly straightforward and well understood in consumer-purchase situations.

In processing situations the decisions become much more complicated because of the expanded alternatives available. A variation of the cost-per-unit-product rate equation is often adopted in situations where a surface transfer is critical. This variation is to consider the available-energy cost as proportional to the area and conductance combined, so that the specific cost is on the basis of cost per unit conductance instead of cost per unit area. This variation allows one to select components such as heat exchangers based not on their efficiency but on their specific cost per unit conductance delivered.

This provides a combined technical and economic measure of the effectiveness of components in a process. Alternative exchangers, converters, and so on can then be designed and evaluated on the basis of combined factors. In the design of heat exchangers, the measure of effectiveness has often been the conductance alone, without regard to cost or to the impact on the energy required to produce that conductance. A simple example is the design of internally finned heat-exchanger tubes. The gain in conductance obtainable is acquired at the increased expense of providing more energy to force the fluid through the tube. The total cost may thus be increased by optimizing a subparameter, conductance in this case, without regard to the economic factors.

Returning to the cost equation in terms of the available energy,

$$C_T^* = C_p^* \dot{p} = C_c^* \frac{J^2}{C\dot{b}} + C_v^* \dot{b} + C_m^*. \qquad (9.7)$$

One can make some simple observations with respect to minimization of cost and energy conservation. The first is that the least-cost solution is not the least-available-energy-use solution, since the cost as a function of available-energy use rate appears a function of $\dot{b}$, as shown in figure 9–1.

Minimum $\dot{b}$ use corresponds to the infinite-capital-cost solution. Similar

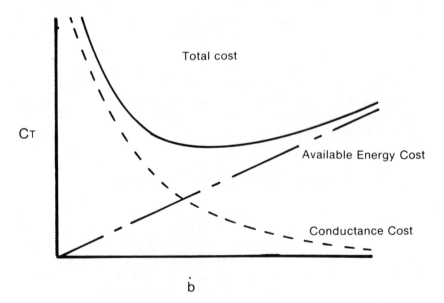

Figure 9–1. Illustration of the Optimization of Systems with Major Available-Energy and Conductance Costs

proposals that suggest selecting a solution with minimum energy use per unit dollar cost are again faced with the minimization of $b \rightarrow 0$ and a capital cost approaching infinity.

In summary, this approach has proved useful for comparing alternative methods of processing to reach minimum-cost solutions. Environmental and energy-conservation factors are not present for the formulation as it stands.

Berry's comment is quite to the point of energy as a currency. In fuel terms, a petrodollar economy would be an economy in which available energy rather than gold was used as a monetary standard.

In order to reach a theory in which this is analyzable, we hypothesize a *capital available energy*. In the minimum-cost solution to the electrical-wire problem, we considered the variable available energy dissipated in supplying power. To this should be added the available energy of the wire itself, allocated over its lifetime. Instead of this we consider the available energy of the wire as capital available energy. To this available energy of the wire itself is then added the available energy expended to produce the wire from its datum state. This includes the fuel energy for extraction, electrolytic refinement, shaping, transport, and so on, and also the distribution of the available energy utilized to construct the processing machines, buildings, and so forth (that is, the capital investment allocated to the processing of this particular product).

This may sound very heuristic, but it is not much more difficult than the energy accounting done by Pimentel et al. [4] and Steinhart and Steinhart [5] with respect to food production, or Berry [6] and others [10] with respect to how much energy is required for an automobile (see chapter 2). The advantage of an available-energy accounting is that it identifies the value of the waste streams and assists in developing combined systems with greater conservation. It also provides a basis of value that a unit of energy does not represent.

Thus for optimization with respect to choices and designs, an available-energy analysis can provide information not obtainable by an energy accounting. An initial national accounting in available-energy terms has been attempted by Riestad [11]. Figure 9–2 shows the results of his analysis, which can be compared with an energy analysis in figure 9–3.

Available-Energy Implications of Solar Energy Systems

In solar energy conversion it is the intercepting area and its properties that determine the conversion to thermal modes in the receiver. In the engineering analysis of solar energy systems, the energy input is usually considered free, and the cost of these systems is associated with the expense of the materials

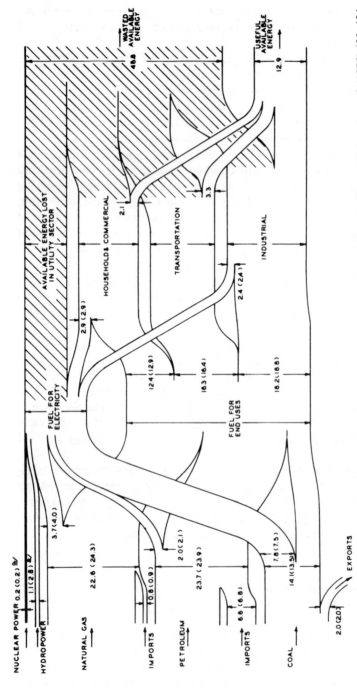

Source: G.J. Riestad, "Available Energy Conversion and Utilization in the United States," *J. of Engr. for Power* 97, Sec. A, no. 3 (1975):429–434. Reprinted with permission.

Note: All values reported are 10^{15} Btu ($1{,}055 \times 10^{18}$J).

[a]All in parentheses are energy; other values reported are available energy.

[b]The energy value for hydropower is reported in the usual manner of coal equivalent: actual energy is 1.12×10^{15} Btu [11].

Figure 9–2. Flow of Available Energy in the United States

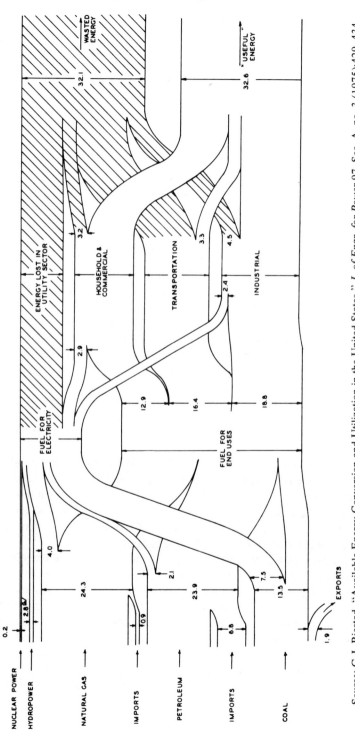

Figure 9-3. Flow of Energy in the United States

Source: G.J. Riestad, "Available Energy Conversion and Utilization in the United States," *J. of Engr. for Power* 97, Sec. A, no. 3 (1975):429–434. Reprinted with permission.

Note: All values reported are 10^{15} Btu (1.055×10^{18}J).

[a]The energy value for hydropower is reported in the usual manner of coal equivalent [11].

to provide the collecting area. The solar radiation on the earth's surface might also be considered as a resource associated with a given surface area, in the sense that the equatorial areas are more valuable since they have greater solar-energy potential. In photosynthetic-potential studies areas of the earth have been mapped according to their potential for photosynthesis based on solar-energy input, cloud cover, growing season, weather, and temperature (see figure 9–4.) Desert regions have been proposed for solar-energy-conversion regions not only on the basis of their sunlight-input potential but also because of their low area value for other alternative economic uses.

Freeman Dyson [7] hypothesized that advanced civilizations in the universe might be located by their infrared emissions. This concept was based on the idea that a very advanced civilization would require more surface area to emit waste thermal energy into space than a planet itself would provide. To accomplish this, the inhabitants would collect space fragments or other materials and build a surface larger than the planet itself (a Dyson sphere). This civilization would then be detectable from its high infrared emission.

If nuclear-waste-disposal problems are not solved, the construction of such a sphere might also be required to intercept a greater quantity of solar energy. In this light, land surface may be valued according to its solar-energy potential. In fact, the general distinction between developed and under-developed countries can be determined by observation of the photosynthetic potential [8], as shown in figure 9–4. (Exceptions are those countries with large fossil-fuel resources.)

In solar-energy-conversion systems, the crucial factor is the collector capital cost per unit area. For example, first consider a simple photoelectric device. The input is sunlight, and the outputs are electrical energy and waste thermal energy.

Photoelectric cells operate more efficiently at lower temperatures because of the reduced interaction between photo quantum transition states and thermal states at low temperatures. The low-energy photons in the solar spectrum are thus detrimental and reduce the efficiency of these devices. It is in fact beneficial to utilize some of the electrical-energy output to drive fans or pumps and maintain the photo cells at a lower temperature.

Let us ask how such a system is to be evaluated. Clearly, no energy inputs have economic value that are charged on an energy basis (that is, dollars/kilowatt-hour). Efficiency has a technological value in terms of useful energy out, but the decisions on the utility of a system are mainly the cost to convert the energy; the energy input has no valued cost. This is, of course, the situation with all extraction process (unless a depletion allowance might be considered as this cost). What is the cost of coal to the coal-mining company? of crude oil to the oil company? of copper to the copper-mining company? of air to the ammonia-processing plant? Often only in secondary processing or

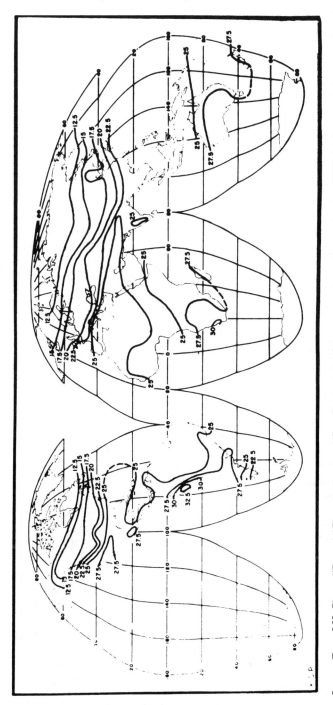

Source: From J.H. Chang, "Potential Photosynthesis and Crop Productivity," *Ann. Assoc. Amer. Geographers* 60(1970):96. Reprinted with permission of Association of American Geographers, 1710 16th St. NW, Washington, D.C. 20009.

Figure 9-4. Average Annual Potential Photosynthesis in Grams per Square Meter per Day

where alternative economic uses for a resource are available is there an economic cost of the input. Although it may be argued that a mineral resource in the ground has value in cases such as copper mines, gold mines, oil fields, and so forth, in reality the major value resides in the technology to extract these resources. The location of, for example, Dow Chemical Corporation in Midland, Michigan, is predicated on a large brine deposit. This deposit extends from western New York through central Michigan, so that although the brine has value with respect to the average earth composition, the land resource is negligible. The value of oil and gas leases is about 12 percent of the wellhead price.

In this book local high concentrations of valuable minerals are taken to have economic value that can be associated with its available energy with respect to a datum or average earth concentration. The cost of available energy from a solar energy system is then a capital cost, which is a function of the conversion area A and the technology, often expressed as a conductance C:

$$\text{Capital cost} = C_C^* \frac{A}{C}$$

In this book we consider the cost as the cost per unit available energy. This can be written in terms of the available-energy input. If the effectiveness is considered as the ratio of the available energy of the output to the available energy of the input,

$$E \equiv \frac{b_{out}}{b_{in}},$$

then effectiveness depends on the technology. It depends specifically on the economic cost of the construction and materials that are required to provide the conversion. The output of a solar-collection system depends on the collection area, which controls the quantity but not the quality of the input. The specific available energy of the input is not controllable. The quantity, however, is—through the interception area size and the absorptivity, conduction, and so forth, as discussed in chapter 14. The output-energy quantity and quality are closely interrelated, however. In a simple solar water-heating collector system, as shown in figure 9–5, a high-temperature, high-quality energy is obtainable by lowering the water flow rate. The total energy loss by convection of heat to the atmosphere is higher for a high temperature, so that quality is gained at the expense of quantity.

For the system shown the input is the solar intensity I times the collector area A. The available-energy output is the heat flow to the fluid q times the Carnot factor $[1 - (T_0/T_c)]$, where T_0 is the environment temperature and T_c

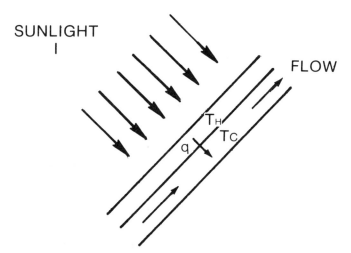

Figure 9–5. Schematic of Energy Flow in a Solar Water-Heating Collector

is the collector-fluid temperature. The energy absorbed by the fluid is then

$$\dot{b}_{out} = q\left(1 - \frac{T_0}{T_c} \right) = CA(T_H - T_C)\left(1 - \frac{T_0}{T_c} \right).$$

The specific cost of available energy decreases with increased conductance C. The technological factor is the improvement of the conductance. The cost may then be related extensively to the area and intensively to the conductance. The effectiveness,

$$\frac{\dot{b}_{out}}{\dot{b}_{in}} = \frac{C(T_H - T_C)}{I}\left(1 - \frac{T_0}{T_H} \right),$$

does not depend on the area. This would indicate no energy economy of scale in solar heating systems. It would also indicate that increased emphasis be placed on the development of individual-home solar energy systems, as opposed to large-scale, centralized systems.

The cost can be assumed to be the cost of the processing area A and the cost of the conductance C. The total cost can be expressed as

$$C_T^* = C_c^* \frac{A}{C} + C_a^* \dot{b}_{input}.$$

$C_a^* \dot{b}_{input}$ is the cost of pumping fluid through the collector or other energy inputs required to produce the energy exchange.

Availability Analysis and the Consumer

The behavior of the majority of consumers in making purchases is based on minimum initial cost. In the absence of information on operating costs, this is a rational criterion. An availability energy-cost analysis can help to establish a basis for estimating both the total cost and the total energy utilized. A prime example to be considered will be air conditioners. The initial cost of an air conditioner can be reduced by decreasing the evaporator and condenser sizes. This reduces initial cost, but at the expense of a higher-compression-ratio compressor that operates less efficiently. Other examples readily examined are humidifiers, refrigerators, washers, driers, and home furnaces.

Furnace-combustion-efficiency checks and criteria for adjustments and cleaning can also be developed on a more scientific basis than yearly inspection. Consideration must be given to the cost of inspection, adjustment uncertainties, and available-energy requirements for improved furnace designs.

New innovations such as inflatable structures will require an available-energy analysis for proper technical assessment of their total energy requirements. One usually assumes that a structure that requires energy for maintaining its shape must be more energy expensive than a rigid structure. However, this may not be the case if the amounts of energy required for material processing, construction, and maintenance are all considered. Similarly, heat-pump and solar-energy-system considerations must be technologically assessed by an availability analysis. Any technological assessment of consumer product choices must include consideration of the pricing policy of energy vendors.

The difficulty of providing simple rules of thumb for decisions on selection of materials can be illustrated by a short example. This problem, considered almost a classical enconomic-choice example, is shown to involve other principles and a different approach if energy assessment is to be included.

Example: Electrical-Wire Analysis

In this example we will consider the question of specifying the wire size and wire material to minimize the total cost and also the minimum total available-energy loss. Consideration will be given to the cost and energy dissipation in the wire and to the cost and energy availability of the wire material. The choice between copper and aluminum wire will be discussed in this example.

Take a simple problem where the power load is fixed, as shown in the sketch (figure 9-6). The power supplied is:

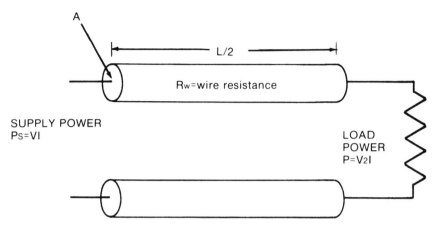

Figure 9–6. Electrical-Wire Nomenclature for Thermoeconomic Analysis

the rate of available energy flow. $\dot{b} = V_2 I + I^2 R_w,$ (9.8)

The available energy of the wire, which is taken as the available energy required for manufacture and extraction of the metal, is taken as:

$$b_c = (b_w)ALd, \qquad (9.9)$$

where b_w = available energy/lb.

A = cross-sectional area.

L = length.

d = density of the wire.

The wire resistance is

$$R_w = \rho \frac{L}{A},$$

where ρ is the resistivity. The total available-energy cost is then

$$b_T = b_c + \dot{b}(t), \qquad (9.10)$$

where b_c is the capital energy investment.

$\dot{b}$ is the rate of energy use.

t is the time over which the wire is to be amortized or the time it is used (continuous operation is assumed here).

Substituting from equations 9.8 and 9.9 in equation 9.10 gives

$$b_T = (I^2 \rho \frac{L}{A} + V_2 I)(t) + b_w A L d. \tag{9.11}$$

If the cost per pound is C^* of the wire, and C_b^* is the cost per unit of power, then the total cost is

$$C = C^* d A L + C_b^* \dot{b} t. \tag{9.12}$$

If the load power is fixed, then $IV_2 = P$ is constant; then equations 9.11 and 9.12 become

$$b_T = \left(\frac{P^2}{V_2} \rho \frac{L}{A} + P \right) t + b_w A L d. \tag{9.11a}$$

$$C = C^* d A L + \left(\frac{\rho L}{A} \frac{P^2}{V_2} + P \right) C_b^*(t). \tag{9.12a}$$

For a minimum cost to be achieved by selection of wire size (area A), then a simple minimization gives, setting

$$\frac{\partial C}{\partial A} = 0,$$

$$A^2 = \frac{P^2}{V_2} \rho \frac{C_b^* t}{C^* d}. \tag{9.13}$$

For minimum available-energy cost, similarly,

$$A^2 = \frac{P \rho t}{V_2 b_w d} \tag{9.14}$$

A minimum-cost solution corresponds to a minimum available-energy cost only if $C^* = C_b^* b_w$ or the cost per pound of material is equal to the energy-rate cost times the available energy required for processing material. There is reason this equality should hold only if energy cost is the major cost of the wire; ultimately, this may be the case, as Berry remarks [6]. Note that inclusion of the cost of money (depreciation, inflation, and so on) can be made but does not alter the analysis, only the cost, C^* value.

A simple analysis for the decision situation considering copper and aluminum materials with the parameters shown in table 9–1 is now illustrated.

Table 9-1
Wire-Example Cost Factors

	Copper	Aluminum
b_w	10 kW-hr/lb	33kW-hr/lb
C^*	$0.56/lb	$0.45/lb
ρ	1.073×10^{-6} ohm-cm	2.6548×10^{-6} ohm-cm
d	556 lb/ft	165 lb/ft3

Note: Electrical energy cost is taken as $0.02/kW-hr, and the time is taken as twenty years.

Relative available-energy-cost curves are shown in figure 9-7. As expected, the area at which a minimum occurs for aluminum is higher than for copper. What is unexpected is the relative shape of the curves. The copper minimum is much sharper than the aluminum minimum, indicating a more critical decision process in the copper case. Also indicated on these curves are the areas at which minimum economic cost occurs. It is notable that in the copper situation a larger wire size than economically optimum is best in terms of energy, but in the aluminum situation a smaller wire is optimum in terms of energy although this is not indicated economically.

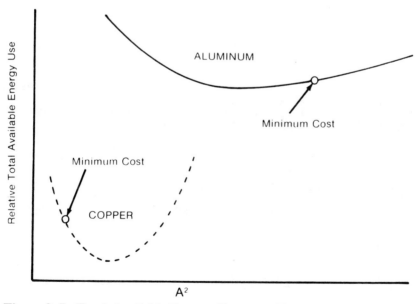

Figure 9-7. Total Available-Energy Use as a Function of Wire Cross-Section Squared A^2

Certainly a rule of thumb that would indicate logically that increasing the size will reduce energy losses is not sufficient for this simple choice situation. The energy cost of switching from copper to aluminum is also indicated in this figure.

Figure 9–8 shows the relative economic-cost curves for the same situation. Here it is shown that economically the choice is aluminum rather than copper, but with a much larger wire. Because of the flatness of the aluminum energy and cost curves, the decision about size is not critical. In the copper case the choice is very critical.

Also shown on this figure are the societal-cost penalty associated with choosing a minimum available-energy size rather than least cost, and also in selecting aluminum rather than copper wire, as is now the trend in the construction industry.

This illustration indicates the value of including available-energy analysis in decision making but does not yet point to the solution when the minimum of energy and economy do not coincide. Resolution of this question will require consideration of future projections on cost of materials, energy, and impacts on the economy and the environment. This consideration adds a new dimension to environmental and energy decision making.

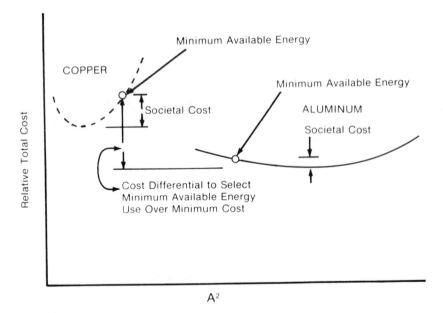

Figure 9–8. Total Economic Cost as a Function of Wire Cross-Section Squared A^2

Available-Energy Accounting: Desiderata and Criteria

The procedures of thermoeconomics require careful consideration of many elements in their development. In this section I have attempted to outline how the elements might be selected and formulated.

To be valid, an accounting system must be useful. To this end, certain desiderata and criteria will be applied to the categorization of the accounting elements. (A summary of the analysis steps is given in appendix 9A.)

1. A standard of effectiveness needs to be computed. This indicates the need to define as an element the reversible available energy required to produce the product from the input resource factors. This is only theoretically attainable with infinitely large capital investment and infinitely small process rates. Thus in practice it is never attained but needs to be available for merit considerations.

2. Waste-stream available energies and dissipation losses also need to be computed. The dissipation losses in available energy will usually be computed as the difference between the input and output available-energy flow rates, with storage taken into account when applicable.

3. The storage factors or available energy stored as capital investment need to be accounted. The capital investment in a large furnace or turbine, for example, are accumulated available-energy stores that are required for processing. In one sense they have a component that is recoverable, representing the actual available energy of the component. This is directly utilized, for example, in the replacement of an open-hearth furnace by an electric-arc funace when the old furnace is fed into the new one. There is a component of the available energy that goes into producing a system that is recoverable, and this should be identified separately from the available energy dissipated in the construction.

4. The speed of processing must be a factor that is resolvable in terms of the dissipation function. For example, a large compressor (or other large machine) operating at slow speed is more efficient than a small compressor operating at high speed.

5. Convergence of dollar accounting and available-energy accounting permits common discussion of such factors as economy of scale, resource value, discounting, depreciation, and obsolescence.

6. Local conditions with respect to resource availability and environmental factors such as weather, climate, geography, and so on are included in the analysis. This is included in the datum or reference-state definition and the transport requirements. The changes in these factors with time as the resources are depleted and the environment changes are important factors not considered in any present analyses.

Let us illustrate the accounting by discussing the wire example for which the cost optimum was determined. This will require comparison of copper and aluminum resource materials, beginning with the ore-extraction phase and ending with the final wire form. Estimates will be required in arriving at numbers for each of the categories of available energy just described.

1. The theoretical available energy of extraction of the metal from the ore can be estimated in two steps. The first is the theoretical available energy to extract the material from its chemically found form. In the case of aluminum, ore in the form of bauxite will be assumed, which is 50 percent Al_2O_3. Then approximately 4 pounds of ore are required to give 1 pound of aluminum. The available energy required for extraction of aluminum from Al_2O_3 is approximately 3.30 kW-hr/lb, to which is added the concentration available energy of approximately 0.01 kW-hr/lb, giving a total of approximately 3.04 kW-hr/lb. This is the reversible available energy required to extract the metal. For copper, assuming Cu_2S as the ore form and a concentration of 0.1 percent, in many currently mined deposits gives 0.11 kW-hr/lb for chemical processing, plus approximately 0.05 kW-hr/lb for concentrating, or a total of 0.16 kW-hr/lb theoretical reversible available energy required.

Note that this computation requires knowledge of the chemical form of the ore and the concentration in the resource deposit. The concentration changes with depletion, and so also may the chemical composition as one moves from, say, bauxite to red clays (kaolinite) for aluminum processing. If the aluminum were extracted from the earth crust with an average of 8 percent aluminum, the concentration available energy would increase to 0.03 kW-hr/lb, so that an additional 0.02 kW-hr/lb would theoretically be required if the Al_2O_3 were uniformly distributed in the crust as an average environment. The copper would increase to 0.07 kW-hr/lb concentration energy. Of course, the value of a given ore deposit is a combination of a loosely formed chemical bond of the ore and the concentration. In the case of copper it ordinarily oxidizes under atmospheric weathering conditions to copper sulfate, with a loss of available energy of about 1.5 kW-hr/lb. This loss is then representative of the available energy lost because of deterioration, which leaves the copper in a more degraded form. This energy is required to extract the copper if the copper is not recycled before environmental deterioration occurs. Aluminum, on the other hand, oxidizes to Al_2O_3, which is then extractable with energy requirements close to those of the original ore.

2. Waste-stream available energies are not currently available from published materials. In the aluminum case this would be represented by waste electrolyte that is not recycled, CO_2, and water vapor, as well as any fluorine gas that escapes to the atmosphere in the processing. Secondary waste streams would be the CO_2 and SO_2 emitted by the power plant supplying the electrical energy. In the copper case, sulfur compounds would be produced in the Cu_2S separation process, as well as CO_2 and SO_2 from smelting operations.

The inventories of aluminum and copper are not usually considered in the estimates of actual energy required per pound of metal produced. Data from Department of Commerce *Census of Manufacturers* reports notes production and energy purchased on a yearly basis. In many cases these represent energy purchased during the year without regard to storage, and metal sold rather than actually manufactured. Inventory records would be required to estimate the storage factors.

3. As noted in the previous paragraph, storage factors are available but not counted in the usual energy accounting. Hirst [14] estimates the capital investment of total energy based on a requirement of one pound of machinery for each pound of metal product produced. He depreciates heating processes over thirty years and electrical equipment over twenty years. This results in a capital investment of 0.5 kW-hr/lb for aluminum production per year and 0.4 kW-hr/lb for copper production.

This energy accounting for machinery or capital investment is very tenuous at present, particularly in regard to nonferrous metals, since machinery is principally of iron extraction on a weight basis. The scrap after the depreciation period is recoverable. If this iron is recycled, it requires about one-third the energy cost of new material. If it is aluminum, its recycled energy cost is about 4 percent that of virgin-material refining. In an aluminum smelter, carbon electrodes are consumed at a very high rate, converted to CO and CO_2, and released to the atmosphere. Typically, 0.7 pound of carbon is consumed per pound of aluminum smelted. The energy required to produce this carbon from coke is approximately 5 kW-hr/lb (of electricity) carbon or 13.2 kW-hr thermal energy at the electrical-power plant. This energy requirement is included in an aluminum industry's reporting of energy use if the carbon plant is owned by the industry. This energy requirement is not small, representing approximately one-third of the reported 35 kW-hr/lb aluminum required in the production of aluminum metal.

The consumption of the carbon electrodes is required in the reduction of the aluminum oxide to aluminum metal in the reaction:

$$Al_2O_3 + 3C \rightleftharpoons 2Al + 3CO,$$

which theoretically requires 3 pounds of carbon for each 2 pounds of aluminum produced.

4. The energy cost of a rapid processing rate appears in the high current density used in the electrolysis cell produced by a high voltage. This high current density is used to provide electrical heating of the electrolyte to keep it in a molten state. In this conversion process approximately 12 kW-hr/lb of aluminum are required, instead of the theoretical 3.04 kW-hr/lb. This difference, 9 kW-hr/lb, is a rate loss that could be reduced by direct thermal

heating and lowered current density. The 12 kW-hr/lb of electrical energy represents about 31.6 kW-hr/lb of coal energy.

An interesting economic study of the differences between private costs and public costs is summarized in figure 9–9. This study examines the cost of paper production as a function of the reuse or recycle of paper waste. The minimum cost for the paper industry occurs at a lower reuse ratio than the reuse ratio that minimizes the total cost including the public cost to treat wastes and reduce external damages [17].

Example: Electrolytic Aluminum Production

Aluminum is produced from alumina (Al_2O_3) by electrolysis in a reduction cell. This process, termed the Hall or Heroult process, requires large quantities of electrical energy. In this process, alumina at a concentration of 2–20 percent in a bath of fused cryolite at about 900°C is subjected to an electrical voltage of 6–7 volts at a current density of about 1,000 amp/ft^2 at the cathode. At the cathode nearly pure aluminum is deposited as a liquid. At the anode the oxygen is released as a gas. The reaction is:

$$Al_2O_3(s) \rightleftharpoons 2Al(\ell) + O_2(g).$$

The Gibbs free energy for this reaction is $\Delta G = 401{,}400 - 77T$ cal/mole. The reversible cell potential required for this reaction at 900°C is found by equating this free energy to the valence n times the potential V_o times the Faraday constant $\mathscr{F} = 23{,}066$ cal/volt equivalent; $\Delta G = nV_o\mathscr{F}$. For aluminum the valence n is 3; therefore, the reversible (minimum)potential required is $V_o = 2.24$ volts. The voltage actually used to produce a reasonable production rate is higher than this (6 volts).

The available-energy use per unit mass-deposition rate may be determined by considering the following illustration. The available energy required is

$$b = IV_o + I^2R,$$

where I is the current and R the resistance of the electrolyte. The first term on the right represents the minimum available-energy rate required. The second term I^2R represents the energy dissipated to achieve a production rate. The mass rate of deposition of a material electrolytically is fixed by Faraday's law as:

$$\dot{m} = \frac{I}{n\mathscr{F}}$$

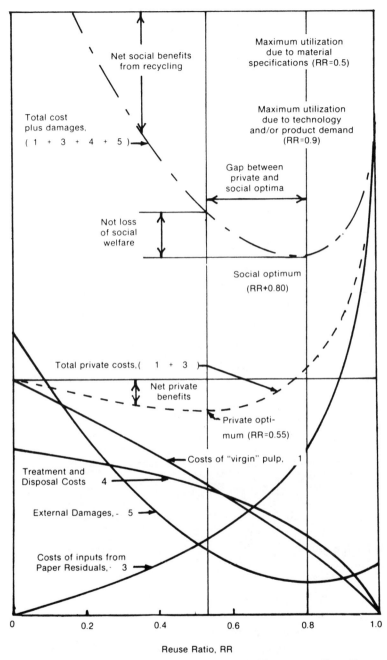

Source: Redrawn from W.D. Spofford, Jr., "Solid Residuals Management: Some Economic Considerations." Reprinted with permission from 11 *Natural Resources Journal* 585 (1971) published by the University of New Mexico School of Law, Albuquerque, NM 87131.

Figure 9–9. Components of Cost in the Use of Paper Residuals

at 100 percent current efficiency. For aluminum with $n = 3$,

$$\frac{\dot{m}}{I} = 0.336 \text{ gm/amp-hr.}$$

Dividing b by $\dot{m}$ gives

$$\frac{b}{\dot{m}} = \frac{IV_o}{\dot{m}} + \frac{I^2 R}{\dot{m}} = \frac{V_o}{0.336} + \frac{IR}{0.336} \quad \frac{\text{watt-hr}}{\text{gm}}.$$

The minimum energy per gm is then

$$\frac{b}{\dot{m}} = \frac{V_o}{0.336} = \frac{2.24}{0.336} = 6.67 \frac{\text{watt-hr}}{\text{gm}}.$$

Cost enters with consideration of the capital costs involved in the construction of the electrolyte cell. If we take these costs as proportional to the electrode area as $C_A^* A$, then the total-cost equation may be approximated as:

$$C^* = C_A^* A + C_b^* IV_o + C_b^* I^2 R.$$

Since the resistance is proportional to the resistivity ρ and the distance between electrodes L and varies inversely with the electrode area:

$$R = \frac{\rho L}{A}$$

The cost per unit aluminum deposition can be written as

$$\frac{C^*}{\dot{m}} = \frac{C_A^*}{0.336}\left(\frac{A}{I}\right) + \frac{C_b^* V_o}{0.336} + \frac{C_c^* \rho L}{0.336}\left(\frac{I}{A}\right),$$

using the relation

$$\frac{\dot{m}}{I} = 0.336.$$

This cost equation illustrates the trade-off of current-density variation, which is used to minimize cost. At high current density I/A, the capital cost is low but the available energy cost is high, and a trade-off is required to minimize cost. Minimum available-energy use per unit of material deposited is at

$I/A \to 0$, but this requires large electrodes and associated high costs. Typical current densities in actual cells are 100 amp/m^2 in locations where electrical-energy costs are low. With future increased electrical-energy costs, the optimum-cost current densities would be reduced. Extra electrolyte cells would need to be constructed to maintain production rates in these conditions.

In actual cells the chemical reaction of alumina producing oxygen at the anode is accompanied by immediate reaction of this oxygen with the carbon electrode:

$$C + \frac{1}{2} O_2 \rightleftharpoons CO.$$

The Gibbs free energy of this reaction is

$$\Delta G = -28,200 - 20.16T \text{ cal/mole.}$$

Carbon monoxide, a strong reducing agent, is then a product of the electrolytic cell. This CO has value in such processes as pyrolysis for producing fuels from cellulose. The available energy actually required for the production of aluminum also goes into producing a CO product that has value. If this product is used, the available energy of this CO product should be credited to the process. If it is not used, it must be credited against the process as a pollutant.

Considered from the overall chemical balance of

$$Al_2O_3 + 2C \rightleftharpoons 2Al + 3CO,$$

the minimum energy required at 900°C should be 3.36 W-hr/gm Al, the difference being the energy released in the spontaneous formation of CO from the oxygen and carbon electrode material. The total quantity of available energy lost is

$$A_1 = 6.67 - 3.36 = 3.31 \text{ W-hr/gm.}$$

Of this available energy lost,

$$q \cdot \frac{T - T_0}{T} = 3.37 \left(\frac{1,173 - 300}{1,173} \right) = 2.48 \text{ W-hr/gm}$$

is the available energy lost by the materials if they leave the process region at 900°C and no effort is made to recover its thermal energy.

A more detailed cost and available-energy analysis would include the consumption of carbon electrode materials, the formation of CO_2 as well as CO, and the capital available energy of the cell walls and other required structures amortized over the lifetime of the processing plant.

The Heat-Exchanger Problem and Available Energy

One of the basic loss mechanisms in energy conversion or chemical-processing systems is the heat exchanger. In nuclear power plants it serves a separation function, isolating the primary reactor-core cooling loop from the power-turbine fluid loop. In figure 9–10 this heat exchanger is shown as the steam generator. In fossil-fuel power plants it is an interface between the combustion-product gases and the turbine steam or mercury-cycle fluid. In chemical processing it serves to isolate chemicals in which heating is required to increase reaction rates or provide exothermic energy for a reaction. There are many examples, ranging from steam or electrical heating in brewing, to home heating, to steel making. The loss of available energy in heat-exchange processes accounts for a major available-energy loss in most energy-conversion systems. The avoidance of this loss by some other technical means, such as direct use of combustion gases in turbines or endothermic-reaction chemistry, often produces significant energy-conservation results. Heat-exchanger optimization is a major research effort of the power and heating–air-conditioning business. unfortunately, the parameter usually optimized is the heat transfer per unit exchanger area, without regard to the available-energy loss.

Heat exchangers viewed as input-output energy devices are 100 percent efficient, since the total energy "lost" by the hot-side fluid is "gained" by the cold-side fluid. Efficiency defined as output energy divided by input energy thus is not a useful performance measure. There are two basic problems in the available-energy analysis of heat exchangers. (1) the available-energy requirement to supply high flow rates that increase the heat-transfer coefficients and hence minimize the available-energy difference between the hot and cold side of the exchanger; and (2) the need to reduce the heat-exchanger area to reduce the capital cost of the exchanger, which leads to reduced heat flux. Examination of a simple exchanger will help clarify this problem. This will be formulated as an available-energy transfer problem.

Let us take the simple concurrent heat exchanger shown in figure 9–11. The heat transfer between the two streams at any point may be represented as

$$q = UA(\Delta T)_{1m}.$$

U is the heat-transfer coefficient, and $(\Delta T)_{1m}$ the logarithmic-mean temperature difference between the hot and cold sides [12].

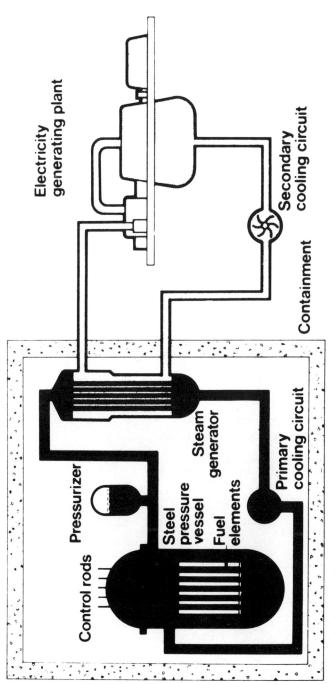

Figure 9–10. Nuclear Power Plant Showing the Primary Cooling Loop with Circulation through the Steam Generator, Which Is the Main Heat-Exchanger Element of the System

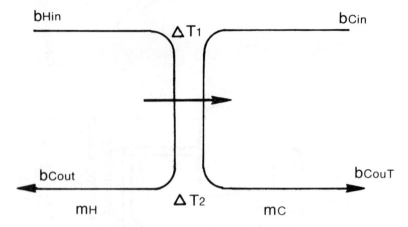

Figure 9–11. Concurrent-Flow Heat-Exchanger Schematic

$$\Delta T_{1m} \equiv \frac{\Delta T_1 - \Delta T_2}{\ln(\Delta T_1/\Delta T_2)} \, ,$$

where $\Delta T_2 =$ the outlet temperature difference.

$\Delta T_1 =$ the inlet temperature difference.

U may be increased by careful design of heat-exchange geometry and by design utilizing input power flow to increase the thermal-exchange coefficient on both input and output sides.

The available-energy decrease of the hot side is

$$\Delta b_H = \dot{m}_H(h_{H_{in}} - T_0 s_{H_{in}}) - \dot{m}_H(h_{H_{out}} - T_0 s_{H_{out}}).$$

The available-energy increase of the cold side is

$$\Delta b_C = \dot{m}_C(h_{C_{out}} - T_0 s_{C_{out}}) - \dot{m}_C(h_{C_{in}} - T_0 s_{C_{in}}),$$

With $$b = h - T_0 s.$$

Balance of available energy

$$\dot{m}_H(b_{H_{in}}) + \dot{m}_C(b_{C_{in}}) = \dot{m}_H(b_{H_{out}}) + \dot{m}_C(b_{C_{out}}) + \text{Loss},$$

or

$$\dot{m}_H(b_{H_{in}} - b_{H_{out}}) = \dot{m}_C(b_{C_{out}} - b_{C_{in}}) + \text{Loss}.$$

Energy balance

$$\dot{m}_H(h_{H_{in}} - h_{H_{out}}) = q = \dot{m}_C(h_{C_{out}} - h_{C_{in}}).$$

From availability,

$$\dot{m}_H[h_{H_{in}} - h_{H_{out}} + T_0(s_{H_{out}} - s_{H_{in}})] = \dot{m}_C[h_{C_{out}} - h_{C_{in}} + T_0(s_{C_{in}} - s_{C_{out}})] + \text{Loss.}$$

Using the energy balance, this reduces to

$$\dot{m}_H(T_0)(s_{H_{out}} - s_{H_{in}}) = \dot{m}_C T_0(s_{C_{in}} - s_{C_{out}}) + \text{Loss.}$$

$$\text{Loss} = T_0[\dot{m}_H(s_{H_{out,}} - s_{H_{in}}) - \dot{m}_C(s_{C_{in}} - s_{C_{out}})].$$

It is this loss in available energy that is inherent in the use of a heat exchanger.

Let us next examine some specific heat-exchanger processes and materials to help describe how the entropy change is found. For most heat-exchanger systems, the work forms are pressure-volume forms used in the process to supply the flow of fluid through the exchangers. Gibbs equation in the form

$$T\,ds = dh - v\,dp$$

is then applicable.

Rearranging for calculation of the entropy changes gives

$$ds = \frac{dh}{T} - \frac{v}{T}\,dp.$$

Then

$$\int_{s_{H_{in}}}^{s_{H_{out}}} ds = \int_{T_{H_{in}}}^{T_{H_{out}}} \frac{dh}{T} - \int_{P_{H_{in}}}^{P_{H_{out}}} \frac{v}{T}\,dp.$$

The first term on the right represents the entropy change due to the heat exchange, and the second term represents the pumping-energy entropy effect.

Two particular cases are readily solved for the entropy change. One is the perfect gas case, where

$$dh = c_p\,dT \qquad \text{and} \qquad \frac{v}{T} = \frac{R}{p}.$$

Then

$$\int_{s_{Hin}}^{s_{Hout}} ds = s_{Hout} - s_{Hin} = c_p \ln \frac{T_{Hout}}{T_{Hin}} - R \ln \frac{P_{Hout}}{P_{Hin}}.$$

Then the loss of available energy is

$$\text{Loss} = \dot{m}_H T_0 \left(c_p \ln \frac{T_{Hout}}{T_{Hin}} - R \ln \frac{P_{Hout}}{P_{Hin}} \right)$$

$$- \dot{m}_C T_0 \left(c_p \ln \frac{T_{Cin}}{T_{Cout}} - R \ln \frac{P_{Cout}}{P_{Cin}} \right).$$

The other case is the situation where a change of phase at constant T and P occurs; then,

$$\int_{s_{Hin}}^{s_{Hout}} ds = -\frac{h_{fgH}}{T_H} \quad \text{and} \quad \int_{s_{Cout}}^{s_{Cin}} ds = \frac{h_{fgC}}{T_C}.$$

The loss in this case is

$$\text{Loss} = -\dot{m}_H T_0 \frac{h_{fgH}}{T_H} + \dot{m}_C T_0 \frac{h_{fgC}}{T_C}.$$

Since $\dot{m}_H h_{fgH} = \dot{m}_C h_{fgC} = q$ from energy conservation,

$$\text{Loss} = -q T_0 \left(\frac{1}{T_H} - \frac{1}{T_C} \right).$$

If $T_0 = T_C$ (that is, the available energy in the incoming stream is all lost),

$$\text{Loss} = q \left(1 - \frac{T_C}{T_H} \right),$$

the familiar Carnot relation. This represents the loss in available energy in the heat-transfer process from the fluid at T_H to a fluid at T_C.

In heat-exchanger design, one of the objectives is to increase the heat-transfer coefficient U in the conductance equation

$$q = UA\Delta T$$

so that the temperature difference ΔT is minimized. In the change of phase case, for example,

$$q = UA(T_H - T_C)$$

for a given heat transfer, the loss is minimized if $T_H \rightarrow T_C$, which means that a U or A increase is designed to reduce the available-energy loss, or, from the loss equation and the heat-transfer equation,

$$\text{Loss} = \frac{q^2 T_0}{UA T_H T_C}.$$

This equation illustrates the advantage of designing the heat exchanger to operate at as high a temperature level as possible: the available energy per unit of thermal energy is higher.

The cost incurred to convert the available energy on the input side of the exchanger to available energy on the output side is the cost of available energy times the loss incurred plus the cost of the equipment to provide this conversion.

$$C_T = C_{b*}(\text{Loss}) + C_{A*} UA.$$

Here the equipment cost of the exchanger is assumed proportional to the exchanger area A times the conductance U per unit area. This takes into account the increased cost of a heat exchanger if it is better designed to give a higher conductance. Substituting the loss expression then gives the total cost,

$$C_T = C_{b*} \frac{q^2 T_0}{UA(T_H)(T_C)} + C_{A*} UA.$$

For a fixed heat flux q, then, the minimum cost is found for an area,

$$A = \frac{q}{U}\left(\frac{T_0 C_{b*}}{T_H T_C C_{A*}}\right)^{1/2} = \left(\frac{\text{Loss}}{U^2} \frac{C_{b*}}{C_{A*}}\right)^{1/2},$$

so that the minimum cost is:

$$C_T = \left[(C_{A*} C_{b*})^{1/2} \, 2q\left(\frac{T_0}{T_H T_C}\right)^{1/2} \right].$$

This analysis illustrates that, in an economy where the cost of available energy C_{b*} is increasing, the size of heat exchangers should be increased in proportion to the square root of the cost of available energy.

The total cost, however, increases proportionally to the square root of the available energy cost, not directly. Thus, if a heat exchanger is used in an energy-conversion system, the cost increase to operate this exchanger increases less than proportionally if the exchanger is optimized.

In times of high fuel costs, one should expect that the size of heat exchanges, and hence power plant sizes, will be large. Similarly, as is the case for nuclear power plants, the fuel cost is low and the optimum areas of heat exchangers are small. The small-size heat exchangers in nuclear power plants, which reduce the cost, also decrease the thermal efficiency, as will be shown in the next section. This economic minimization then increases the thermal discharge for a nuclear power plant compared with that for an equivalent-power-output fossil-fuel plant. It is sometimes noted that the efficiency of atomic power plants is lower than that of fossil-fuel power plants because of a lower-temperature medium on the hot side of the primary heat exchanger. This means, in the loss equation, that T_H is lower, thus increasing the loss. Although it is certainly true that this temperature is lower for reasons for safety, the incentive to do this is less in fossil-fuel plants. The economic optimization, however, in the final analysis leads to the lower efficiency of these plants because fuel costs are low compared with capital costs. This will change as the cost of uranium increases with diminishing supply and increased demand.

Consideration of the technical evaluation of heat exchangers can also be made on the basis of heat-exchanger effectiveness. The effectiveness of a heat exchanger is defined as the available-energy output per unit of available-energy input:

$$E = \frac{b_{C_{out}}}{b_{H_{in}}} = \frac{b_{H_{in}} - \text{Loss}}{b_{H_{in}}} .$$

Now

$$b_{H_{in}} = q\left(1 - \frac{T_0}{T_H}\right) ,$$

if energy in the fluid leaving the hot-side heat exchanger is lost.
Then

$$E = 1 - \frac{q T_0}{UAT_C(T_H - T_0)} .$$

From this equation the effectiveness is increased by increasing the area A and the heat-transfer coefficient U. The effectiveness is a maximum for the largest area possible with U fixed.

From consideration of the cost and effectiveness equations, it is clear that if cost is to be minimized, then for a system operating with lower-cost fuel the heat-exchanger area will be smaller and the effectiveness smaller for a given available energy out of the heat exchanger. Thus nuclear power plants with lower fuel costs C_{b*} and higher capital costs C_{A*} are built with smaller heat exchangers, resulting in efficiency lower than for conventional fossil-fuel plants. These considerations hold with respect to the heat-exchanger processes, since (1) this is the major available-energy loss in a power plant, and (2) this is the major difference in the power cycle between the two systems. This lowered efficiency in turn requires that a larger loss be incurred, resulting in greater heat loss to the environment. Nuclear plants thus require greater amounts of cooling water or larger cooling towers. From the thermal-environment point of view, fossil-fuel power plants are thus to be preferred.

Note that if cooling-tower costs are included (as a capital cost) proportional to the loss, then this will increase C_{b*}. This strategy should be followed, rather than including cooling-tower costs as a separate capital cost independent of the rest of the plant. This procedure should result in increased effectiveness even with increased total cost.

A note should be added here regarding concurrent and counterflow heat exchangers. From the available-energy point of view, the counterflow exchanger allows the fluid on the cold side to leave at a higher temperature than does a concurrent flow system. This effectiveness advantage is often not utilized because of the greater heat exchange per unit area achievable in a concurrent-flow device since the average ΔT can be maintained at a higher level. The use of phase-change fluids is important for maintaining a high ΔT uniformly across the heat exchanger.

In a typical fossil-fuel electrical-power plant, the effectiveness of the boiler system is about 0.60. An efficiency of 0.9 is representative of boilers, indicative that 90 percent of the energy in the combustion products is converted to energy in the steam. Of the available energy in the combustion gas, however, only 60 percent is converted to available energy in the steam. This conversion loss represents the greatest loss in available energy in the power-plant cycle.

In a nuclear plant the primary heat-exchange-cycle temperatures range in the order of 1,500°F, compared with 2,000°F for combustion temperatures. This lowered-temperature hot-side fluid-exchange process also limits the efficiency, if heat-exchanger areas are to be kept to reasonable size.

The differences in this primary heat exchanger or boiler preclude the replacement of the combustion apparatus by a nuclear reactor, even in a gas-cooled reactor. Thus replacement of the primary fuel system is not feasible.

Table 9–2 indicates the cost breakdown of nuclear and coal-fired power plants. Although the total cost of electric power has increased over the ten years indicated, the fuel cost as a percentage of the total cost has changed differently for thèse plants. With increased cost of nuclear fuels, more effort has gone into improvement of their efficiency. In the case of coal plants, the same capital-cost increase has occurred and has reduced even more the relative cost of fuel. Movement has been toward equal capital and fuel costs, as would be expected for optimized cost based on the available-energy analysis.

The analysis indicated previously neglects the additional loss that occurs if the hot-side medium is exhausted to the atmosphere as it is in a combustion process—for example, a fossil-fuel power plant operating on a Rankine cycle. Power systems tht exhaust combustion products without preheating combustion air incur this loss. Automotive engines with the combustion air intake preheated by the exhaust manifold demonstrate this increased effectiveness.

Renewable Energy Systems

Renewable energy systems are fundamentally different from nonrenewable fossil and nuclear energy systems. The difference can be attributed to the fact that renewable energy sources are inexhaustible and are of low or zero economic value before conversion to a useful form. Their inexhaustibility is due to their origin in the stream of energy from the sun, whether the effect is

Table 9–2
Generation Costs for Nuclear and Coal-Fired Plants Compared as Percentages of Total Costs of Generation

	Percentage of Total	
Nuclear	1965[a]	1975
Fuel cost	27	34
Capital cost (reactor and associated plant;	33	
heat exchanger and turbo alternator		49
and associated plant)	27	
Operating cost	13	17
Total	100	100
Coal Fired		
Fuel cost	64	44
Capital cost (boiler plus heat exchanger;	20	
turbo alternator)	7	38
Operating cost	9	18
Total	100	100

[a]1965 data from K.H. Spring, "The Thermodynamic and Economic Base," in *Direct Conversion of Energy* (London: Academic Press, 1965).

directly solar or indirectly solar as with wind, tides, hydro, and so forth. The radiation from the sun is absorbed and stored in the earth system in various mechanisms before being reradiated to space (see chapter 17). The economic value is usually assumed to be zero since there is no material value associated with its use. In fact, the economic value is in the cost of the "net" used to collect the energy. The basic cost of the net is the collector surface value. This is an area value, whether it be that of a land area or an ocean area. Each of these has a capital cost that may be amortized or considered as rent. This cost has been assumed small in past energy analyses. This value can be associated with savings in heating costs at southern latitudes. The shift in industry to these latitudes in the United States can be attributed in part to reduced heating costs where solar heating (natural or technical) is more economic.

In terms of the discussion earlier in this chapter, renewable resources have a capital available energy of equipment but no net variable energy cost for operation. This is true because, although energy may be required to operate the system, the net energy produced is greater than that used. The capital energy required to process the materials of the collector and build the system are usually high. Renewable energy systems are often net energy consumers over a considerable time before they pay back the energy required for their construction.

The net energy curves for renewable energy systems are shown in figure 9–12 compared to nonrenewable systems. There is a basic difference between nonrenewable and renewable systems that is not indicated in this figure—namely, that once the renewable system reaches the crossover point to net positive energy production, the energy is resource energy: it is energy that adds to the energy resources. In comparison, the nonrenewable-energy crossover point only indicates the point at which resources being used (converted) give a net useful effect.

A better description of the difference is shown in figure 9–13. In the nonrenewable system, the on-line point indicates the point at which resources are converted to useful forms at a more rapid pace, but net energy is not "produced." In the renewable-resource on-line point, the system starts to pay back the energy invested; and, once it is paid off, the energy is in fact "produced" energy.

Energy invested in renewable resources extends the resources of the earth. Energy invested in nonrenewable resources accelerates the depletion of the earth's resources. This difference is crucial in the analysis of energy-system alternatives. Slesser and Lewis [13] have termed these renewable energy systems *energy breeders*. They are different from the breeder reactor, which converts the element U_{235} to plutonium. Breeder reactors do not produce energy; they release energy potential which, once used, is exhausted.

The effectiveness of renewable energy systems needs to be based on

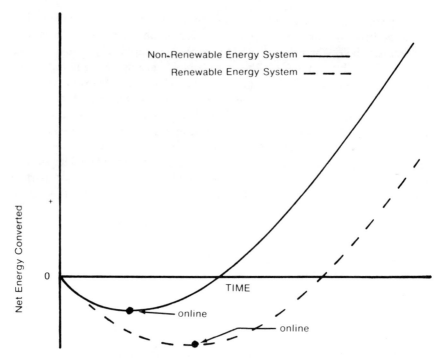

Net Energy Converted by Energy Systems as a Function of
Time from Start of Construction.

Figure 9–12. Net Energy Converted by Renewable and Nonrenewable
Energy Systems as a Function of Time from the Start of
Construction

amortization of the capital available energy over the lifetime of the system
and subsequent inclusion of this as processing energy. The available energy
of the main energy flow must also be carefully defined. If the input is direct
solar energy, the spectral and spatial characteristics outlined in chapter 17
must be used to calculate the available energy of this input.

The effectiveness of present renewable energy systems is low compared
with that of nonrenewable energy systems. Nonrenewable energy systems
used to produce special fuels such as coal to gas or coal to gasoline systems
have lower effectiveness than does direct coal burning. Effectiveness
comparisons of these systems must be based on identical input of the main
flow and the same quality outputs. Table 9–3 summarizes energy-
conversion-to-investment ratios for alternative major energy-conversion
systems.

The basic way that energy systems are now compared is in terms of dollar
cost per unit of energy produced in useful form. With this accounting system,

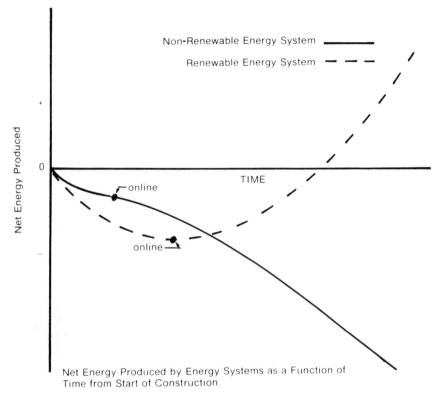

Net Energy Produced by Energy Systems as a Function of
Time from Start of Construction.

Figure 9–13. Net Energy Produced by Renewable and Nonrenewable
Energy Systems as a Function of Time from the Start of
Construction

Table 9–3
Net Energy Converted Compared with Energy Invested in Major Energy
Systems

	Net Energy Converted / Energy Invested
Solar thermal	25
Natural gas and oil	20
SNG and oil from coal	15
Nuclear fission	10
Solar electric	3

Sources: M. Slesser and I. Hounam, "Solar Energy Breeders," *Nature* 262(1976):244; H.
Koenig, personal communication (1980).

the future value of nonrenewable resources is discounted. This method essentially says that, as these resources are depleted, new technology or new resources will be discovered. The faith in technology to extend resources is crucial. If the large energy investment required for present renewable energy systems can be reduced, this dependency can be diminished. If improvement in the technology of renewable energy systems can be accomplished at a rate that balances the resource-depletion rate, this would prove a logical strategy.

Nuclear-energy advocates note that the radioactive elements like uranium and thorium have no other use than as fuels. They then argue that it is preferable to use them rather than fossil fuels, which have chemical-material value. Nuclear reactors operated in the breeder mode have enormous energy potential, as noted in figure 9–14. The term *breeder* is misleading in that new

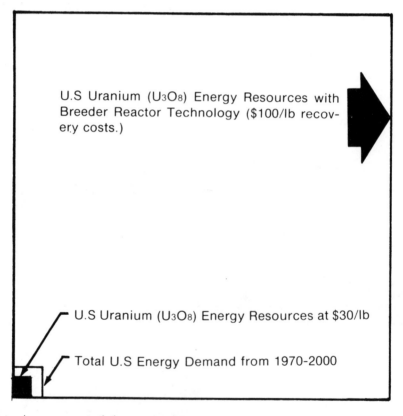

Note: Areas represent relative energy values.

Figure 9–14. Uranium, U_3O_8, Energy Resources with Breeder-Reactor Technology Compared with Conventional Reactor Technology

fuel is not created, any more that it is in the process to concentrate uranium from its oxides or to re-form pentane to natural gas. It is unique in that it produces a new element, plutonium, which has fuel value in a nuclear reactor.

In the analysis of complex industrial processes where multiple energy flows between units occurs, it is often useful to determine separate economic values for separate available-energy inputs. This method recognizes the fact that the technology required to utilize the available energy of different energy streams is different. As an example, the technology required to utilize low-pressure steam to remove sulfur in a coal-refining process is different than that utilizing a limestone slurry.

In cogeneration systems that produce electricity and process steam, the decisions required in their pricing must take into account alternate source costs. The integration of available-energy accounting with industrial-economic accounting must generally be done with consideration of the value of resources and wastes to each industry [18].

References

1. Tribus, M., and Evans, R.B. "Thermoeconomics of Saline Water Conversion." *Ind. Eng. Chem., Process Des. and Devel.* 4(1965):195.
2. El-Sayed, U.M., and Evans, R.B. "Thermoeconomics and the Design of Heat Systems." *J. of Eng. for Power* (trans. ASME, ser. A) 92(January 1970):27.
3. Evans, R.B.; Tribus, M.; and Crellin, G. "Thermoeconomics Considerations of Seawater Demineralization." In *Principles of Desalination*, 2nd ed., K.S. Speigler. New York: Academic Press, 1979.
4. Pimentel, D.; Hurd, L.E.; Bellotti, A.C.; Forster, M.J.; Oka, I.N.; Sholes, O.D.; and Whitman, R.J. "Food Production and the Energy Crisis." *Science* 182(1973):443.
5. Steinhart, C., and Steinhart, J. *Energy: Sources, Use and Role in Human Affairs*. North Scituate, Mass.: Duxbury Press, 1979.
6. Berry, R.S. "Recycling Thermodynamics and Environmental Thrift." *Bull. At. Scient.* 28(May 1972):8.
7. Dyson, F. Personal communication, 1973.
8. Chang, J.H. "Potential Photosynthesis and Crop Productivity." *Ann. Assoc. Amer. Geographers* 60(March 1970):92.
9. Spring, K.H. "The Thermodynamic and Economic Base." In *Direct Conversion of Energy*. London: Academic Press, 1965.
10. Ballard, C., and Herendeen, R. "Energy Costs of Goods and Services, 1963 and 1967." Document 140, Center for Advanced Computation, University of Illinois, Urbana, 1975.

11. Riestad, G.J. "Available Energy Conversion and Utilization in the United States." *J. Eng for Power* (Trans. ASME, ser. A) 97(July 1975):429.
12. Faust, A.S.; Wenzel, L.A.; Clump, C.W.; Mans, L.; and Anderson, L.B. *Principles of Unit Operations*. New York: Wiley, 1960.
13. Slesser, M., and Lewis, C. *Biological Energy Resources*. London: E. and F.N. Spon, 1979.
14. Hirst, E. "How Much Overall Energy Does the Automobile Require?" *Auto. Eng.* 80, no 7 (1972):35.
15. Slesser, M., and Hounam, I. "Solar Energy Breeders." *Nature* 262(1976):244.
16. Koenig, H. Personal communication, 1980.
17. Spofford, W.O. "Solid Residuals Management: Some Economic Considerations." *Nat. Res. J.* 11(1971):561.
18. Reistad, G.M. and Gaggioli, R.A. "Available Energy Costing." In *Thermodynamics: Second Law Analysis*, ed. R.A. Gaggioli. Washington, D.C.: American Chemical Society, 1980.

Appendix 9A:
Available-Energy-
Analysis Steps

1. Identify inputs and outputs of energy and materials in both type and quantity.
2. Determine basic processing rates at the production region (in reverse osmosis processes, the membrane; in power plants, the heat exchanger surfaces, the turbines, and so forth.)
3. Determine the minimum available energy required for the process.
4. Examine alternative technologies for these basic production regions.
5. Determine available-energy values of outputs including wastes with respect to recycling or as inputs to other processes.
6. Determine capital investment in available energy of the processing system.
7. Determine the limitations on and range of control variables to maintain satisfactory outputs of both products and wastes.
8. Estimate environmental effects and hazards.
9. Minimize available-energy use in the process.
10. Minimize cost for the process.
11. Apply decision criteria based on considerations of available energy, cost, and social and environmental factors.

10

Chemical-Engineering Concepts

They have invented a term "energy" and the term has been enormously fruitful because it also creates a law by eliminating exceptions because it gives names to things which differ in matter but are similar in form.
—H. Poincaré, *Science and Method* (1913)

Chemical-Engineering Preliminaries

This chapter introduces concepts and methods from chemical-process engineering. The objective of this introduction is to provide background for the handling of resource-processing, waste-disposal, and energy-conversion systems on a common basis. This basis will be expanded to available-energy methods in a later chapter. Particular emphasis will be placed on the principles of recycle, illustrated in simple chemical-reactor examples and extended to waste-treatment and environmental problems. For reference, the principal results are as follows.

1. *Conservation of mass*: Mass entering a stage must either be stored in the stage or leave the stage. Even with recycle, inert material entering a process stage must leave the process stage. Separated materials may be used for recycle to reduce input requirements. Recycle improves the ratio of output product to input resources used.
2. *Conversion of reactants to products*: Conversion depends on the capital investment for processing. It depends on the quality of reactants and decline as the input quality deteriorates. A separation process is usually required to separate the product from the carrier or reactant materials.
3. *Recycle costs*: The reactor size (capital investment) must be increased if the output is to remain the same with recycle as without it. Complete recycle is theoretically possible, but only with an energy cost that increases with the completeness of recycle.

Batch Processing versus Continuous Processing

Historically, most chemical processes have been developed in the test stages as batch processes. In this mode the components are introduced into the reactor and allowed to react. Then the final products are separated and

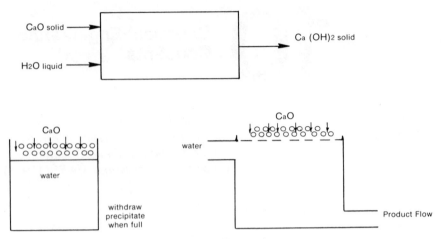

Note: Batch and continuous flow processes are illustrated.

Figure 10–1. Diagrams Representing the Solvay Process for the Slaking of Lime

removed. As a simple example, consider the homogeneous-product case illustrated in figure 10–1). This is part of the Solvay process in the slaking of lime. The solid and the liquid could be mixed until the resultant product was obtained, then dumped out into a truck. In a continuous-flow process, the CaO could be poured into the water and the precipitate removed continuously. Makeup water would be required, but a continuous-flow system is practical. Continuous flow in this case reduces the equipment volume required for a given throughput, since no time is needed to fill and empty the reactor. The stream flow also may not require energy for mixing, as would a batch process.

The batch process is also slower since the water required for assuring complete reaction without mixing is much larger than in the continuous-flow system. The reaction rate is important since with a slow reaction rate the CaO must fall further to react completely. The mixing energy reduces the size of the reactor required so that a trade-off must be made between reactor size (investment) and stirring energy (operating cost).

Chemical-process methods can be extended to environmental problems at several levels. Perhaps the largest scale is the development of concepts of the earth as a chemical reactor [1], as illustrated in figure 10–2. Geological and biological-ecological systems are often analyzed as chemical-reactor systems, utilizing conservation-of-mass and -energy principles. Particular examples include nutrient cycles (nitrogen or phosphorus), photosynthesis and respiration systems (oxygen and CO_2 cycles), and closed food-chain systems (plant-herbivore-carnivore-bacteria). These principles also apply to

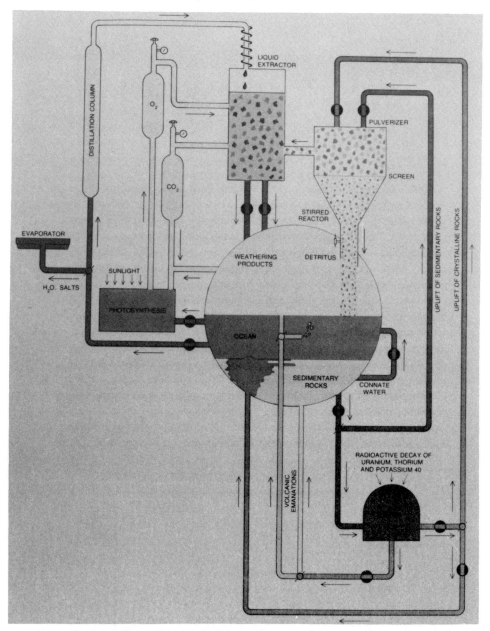

Figure 10–2. The Geochemical System of the Earth Portrayed as a Giant Chemical-Engineering Plant

the analysis of materials-recycling systems involving solid-waste recovery and energy recovery.

A block diagram of a simple chemical-processing system is indicated in figure 10–3. Conservation of mass applied to this system indicates that the rate of change of the mass within the system, dn_s/dt, is equal to the rate of flow into the system, n_i, minus the rate of flow out, n_o:

$$\frac{dn_s}{dt} = n_i - n_o.$$

Mass storage may occur in a system during the charging phase of a chemical reator, and mass loss during the discharge phase. In batch reactors the charging and discharging rates may limit the total output of a reactor as much as do the chemical-reaction rates. Batch reactors are intermittent operations in which both n_i and n_o are zero during the reaction. These reactors require larger storage and higher-capacity pumping or transport systems than do comparable-output continuous-reaction systems. Batch and continuous reactors are illustrated in figure 10–4. Consideration of these illustrations will indicate that the definition of continuous or batch flow system depends to a certain extent on the time scale used and the definition of the system boundary. An automobile engine, looked at on a long time scale, appears to process fuel and air on a continuous basis. If examined in a short time scale, however, it is a series of batch combustion processes. Similarly, a power plant, if examined from a boundary outside the coal storage, appears as a batch instead of a continuous system.

The storage aspect is often handled by inventory-control methods so that production can be adjusted to a changing demand. In the petroleum industry

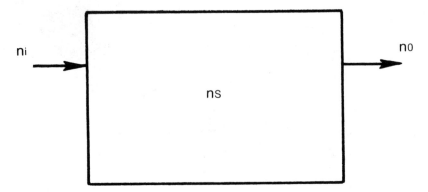

Figure 10–3. Block Diagram of a Simple Chemical-Processing System Showing Nomenclature

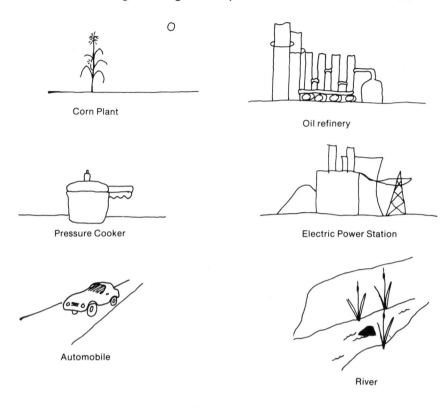

BATCH REACTORS   CONTINUOUS FLOW REACTORS

Figure 10–4. Typical Batch and Continuous-Flow Chemical Reactors

the costs of storage are a significant percentage of the product costs. In hydroelectric systems the major cost is the storage cost (the cost of the impoundment dam facility). Development of new technology for energy storage would significantly reduce electric-energy costs, since the excess power-generation capacity that is now used to meet peak power demands could be reduced.

In chemical reactors and in environmental-biological systems, a storage volume is provided to allow the constituents time to react. Chemical reactions require mixing to produce interaction of reactant molecules. Diffusion processes are generally slow, and most chemical reactors are designed to provide augmented mixing or contact between reacting species. In the environment the mixing is provided by atmospheric circulation and by ocean currents. Sewage-treatment plants use aeration mixing systems to increase the rate of growth of bacteria and hence reduce the size of the aeration tanks required. This mixing requires energy, so that an optimization between the energy cost and the tank-construction cost must be made in the design. Greater capacity of a treatment or reactor system can be obtained by applying more mixing energy. This is often the means by which the production rate of a process is controlled to meet changing demands. In the next section, conventional chemical-engineering nomenclature and methods will be outlined preliminary to the introduction of environmental factors.

Yield and Conversion in Environmental and Chemical Engineering

Yield is classically defined in terms of the product produced in a reactor relative to the total reactant introduced into the system that is not in excess [2–6]. In terms of a simple reaction,

$$v_A A + v_B B \rightleftharpoons v_C C,$$

where v_C refer to stoichiometric coefficients. The yield is the total moles of C formed divided by the total moles of A introduced into the reactor.

We would like to maintain a measure of yield that equals one when complete reaction occurs and zero when no reaction occurs. The yield is defined as

$$Y \equiv \frac{v_A}{v_C} \cdot \frac{C}{A}$$

If v_A moles of A are introduced and v_C moles of C are formed, the reaction is complete and the yield is one. *Conversion*, on the other hand, refers to the ratio of a reactant converted A_r relative to the reactant introduced A. In the foregoing example the conversion is defined as

$$\text{Conversion} \equiv \frac{A_r}{A}.$$

If only one product is formed, these reduce to the same quantity. In reactors where many reactions occur, the reactant converted may exceed that which appears in the product. In these cases the conversion is greater than the yield. From an environmental point of view, the objective is to react a certain species, and the conversion is a more appropriate measure of effectiveness. In chemical production the yield is a better measure of effectiveness. If an environmental cleanup system is to be effective on a long-term basis, the product must have a value for use; thus the yield factor seems to be the better measure.

In some literature the conversion is used to indicate the yield on a single pass, whereas the yield term designates the total yield with recycling utilized. For example, in the production of ammonia the conversion without recycle is about 38 percent, whereas with recycle it reaches 90 percent. The total 90 percent conversion is then termed the yield.

Another useful measure in environmental problems is the unconverted fraction of that pollutant input which appears as output in the waste stream. This is defined as

$$U \equiv \frac{A - A_r}{A} = 1 - \text{Conversion.}$$

Chemical-Reactor Theory in an Environmental-Engineering Context

Let us begin with a simple analysis of a continuous-flow chemical reactor that may be considered as a steady-state system. It will be assumed that there is no accumulation or depletion of material within the reactor. An aeration tank for the treatment of sewage (the growth of bacteria on the organic matter in wastewater) will serve for illustration. Consider figure 10-5, where the input of wastewater n consists of the components of water and organic waste A. In steady flow the total flow in is equal to the total flow out, so that $n_{in} = n_{out}$. This is only an approximation because the oxygen contribution from the air is not insignificant. If the concentration of waste entering is C_{A_i}, then a mass balance will give, for a separation system,

$$\text{Water balance } (1 - C_{A_i})n = (1 - C_{A_i})n. \qquad (10.1)$$

$$\text{Organic balance } C_{A_i}n = C_p n + C_A n. \qquad (10.2)$$

$C_p n$ represents the bacteria converted to dried organic matter, which is separated to be used as fertilizer or for conversion to fuel or food. The purpose of the bacterial action is to remove dissolved organic materials from the input waste stream. The output waste stream will include some dissolved organics, and the quantity of organics leaving in this waste stream will depend on the effectiveness of the bacterial action. The bacterial action and completeness of the conversion depend on the contact of the bacteria with the dissolved organics and oxygen. This is determined by the residence time of the dissolved organics in the reactor and the mixing and aeration rate. If the mixing and aeration is fixed, then the degree of completion depends on the

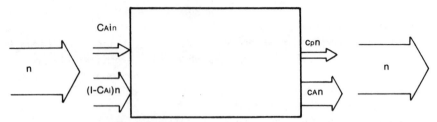

Figure 10–5. Schematic of an Aeration Tank as a Continuous-Flow Chemical Reactor

residence time, which is proportional to the volume and inversely proportional to the flow rate.

$$\tau = K_s \frac{V}{n}. \tag{10.3}$$

If the reaction rate is proportional to the concentration of organics, $r = K_r C_A$ (a first-order reaction), then the conversion of organics to bacterial products can be approximated in a linear manner, so that the output-product flow is $C_p n = \tau r n$;

$$C_p n = K_s \frac{V}{n} K_r C_A n = KVC_A . \tag{10.4}$$

Increasing the volume of the reactor and the concentration of organics increases the bacterial output $C_p n$. If this is substituted into the organic balance equation, the relation between the product flow and the input organic-concentration flow and reactor volume is:

$$C_p n = \frac{C_{A_i} n}{1 + \dfrac{n}{KV}} . \tag{10.5}$$

As expected, the maximum conversion of all input to product occurs at infinite reactor volume. The output-product flow also increases with total flow. In an environmental situation a constraint is usually placed on the concentration of waste in the effluent waste stream. In this case

$$C_A = \frac{C_{A_i}}{1 + \dfrac{KV}{n}} . \tag{10.6}$$

Increasing the volume reduces the concentration of waste in the effluent; increasing the total flow n increases the concentration of waste in the effluent.

The cost of the operation of such a treatment process involves the capital cost, which is proportional to the reactor volume V. The benefit is of course proportional to the conversion to bacterial forms. If this output has value, then the profit is proportion to the conversion. In the sewage-treatment business, the objective is to minimize the cost, since the profit in dollar terms is negative in present systems. The constraint is the output-waste organic concentration that is acceptable for further water use or environmental impact. For comparison purposes the economic benefit will be taken as:

$$B = I^*C_p n - C^*V$$

$$= I^*C_{A_i} n \left[1/\left(1 + \frac{n}{KV} \right) \right] - C^*V. \qquad (10.7)$$

where I^* is the income per unit product.

C^* is the cost per unit reactor volume per unit time amortized over the reactor life.

For a fixed input flow n this becomes a simple maximization problem with respect to the reactor volume

$$V_{max} = n \left(\frac{I^*}{C^*} \frac{C_{A_i}}{K} - \frac{1}{K} \right)^{1/2}. \qquad (10.8)$$

If the concentration of the waste output

$$C_A = \frac{C_{A_i}}{1 + \dfrac{KV_{max}}{n}}$$

is within limits, then this would be the design volume for the reactor. If this is outside the limit of allowable concentration C_{A_L}, then the volume must be greater than V_{max} at the volume corresponding to the limiting concentration

$$V' = \frac{n}{KC_A} (C_{A_i} - C_{A_L}).$$

The benefit is then the value of B with this volume, V'.

In environmental work, new information on the effects of waste-flow impact often requires a reduction in the waste-flow concentration C_A or the total waste flow $C_A n$. Once the reactor has been constructed, the volume V is fixed and difficult to alter. The designer has several alternatives for meeting the additional requirements:

1. recycling of waste effluent;
2. addition of a second reactor;
3. reduced product;
4. preprocessing of input flows;
5. postprocessing of output flows;
6. reduction in flow rate;
7. technological changes.

Emphasis will be placed here on the recycling implications and changes required for optimization with recycle.

Recycle of Waste Effluent

Figure 10–6 indicates the alternative of recycling a portion y of the effluent back into the reactor. The water-flow balance remains the same, since it is an inert dilutant in this problem. The organic balance becomes

$$C_{A_i}n + yC_A n = C_p n + C_A n. \qquad (10.9)$$

Again, $C_p n = KVC_A$. Combining these gives, for the product flow,

$$C_p n = \frac{KVC_{A_i}}{(1 - y) + KV/n}. \qquad (10.10)$$

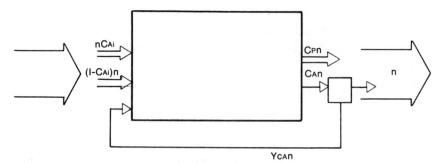

Figure 10–6. Schematic of a Continuous-Flow Reactor with Simple Recycle of Effluent

If effluent is simply recycled without treatment, the through flow n must be reduced by the recycle to $n' = n(1 - y)$. Then the product flow is

$$C_p n(1 - y) = \frac{KVC_{A_i}(1 - y)}{1 - y + \dfrac{KV}{n}}.$$

The effluent concentration is increased, since less product is formed; but the total effluent of A is decreased to

$$C_A(1 - y)n = \frac{C_{A_i}}{1 + \dfrac{KV}{n(1 - y)}}\, n. \qquad (10.11)$$

The use of simple recycle without processing then can usually be used to advantage in reducing total effluent. The input flow rate to the system must usually be reduced to accommodate the recycle flow, thus reducing the overall plant capacity to treat wastes. Therefore, recycle usually implies the addition of some treatment to the waste before recycling. This treatment requires both energy and materials and thus increases costs.

A simple alternative is to add a second reactor in series with the first reactor. This can be used to treat the effluent in an attempt to react more waste and meet effluent standards while producing more product. The optimization of the second reactor is identical to that of the first reactor. The cost of the two reactors is more than the cost a single optimized reactor.

Let us then consider the case of partial concentration of the leaving waste flow for recycle (figure 10-7). C_{A_y} is the concentration of the A component in the recycled stream. The mass conservation for components then becomes

$$\text{Component } A: \ C_{A_i} n + y C_{A_y} n_o = C_p n_0 + C_A n_0.$$

$$\text{Solution: } (1 - C_{A_1})n + y(1 + C_{A_y})n_o = (1 - C_A - C_p)n_o.$$

The product rate is $C_p n_o = KVC_A$. An overall mass balance gives $n = C_p n_o + (1 - y)n_o$.

$$C_p n_o = \frac{C_{A_i} n + y C_{A_y} n_o}{1 + (n_o/KV)}. \qquad (10.12)$$

Substituting and solving for n as a function of the n_o flow gives

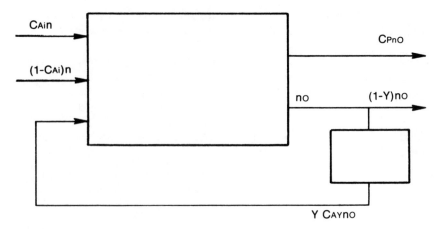

CAin

CPnO

(1-CAi)n

no   (1-Y)no

Y CAYno

Figure 10–7. Schematic of a Continuous-Flow Reactor with Concentration of Effluent before Recycle

$$n = \{1 + \frac{n_o}{KV} - y\left(1 + \frac{n_o}{KV} - C_{A_y}\right)n_o\}/\left(1 + \frac{n_o}{KV} - C_{A_i}\right).$$

$$(10.13)$$

For n_o fixed by the reactor size, the flux of new material into the reactor must be reduced as the recycle increases. In the limit of $y = 1$, complete recycle $n = n_o$, and all A component goes into product. In this limit the input flow

$$n \to 0 \text{ as } C_p \to \frac{C_{A_i} n}{n_o}.$$

The input of new material is decreased with increased recycle, as is the effluent flow.

The recycle problem is related to the cost of the recycle. This cost involves two factors: the cost simply to recycle (that is, increase y) and the cost to increase the concentration of the recycled material C_{A_y}. The thermodynamic cost to concentrate the recycle can then be taken as proportional to the Gibbs free energy required (see chapter 5):

$$\Delta G = k \ln \frac{C_{A_y}}{C_A}.$$

$$(5.27)$$

Returning to the benefit equation, this recycle parameter can be introduced as a cost of recycle, and concentration as a cost

$$R = C_r^* y \, \ln \frac{C_{Ay}}{C_A} . \qquad (10.14)$$

The net benefit is then

$$B = I^* C_p n_o - C^* V - C_r^* y \, \ln \frac{C_{Ay}}{C_A}$$

$$= I^* \left[\frac{C_{A_i} n + y C_{A_y} n_o}{1 + n_o/KV} \right] - C^* V - C_r^* y \, \ln \frac{C_{Ay}}{C_A} . \qquad (10.15)$$

But

$$n = \frac{1 + n_o/KV - y \left\{ 1 + \dfrac{n_o}{KV} - C_{Ay} \right\} n_o}{\{ 1 + n_o/KV - C_{A_i} \}} .$$

With n_o fixed as before by the optimization of the reactor volume initially,

$$n_o = \frac{V_{max}}{\left(\dfrac{I^*}{C^*} \dfrac{C_{A_i}}{K} - \dfrac{1}{K} \right)^{1/2}} .$$

The maximization of B with respect to y or C_{A_y} can then be made. The maximization with respect to the recycle is determined by

$$\frac{\partial B}{\partial y} = 0 = \frac{I^* n_o}{1 + n_o/KV} \left[\frac{C_{Ay} - (1 + n_o/KV) C_{A_i}}{1 - n_o/KV - C_{A_i}} + C_{Ay} \right]$$

$$- C_r^* \left[\ln C_{Ay} - \ln \left[\frac{C_{A_i} n + y C_{A_y} n_o}{n_o + KV} \right] \right.$$

$$\left. + \frac{y C_{A_y} n_o}{C_{A_i} n + y C_{A_y} n_o} \right] .$$

Solving for y will give the maximum or minimum cost with respect to the recycle. Now

$$C_A = \frac{C_{A_i} n + y C_{A_y} n_o}{n_o + KV}.$$

Thus

$$R = C_r^* y \ \ln\left[\frac{C_{A_y}}{C_{A_i}\{1 + n_o/KV\}} \right]$$

$$= C_r^* y \ \ln C_{A_y} - C_r^* y \ \ln\left[\frac{C_{A_i} n + y C_{A_y} n_o}{n_o + KV} \right], \qquad (10.16)$$

and

$$n = \frac{1 + n_o/KV - y\{1 + n_o/KV - C_{A_y}\} n_o}{1 + n_o/KV - C_{A_i}}.$$

A maximization with respect to the concentration of the recycle can also be attempted in a similar manner.

The maximization with respect to both the recycle and the recycle concentration can be accomplished, since for $B = B(y, C_{A_y})$, requires both

$$\left. \frac{\partial B}{\partial y} \right)_{C_{A_y}} \qquad \text{and} \qquad \left. \frac{\partial B}{\partial C_{A_y}} \right)_y$$

to be zero.

Introduction of a following reactor to reduce the effluent concentration to an acceptable level is shown in figure 10–8. The reaction equation for the additional reactor is

$$C_p'\{1 - C_p\} n = K'V'C_A'.$$

The organic balance is

$$C_A n = C_p'(1 - C_p)n + C_A'(1 - C_p)n.$$

Combining these gives the effluent concentration

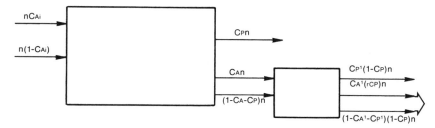

Figure 10–8. Schematic of a Continuous-Flow Reactor with the Addition of a Secondary Reactor to Extract More Product and Reduce Effluent Concentration

$$C'_A = \frac{C_A n}{K'_i V' + (1 - C_p)n}. \qquad (10.17)$$

The additional product flow is then

$$C'_p \{1 - C_p\}n = \frac{C_A n}{1 + \dfrac{(1 - C_p)n}{K'V'}}. \qquad (10.18)$$

The benefit is then

$$B = I * C_p n - C * V + I * C'_p (1 - C_p)n - C * 'V'. \qquad (10.19)$$

Maximizing this benefit with respect to the volume V' gives a volume

$$V'_{max} = (1 - C_p)n \left(\frac{I*}{C*} \frac{C_A}{K'} - \frac{1}{K'} \right)^{1/2}. \qquad (10.20)$$

This volume is smaller than the original reactor volume, since

$$C_A < C_{A_i} \text{ and } 1 - C_p < 1.$$

Again, if C_{A_e} is a limiting parameter of the effluent that must not be exceeded, the second reactor may not be optimized. The second-reactor volume must then be at least

$$V'_e = \frac{n[C_A - C_{A_e}(1 - C_p)]}{C'_{A_e} K'}.$$

The benefit is then

$$B = I*C_p n - C*V + \frac{I*C_A n}{1 + (1 - C_p)nC_{A_e}K'/[K'n\{C_A - C_{A_e}(1 - C_p)\}]}$$

$$- \frac{C*n\{C_A - C_{A_e}(1 - C_p)\}}{C_{A_e}K'} \tag{10.21}$$

If the total flow of the A component in the effluent is limiting instead of the effluent concentration, a similar optimization can be made.

An analysis of the overall system will show that, if the cost per unit volume of the additional and original reactors are the same, the sum of the two reactor volumes will be the same as the volume required of one reactor designed to meet the effluent constraints. Usually the cost per unit volume for adding a second reactor will increase because the surface area is greater for two smaller reactors than for one tank with the same total volume. In any case, the additional piping and controls required will increase the costs. Therefore, there is a penalty associated with failure to anticipate future effluent requirements in the design of a reactor.

References

1. Siever, R. "The Steady State of the Earth's Crust, Atmosphere and Ocean." *Sci. Amer.* 230, no. 6(1974):72.
2. Denbigh, K.G., and Turner, J. *Chemical Reactor Theory*, 2nd ed. Cambridge: Cambridge University Press, 1971.
3. Aris, R. *The Optimal Design of Chemical Reactors*. New York: Academic Press, 1961.
4. Whitwell, J.C., and Toner, R.K. *Conservation of Mass and Energy*. Waltham, Mass.: Blaisdell, 1969.
5. Meissner, H.P. *Process and Systems in Industrial Chemistry*. Englewood Cliffs, N.J.: Prentice-Hall, 1971.
6. Holland, C.D., and Anthony, R.G. *Fundamentals of Chemical Reactor Theory*. Englewood Cliffs, N.J.: Prentice-Hall, 1979.

11

Chemical Thermodynamic Mixing and Environmental Impact

While very little Direct Exchange between material and energy occurs, it is important to note that there are significant trade-offs between these residual streams. For example, an effort to achieve complete recycle with present levels of material flow would require monstrous amounts of energy to overcome entropy.

—A. Kneese, *Analysis of Environmental Pollution* (1971)

Thermodynamics of Recovery of Materials in Mixed States

The question of minimum-available-energy requirements for separation of the components of a mixture as a function of the recovery ratio is discussed in this section.

In processes to remove metal from ores, salts from brines, fresh water from seawater, and so on, an evaluation of technical alternatives must include an estimate of the energy required. This is also crucial to making decisions about research and development priorities in process design. Let us illustrate the relationship between the energy requirement for separation and the recovery ratio for a process. The recovery ratio will be defined as the ratio of the product recovered to the total input of components. In recovery of a metal it is the ratio of the metal output to the total input ore.

As an illustration let us consider the desalinization of water, as shown symbolically in figure 11–1. The input will consist of water and salt, designated by subscripts w and s, respectively. The product stream in desalinization process is not pure water, but water of a low salt concentration that depends on the particular use. The waste stream consists of a high-salt-concentration brine. Mass balances for each component and a total mass balance are as follows:

$$\dot{n}_I = \dot{n}_P + \dot{n}_F, \tag{11.1}$$

$$\dot{n}_I = \dot{n}_{Is} + \dot{n}_{Iw}, \tag{11.2}$$

$$\dot{n}_F = \dot{n}_{Fs} + \dot{n}_{Fw}, \tag{11.3}$$

271

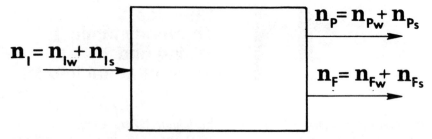

Note: Saline water enters at the left and product water and concentrated brine leave to the right. Each stream consists of water designated with a subscript w and salts with a subscript s.

Figure 11–1. Schematic of Water Purification from Saline Water

$$\dot{n}_P = \dot{n}_{Ps} + \dot{n}_{Pw}, \qquad (11.4)$$

$$\dot{n}_{Iw} = \dot{n}_{Pw} + \dot{n}_{Fw}, \qquad (11.5)$$

$$\dot{n}_{Is} = \dot{n}_{Ps} + \dot{n}_{Fs}. \qquad (11.6)$$

This is a set of six equations with nine unknowns. In order to specify a solution, three additional conditions are required. In seawater-conversion processes, these are usually [1]:

1. the concentration of salt in the input stream:

$$A \equiv \frac{\dot{n}_{Is}}{\dot{n}_I}; \qquad (11.7)$$

2. the concentration of salt required in the product stream:

$$B \equiv \frac{\dot{n}_{Ps}}{\dot{n}_P}; \qquad (11.8)$$

3. the recovery ratio, defined as the ratio of the product stream to the input stream:

$$r \equiv \frac{\dot{n}_P}{\dot{n}_I}. \qquad (11.9)$$

Note that $A + r$ must be less than 1 to maintain the flow directions assumed. Designation of these three factors will determine the minimum available energy required for the processing. For illustration the enthalpy changes will be assumed small. This means that there is no change in the temperature of components or in the chemical form of components. The available energy required is represented by the product of the datum temperature T_0 and the entropy change for each component from input to product or waste stream (see chapter 5).

$$T_0 \Delta S_T = \underbrace{T_0 \Delta S_{Pw} + T_0 \Delta S_{Ps}}_{\text{Product stream}} + \underbrace{T_0 \Delta S_{Fw} + T_0 \Delta S_{Fs}}_{\text{Waste stream}} . \quad (11.10)$$

Product stream:

$$T_0 \Delta S_{Pw} = \dot{n}_{Pw} R T_0 \ln \left(\frac{\dfrac{\dot{n}_{Pw}}{\dot{n}_P}}{\dfrac{\dot{n}_{Iw}}{\dot{n}_I}} \right) = \dot{n}_{Pw} R T_0 \ln \left[\frac{(1-B)}{(1-A)} \right] ,$$

$$T_0 \Delta S_{Ps} = \dot{n}_{Ps} R T_0 \ln \left(\frac{\dfrac{\dot{n}_{Ps}}{\dot{n}_P}}{\dfrac{\dot{n}_{Is}}{\dot{n}_I}} \right) = \dot{n}_{Ps} R T_0 \ln \left(\frac{B}{A} \right) .$$

Waste stream:

$$T_0 \Delta S_{Fw} = \dot{n}_{Fw} R T_0 \ln \left(\frac{\dfrac{\dot{n}_{Fw}}{\dot{n}_F}}{\dfrac{\dot{n}_{Iw}}{\dot{n}_I}} \right) = \dot{n}_{Fw} R T_0 \ln \left[\frac{(1-A)+(B-1)r}{(1-r)(1-A)} \right] ,$$

$$T_0 \Delta S_{Fs} = \dot{n}_{Fs} R T_0 \ln \left(\frac{\dfrac{\dot{n}_{Fs}}{\dot{n}_F}}{\dfrac{\dot{n}_{Is}}{\dot{n}_I}} \right) = \dot{n}_{Fs} R T_0 \ln \left[\frac{A-Br}{A(1-r)} \right] .$$

It is common to represent the available-energy requirement per unit flow of product; in this case n_p. The available energy per unit product flow is then given by

$$\frac{T_0 \Delta S_T}{\dot{n}_P} = RT_0 \left[(1 - B) \ln\left(\frac{1 - B}{1 - A} \right) + B \ln\left(\frac{B}{A} \right) \right.$$

$$+ \left[\frac{(1 - A) - (1 - B)r}{r} \right] \ln\left[\frac{(1 - A) + (B - 1)r}{(1 - r)(1 - A)} \right]$$

$$\left. + \left[\frac{(A - r) + (1 - B)r}{r} \right] \ln\left(\frac{A - Br}{A(1 - r)} \right) \right] \quad (11.11)$$

The first two terms on the right represent the product available-energy change and do not include the recovery ratio.

Let us examine this equation for several limiting cases:

1. Let the recovery ratio approach zero. The water leaving as brine then has the same concentration as the entering seawater.

$$\frac{T_0 \Delta S}{\dot{n}_P} \rightarrow RT_0 \left[(1 - B) \ln\left[\frac{(1 - B)}{(1 - A)} \right] + B \ln\left(\frac{B}{A} \right) \right]. \quad (11.12)$$

This represents the minimum available energy required to produce a product with a property B from an input with property A. This is an impractical limit, as it requires an infinite input flow to give an infinitesimal product flow.

2. Let the recovery ratio approach one. All entering material is converted to product.

$$\frac{T_0 \Delta S}{\dot{n}_P} \rightarrow \infty. \quad (11.13)$$

This represents an impossibility in this example, as all inputs are required to be converted to product.

3. Let the concentration of salt in the product stream approach zero $(B \rightarrow 0)$.

$$\frac{T_0 \Delta S}{\dot{n}_P} = RT_0 \left[\ln\left(\frac{1}{1 - A} \right) + \left[\frac{1 - A}{r} - 1 \right] \right.$$

$$\left. \times \ln\left[\frac{(1 - A) - r}{(1 - r)(1 - A)} \right] + \left(\frac{A}{r} - 1 \right) \ln\left(\frac{A}{1 - r} \right) \right]. \quad (11.14)$$

This represents the available-energy requirement when a pure product is required. It is the maximum available energy per unit product or the maximum available energy that could be extracted from the product stream.

4. Let the concentration of salt in the resource approach zero (that is, $A \to 0$ and $B \to 0$)

$$\frac{T\Delta S}{\dot{n}_P} \to 0.$$

No work is required since no change in the streams is produced.

5. Let the concentration of salt approach pure salt ($B \to 1$).

$$\frac{T_0 \Delta S_T}{\dot{n}_P} = RT_0\left[\left(\frac{1-A}{r} \right) \ln\left(\frac{1}{1-r} \right) + \frac{A-r}{r} \ln\left(\frac{A-r}{A(1-r)} \right) \right.$$

$$\left. - \ln A \right] \qquad\qquad (11.15)$$

or, in terms of the input stream flow,

$$\frac{T_0 \Delta S_T}{\dot{n}_I} = RT_0\left[(1-A) \ln \frac{1}{1-r} + (A-r) \ln\left(\frac{A-r}{A(1-r)} \right) \right.$$

$$\left. - r \ln A \right] \qquad\qquad (11.16)$$

The difference between this result and the limit of $B \to 0$ illustrates the difference between removing salt from water and water from salt.

A plot of the available energy required per unit product for desalinization of seawater is shown in figure 11–2. This indicates how rapidly the energy required for processing increases as the recovery ratio or specification of purity of product is increased.

This analysis is applied in the next section to the case of a waste stream that is mixed with a receiving stream (river). In that context the environmental impact of the waste stream is estimated in quantitative thermodynamic terms for inclusion in economic assessment.

A useful way to consider the environmental input of mixing or adding pollutants with air or water can be examined using result number 5, where $B \to 1$.

Suppose an industry wishes to discharge a pollutant to a stream that already exceeds a desirable concentration. One approach to evaluating the

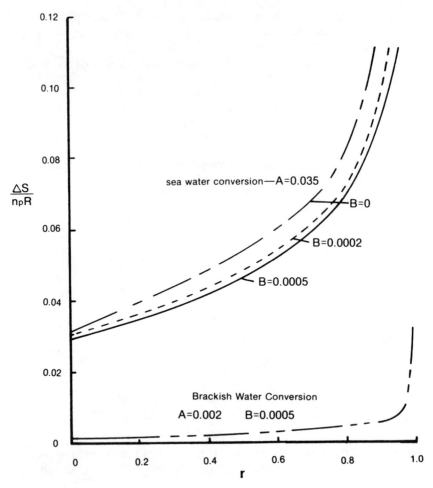

Note: Seawater and brackish-salt concentrations are illustrated.

Figure 11–2. Energy of Extraction per Unit Water Product Flow as a Function of the Recovery Ratio r

value of the discharge is to calculate the available energy required to process the pollutant flow to a salable product. Then compare this with the available energy required to produce the same salable product after the pollutant is mixed with the stream. This difference might then be considered an additional cost of pollution mixing. As an example of this effect, let us take a stream with an initial salt concentration of

$$A = \frac{\dot{n}_{Is}}{\dot{n}_I} = 0.2.$$

If an acceptable concentration of salt is

$$\frac{\dot{n}_{Fs}}{\dot{n}_F} = 0.1,$$

then the recovery ratio for separating the salt as a pure component $B = 1$ is found from

$$\frac{\dot{n}_{Fs}}{\dot{n}_F} = 0.1 = \frac{A - r}{1 - r}$$

as

$$r' = \frac{1}{9} .$$

The available energy of the stream required per unit flow $\dot{n}_I$ is from equation 11.16.

$$\frac{T_0 \Delta S_T}{\dot{n}_I} = 0.211 \, RT_0 .$$

If the pollutant flow has a concentration $A = 0.4$, then the available energy required to produce salt with concentration $B = 1$ and an effluent with acceptable concentration

$$\frac{\dot{n}_{Fs}}{\dot{n}_F} = 0.1$$

is

$$\frac{T_0 \Delta S_T}{\dot{n}_I} = 0.456 \, RT,$$

with the recovery ratio $r = 1/3$.

If two streams have the same flow, then the higher-concentration stream requires more available energy to produce a pure product. Note, however, that the amount of pure product produced is three times that of the dilute concentration $A = 0.2$ stream.

If these two streams have equal flow, then for a total flow $2\dot{n}_I$ the available energy required is

$$T_0 \Delta S_T = RT_0(0.456 + 0.211)\dot{n}_I = 0.667 \, \dot{n}_I RT_0.$$

Now suppose the two equal streams are allowed to mix (the pollutant discharges into the stream). Then $A = 0.3$, and for

$$\frac{\dot{n}_{Fs}}{\dot{n}_F} = 0.1$$

still required with $B = 1$, then $r = 2/9$, and

$$\frac{T_0 \Delta S_T}{\dot{n}_I} = 0.358 \, RT_0 .$$

Now with total flow $2 \, \dot{n}_i$,

$$T_0 \Delta S_T = (0.716) \dot{n}_I RT_0 .$$

The same quantity of salt is extracted in this case, so that the salable product would be the same as before mixing.

The difference in the available energy from that required initially is

$$(0.716 - 0.667) RT_0 \dot{n}_I = 0.049 RT_0 \dot{n}_I .$$

This additional energy is required after mixing.

This additional energy cost should be assigned to the industry for mixing if the energy must eventually be expended by a public water-processing system. This is the extra energy required compared to that needed initially before mixing. The change would then be the original energy required.

$$0.456 \, RT_0 \dot{n}_I$$

plus the additional of

$$0.049 \, RT_0 \dot{n}_I ,$$

or

$$0.505 \, RT_0 \dot{n}_I .$$

Let us now consider the additional change if the receiving stream has a lower initial concentration before the pollutant is added. Let us take $A = 0.1$ instead of 0.2. Then the mixed stream will have

$$A = \frac{0.1 + 0.4}{2} = 0.25 .$$

For $B = 1$,

$$\frac{\dot{n}_{Fs}}{\dot{n}_F} = 0.1,$$

$$r = \frac{1.5}{9},$$

$$\frac{T_0 \Delta S}{\dot{n}_I} = RT_0(0.291).$$

For twice this flow $2\,\dot{n}_I$,

$$T_0 \Delta S = (0.582)\,RT_0\,\dot{n}_I.$$

The increase in the available energy required is then

$$\Delta A = (0.582 - 0.456)\,RT_0\,\dot{n}_I,$$

$$\Delta A = 0.126\,RT_0\,\dot{n}_I.$$

The effect of the available energy of mixing for various receiving stream-pollutant concentrations (A) is summarized in figure 11–3. The pollutant addition is assumed to have a concentration of 0.4, and the acceptable stream concentration is

$$\frac{\dot{n}_{Fs}}{\dot{n}_F} = 0.1.$$

Both the waste and the receiving streams are assumed to have equal total flows.

Curve (a) indicates the available energy required to produce pure pollutant concentration $B = 1$ and acceptable stream concentration for different receiving-stream concentrations. Note that, for receiving concentrations lower than acceptable, the available energy required is negative. This indicates the capacity of this stream to take on waste. Negative values represent available energy lost if salt is added to the stream to reach the limiting acceptable stream concentrations.

Curve (b) represents the recovery ratio or the ratio of the total salt produced per total flow input of both waste and receiving streams. Note that r is the same before mixing as after mixing.

Curve (c) is the available energy required to produce an acceptable output concentration from the two streams before mixing.

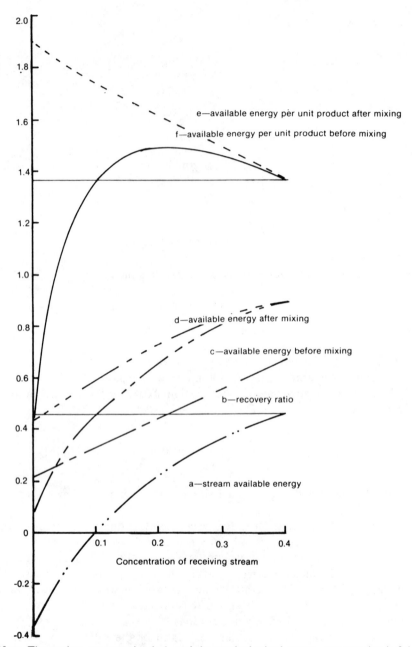

Note: The product concentration is 1, and the required mixed stream concentration is 0.1. Available-energy scale is $\Delta A / n_i R T_o$.

Figure 11–3. Available-Energy Effects of Mixing a Pollutant Stream of Concentration 0.4 with a Receiving Stream of Equal Flow Rate

Curve (d) is the available energy required to produce an acceptable stream if the mixing has occurred. The difference between curves (c) and (d) represents the extra available energy required after mixing has occurred. This is the penalty associated with mixing before separation.

Curves (e) and (f) represent the available energy required per unit of product produced before mixing (e) and after mixing (f).

The difference between curves (e) and (f) represents the extra available energy required per unit product of pollutant if the mixing occurs before separation occurs. Curve (e) indicates that there is a maximum-available-energy requirement for the two steams per unit product.

The effect of adding a polluted stream to one with zero concentration of pollutants is indicated in figure 11–4. Again, equal flows of a polluted stream and a zero-concentration stream are mixed. The available energy to give pure product and stream concentration,

$$\frac{\dot{n}_{Fs}}{\dot{n}_F} = 0.1,$$

is calculated. Curve (a) indicates the sum of the available energies required for the two streams before mixing, and curve (b) for after mixing. The difference shown in curve (c) is the additional available energy required after mixing. Curve (d) represents the total recovery ratio.

Several observations can be made from this figure. The first is that the available-energy loss with mixing increases as the added-pollutant concentration increases. The second is that the concentration of the added-pollutant stream above which available energy is required to meet the stream-concentration limit is 0.3, *not* 0.2 as would be supposed based on simple mixing. The zero-concentration stream can then be used effectively to assimilate more wastes if it is used for purposes other than simple mixing.

Waste-Flow Environmental Impacts

In the formulation of environmental-impact statements [9], both the quantitative effect and the importance of the effect are solicited. When a decision is required on where an effluent is to be released, the capability of the environment to assimilate the wastes is of special concern. One example of how to make this decision is noted by Rich [2] in his discussion of the effect of surface excavations on the sedimentation of streams. Rich states, "However, if streams draining the area generally carry a high sediment load anyway, then the importance of the interaction may be rated low." This reasoning might be restated as follows: the more polluted a stream, the less important it is if more wastes are added. Analogously, the more polluted the atmosphere, the less important is the addition of more pollutants.

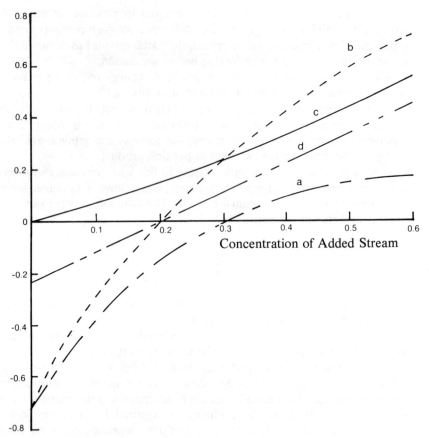

Note: Pure product, $B = 1$, and a mixed concentration of 0.1 are required. Total flows of each stream are equal. Available-energy scale is $\Delta A / n_I R T_0$.

[a]Available energy of unmixed streams.

[b]Available energy of mixed streams.

[c]Difference between the available energy required after mixing and before mixing.

[d]Recovery ratio.

Figure 11–4. Available-Energy Effects of Adding a Polluted Stream to a
          Stream of Zero Concentration

This statement raises some basic environmental concerns. For example, should power plants be located in areas where the air is relatively clean or in areas near other power plants or chemical complexes where the air is already heavily contaminated? Protagonists for dispersion and dilution as environmental solutions argue for the dispersion of industrial and utility enterprises. This argues, in effect, that there is an assimilative environmental capacity.

On the other hand, Rich assumes that once an environmental capacity is exceeded locally, additional loads are not as detrimental.

Both points of view are unattractive to an environmentalist and lead to much misunderstanding and confusion. How can concentration and dispersion both be solutions to environmental problems? A general answer is difficult. One approach is through the thermodynamics of the available energy of the waste stream relative to that of the receiving ecosystem. If the available energy difference is large, the impact is greater. An additional factor—the available energy of the receiving ecosystem relative to that of an acceptable ecosystem—is required, as discussed in the previous section.

The thermodynamic analysis of the first section of this chapter can be recast to estimate the effect of adding a pollutant to an existing environment. Particular examples would be the addition of heavy-metal wastes or sulfite liquors to a stream that is already polluted. In these cases we are interested in the impact of the additional stream on the one it is entering. Figure 11–5 indicates the schematic, which is that of figure 11–1, with arrows reversed.

The same symbols are now used, but they have different meanings: Let

$\dot{n}_p$ = pollutant stream flow.

$\dot{n}_{pw}$ = pollutant water or solvent flow.

$\dot{n}_{ps}$ = pollutant or solute flow.

$\dot{n}_r$ = mainstream flow.

$\dot{n}_{rw}$ = water or solvent flow of mainstream.

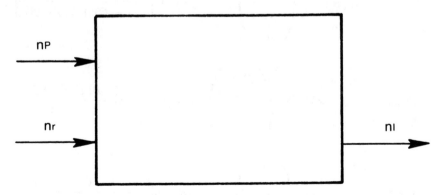

Figure 11–5. Schematic Giving the Nomenclature for the Addition of Pollutants to an Already Polluted Stream

$\dot{n}_{rs}$ = pollutant or solute flow of mainstream.

$\dot{n}_I$ = combined total flow.

$\dot{n}_{Iw}$ = combined water or solvent flow.

$\dot{n}_{Is}$ = combined pollutant or solvent flow.

For discussion purposes, three new combined variables are defined:

1. $B \equiv \dfrac{\dot{n}_{ps}}{\dot{n}_p}$ the concentration of pollutants in the added pollution stream.

2. $C \equiv \dfrac{\dot{n}_{rs}}{\dot{n}_r}$ the concentration of pollutants in the mainstream before added pollutants.

3. $Y \equiv \dfrac{\dot{n}_p}{\dot{n}_r}$ the ratio of the total additional pollutant flow to the mainstream flow.

These three variables are those most easily measured and of interest in discussing pollutant effects in combining streams.

The entropy contribution to the available energy of this mixing is:

$$T_0 \Delta S_T = T_0 \Delta S_{pw} + T_0 \Delta S_{ps} + T_0 \Delta S_{rw} + T_0 \Delta S_{rs},$$

$$T_0 \Delta S_{pw} = \dot{n}_{pw} RT_0 \ln \left(\frac{\frac{\dot{n}_{pw}}{\dot{n}_p}}{\frac{\dot{n}_{Iw}}{\dot{n}_I}} \right) = \dot{n}_{pw} RT_0 \ln \left[\frac{(1-B)(Y+1)}{(1-C)+Y(1-B)} \right],$$

$$T_0 \Delta S_{ps} = \dot{n}_{ps} RT_0 \ln \left(\frac{\frac{\dot{n}_{ps}}{\dot{n}_p}}{\frac{\dot{n}_{Is}}{\dot{n}_I}} \right) = \dot{n}_{ps} RT_0 \ln \left[\frac{B(Y+1)}{C+BY} \right],$$

$$T_0 \Delta S_{rw} = \dot{n}_{rw} RT_0 \ln \left(\frac{\frac{\dot{n}_{rw}}{\dot{n}_r}}{\frac{\dot{n}_{Iw}}{\dot{n}_I}} \right) = \dot{n}_{rw} RT_0 \ln \left[\frac{(1-C)(Y+1)}{(1-C)+Y(1-B)} \right],$$

$$T_0 \Delta S_{rs} = \dot{n}_{rs} R T_0 \ln \left(\frac{\dfrac{\dot{n}_{rs}}{\dot{n}_r}}{\dfrac{\dot{n}_{Is}}{\dot{n}_I}} \right) = \dot{n}_{rs} R T_0 \ln \left[\frac{C(Y+1)}{C+BY} \right].$$

The total change per total additional pollutant flow is compared:

$$\frac{T_0 \Delta S_T}{\dot{n}_p} = R T_0 \left[(1+B) \ln \left[\frac{(1-B)(Y+1)}{(1-C) + Y(1-B)} \right] \right.$$

$$+ B \ln \left[\frac{B(Y+1)}{C+BY} \right] + \frac{(1-C)}{Y}$$

$$\left. \times \ln \left[\frac{(1-C)(Y+1)}{1-C+Y(1-B)} \right] + \frac{C}{Y} \ln \left[\frac{C(Y+1)}{C+BY} \right] \right].$$

$$(11.17)$$

This is a measure of the loss of available energy when the streams are combined. The maximum, of course, occurs if the main stream is not polluted initially (that is, $C = 0$). Then

$$\frac{T_0 \Delta S_T}{\dot{n}_p} = R T_0 \left[(1-B) \ln \left[\frac{(1-B)(Y+1)}{1 + Y(1-B)} \right] \right.$$

$$\left. + B \ln \left[\frac{Y+1}{Y} \right] + \frac{1}{Y} \ln \left[\frac{1+Y}{1 + Y(1-B)} \right] \right].$$

$$(11.18)$$

Another interesting case is the situation in which the mainstream flow is large compared with the pollutant stream (that is, $Y \to 0$). Then

$$\frac{T_0 \Delta S_T}{\dot{n}_p} = R T_0 \left[(1-B) \ln \left[\frac{(1-B)}{(1-C)} \right] + B \ln \left(\frac{B}{C} \right) \right].$$

$$(11.19)$$

Note, however, that the case in which the concentration of pollutants in the added stream is equivalent to that of the mainstream introduces no change in the available energy. This is as expected and indicates that $T_0 \Delta S_T$ alone is

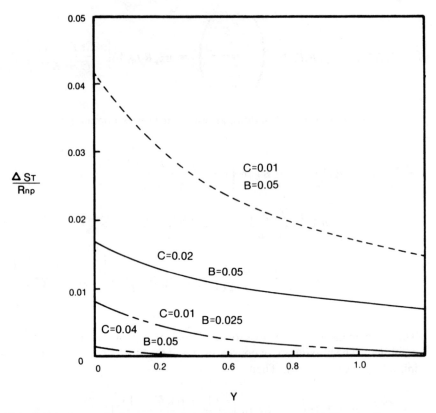

Note: Pollutant concentrations are noted as B and main stream concentrations by C.

Figure 11–6. Entropy Changes per Unit Pollutant Flow for the Mixing of Pollutant and Main Streams as a Function of the Ratio Y of the Pollutant to Main-Stream Flows

not sufficient as a measure of the pollutant effect. The total pollutants in the flow after combination are, of course, greater than in the initial stream. Quality and quantity are required for determination of the impact.

Figure 11–6 shows the relative changes in quality for different stream-flow ratios and quality.

The case in which the added stream is unpolluted is also of interest. In this case $B = 0$ gives:

$$\frac{T_0 \Delta S_T}{\dot{n}_p} = RT_0 \left[\ln \left[\frac{Y + 1}{1 - C + Y} \right] + \frac{(1 - C)}{Y} \right]$$

$$\times \ln\left[\frac{(1 - C)(Y + 1)}{1 - C + Y}\right] + \frac{C}{Y} \ln(Y + 1)\right] . \quad (11.20)$$

Another is when the added stream is all pollutant. Then $B = 1$ gives:

$$\frac{T_0 \Delta S_T}{\dot{n}_p} = RT_0\left[\ln\left[\frac{(Y + 1)}{(C + Y)}\right] + \frac{(1 - C)}{Y} \ln(Y + 1)\right.$$

$$\left. + \frac{C}{Y} \ln\left[\frac{C(Y + 1)}{C + Y}\right]\right] . \quad (11.21)$$

This latter may also be considered as a maximum pollutant effect, which is translated into an overall maximum if the mainstream is unpolluted (that is, $C = 0$). In that case

$$\frac{T_0 \Delta S_T}{\dot{n}_p} = RT_0\left[\ln\left(\frac{Y + 1}{Y}\right) + \frac{1}{Y} \ln(Y + 1)\right] . \quad (11.22)$$

An example is the addition to a stream of sediment as a result of construction activities or washing operations.

Consider the sediment load from a washing operation that carries a stream with a mass concentration of 0.05 and a water flow of 100 gal/min into a creek with a flow of 10×10^3 gal/min and a sediment concentration of 0.01.

If the sediment remains in suspension, then $B = 0.05$, $C = 0.01$, $Y = 0.01$:

$$\frac{T_0 \Delta S_T}{\dot{n}_p} = RT_0\left[0.95 \ln \frac{(0.95)(1.01)}{0.99 + (0.01)(0.95)}\right.$$

$$+ 0.05 \ln \frac{(0.05)(1.01)}{0.01 + (0.05)(0.01)}$$

$$+ \frac{0.99}{0.01} \ln \frac{(0.99)(1.01)}{0.99 + (0.01)(0.95)}$$

$$\left. + \frac{0.01}{0.01} \ln \frac{(0.01)(1.01)}{0.01 + (0.05)(0.01)}\right] .$$

Since no enthalpy change occurs and no reactions are present, the total available-energy change is:

$$T_0 \Delta S_T = 787.3 \, \dot{n}_p = 629.9 \text{ Btu/min.}$$

In fuel equivalents, this is approximately 50 lb of coal/min in thermal terms; or, if an electrical system is used to remove centrifugally, this is approximately 150 lb of coal/min.

This represents the environmental impact of the washing operation and could be viewed as the available energy needed to restore the stream to its condition before washing operations.

The effect of the mixing of streams of different concentrations is illustrated in figure 11-7. In this figure we have assumed that in the washing-operation example the acceptable creek concentration is $C = 0.005$.

The available-energy flow of the creek relative to the acceptable environmental condition is

$$\dot{n}_r A_r = \dot{n}_{rs} RT_0 \ln \frac{0.01}{0.005} + \dot{n}_{rw} RT_0 \ln \frac{0.99}{0.995}$$

$$= RT_0 \dot{n}_w (1.94 \times 10^{-3}).$$

The available energy of the pollution-flow-added stream is

$$\dot{n}_p A_p = \dot{n}_{ps} RT_0 \ln \frac{0.05}{0.005} + \dot{n}_{pw} RT_0 \ln \frac{0.95}{0.995}$$

$$= RT_0 \dot{n}_p (71.16 \times 10^{-3}).$$

The total available-energy flow initially of both streams with $\dot{n}_w = 10 \times 10^3$ gal/min and $\dot{n}_p = 100$ gal/min is

$$\dot{n}_T A_{T_i} = \dot{n}_r A_r + \dot{n}_p A_p$$

$$\dot{n}_T A_{T_i} = 26.5 \, RT_0$$

If the streams are now mixed, the final concentration will be $C' = 1.04 \times 10^{-4}$ and the total flow $n_T = 10,100$ gal/min. The available-energy flow relative to the acceptable concentration $C = 0.005$ is

$$\dot{n}_r AT_F = 22.5 \, RT_0.$$

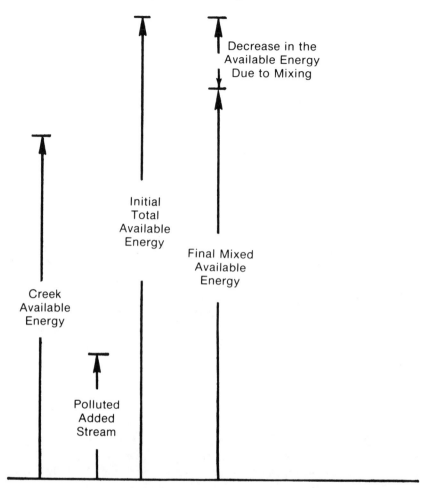

Figure 11–7. Illustration of the Available-Energy Losses Associated with Mixing of Streams with Pollutant Concentrations Higher Than an Acceptable Concentration

The mixing has reduced the available energy of the two streams by

$$A = (26.5 - 22.5)RT_0 = 4RT_0.$$

In some cases an added flow may reduce the available energy required to reach an acceptable environmental level. Examples include the addition of precipitants to a stream or basic salts to an acid stream. In these cases, ΔA would be negative, and the discharge would compensate for a pollutant in the stream. Interpretation of ΔA requires an understanding of the chemistry of many reactions in situations with a myriad of possible catalytic and inhibitory substances. A scheme to carry out this accounting will require comprehensive monitoring and experimentation.

An attempt to determine the available-energy datum state of the atmosphere has been made by Riekert [3] and others [4–8] (see appendix 6B). In the development of the availability function, much effort has been applied to establishing a datum or reference state. Most datum states change hourly, daily, seasonally, and historically. The environmental concern is that the change not be detrimental.

It should be emphasized that the environmental impact does not diminish if the receiving stream becomes more polluted before the pollutant stream enters. This is important to note, since some discussions of environmental-impact statements rate the importance of a pollutant lower if the stream or environment it is entering is already polluted. This is untenable since it leads to the erroneous suggestion that the more polluted an environment is, the more added pollutants it can assimilate. The assimilative capacity is associated with the available energy required for the environment to recover from the pollutant load. This requirement is increased as the total pollutant load increases.

References

1. Evans, R.B., and Tribus, M. "Thermo-Economics of Saline Water Conversion." *Ind. Eng, Chem., Process Des. Develop.* 4(1965):195.
2. Rich, L.G. *Environmental Systems Engineering.* New York: McGraw-Hill, 1973.
3. Riekert, L. "The Conversion of Energy in Chemical Reactions." *Energy Conv.* 15 no. 3–4(1976):81.
4. Evans, R.B. A Proof that Essergy is the Only Consistent Measure of Potential Work." Ph.D. diss., Dartmouth College, 1970.
5. Szargut, J. "Grenzen fuer die Anwendung Smoeglichkeiten des Exergie begriffs." *Brennstoff-Waerm-Kraft* 19, no. 6(1967):309.
6. Keenan, J. "Availability and Irreversibility in Thermodynamics." *Brit. J. Appl. Phys.* 2(1951):183.
7. Riestad, G. "Availability: Concepts and Applications." Ph.D diss., University of Wisconsin, 1970.
8. Obert, E., and Gaggioli, R. *Thermodynamics.* New York: McGraw-Hill, 1963.
9. National Environmental Policy Act, 1969 (P.L. 91–190). Guidelines in *Federal Register*, 1 August 1973.

12 Energy-Conversion Potential from Concentration Differences

They who plough the seas do not carry the wind in their hands.
　　　　　　　　　　　　　　　　—Publilius Syrus (ca 42 B.C.)

Concentration Differences for Energy Sources

In the search for alternative energy sources, several proposals have been made to utilize the available-energy differences between fresh water and seawater. The feasibility and design parameters for these systems can be examined using the methods of available-energy analysis. In this chapter, two systems, the osmotic-power system [3] and the dialytic battery [1], are analyzed in detail. These examples are characteristic of more general ideas for using the available energy of pollutants relative to that of the environment as energy sources.

Available energy is lost when a river of low salt concentration enters the ocean. This energy potential is chemical in nature. Figure 12–1 illustrates a scheme proposed utilizing an osmotic membrane and the osmotic pressure to produce a hydro potential.

The reversible-potential thermodynamic power that is theoretically obtainable can be found from the analysis of the available-energy loss associated with the pollution flow discussed in chapter 11.

The designations of flows are changed to that shown in figure 12–1. Then

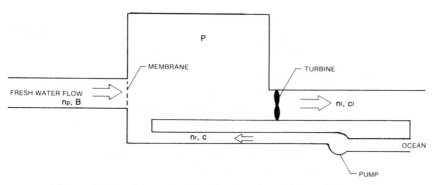

Figure 12–1. Osmotic-Cell Energy-Conversion Schematic

n_p = the freshwater river-flow rate.

n_r = the ocean-flow input to maintain the concentration of salt in the osmotic chamber.

n_I = the total flow = $n_p + n_r$.

Then

B = concentration of salt in the river.

C = concentration of salt in the ocean.

Y = the ratio of river to ocean flow = n_p/n_r.

The available energy per unit of river flow is then, from equation 11.15,

$$\frac{A}{n_p} = RT_0 \left\{ (1 - B)\ln\left[\frac{(1 - B)(Y - 1)}{(1 - C) + Y(1 - B)} \right] + B \ln\left[\frac{B(Y + 1)}{C + BY} \right] \right.$$

$$+ \frac{1 - C}{Y} \ln\left[\frac{(1 - C)(Y + 1)}{1 - C + Y(1 - B)} \right]$$

$$\left. + \frac{C}{Y} \ln\left[\frac{C(Y + 1)}{C + BY} \right] \right\} .$$

When the river flow is water with no salt, $B = 0$ and the maximum potential work is:

$$\frac{\Delta A}{n_p} = RT_0 \left\{ \ln\left[\frac{Y + 1}{1 - C + Y} \right] + \frac{1 - C}{Y} \ln\left[\frac{(1 - C)(Y + 1)}{1 - C + Y} \right] \right.$$

$$\left. + \frac{C}{Y} \ln (Y + 1) \right\} . \qquad (12.1)$$

This work is dependent on the ratio of the river to the ocean-water flow Y. At $Y = 0$, corresponding to high ocean flow, the work is maximum. Design for a high ocean flow will then provide the maximum available energy per unit of river flow. This high ocean-flow rate, however, will require pumping energy, which is accounted for later in this analysis.

If the ocean-water flow is large compared with the river flow ($Y \rightarrow 0$), corresponding to maintenance of seawater concentration on the salt side of the membrane, the available energy per unit river flow is

$$\frac{\Delta A}{n_p} = RT_0 \ln\left(\frac{1}{1-C}\right). \tag{12.3}$$

For $C \ll 1$, this is approximately

$$\frac{\Delta A}{n_p} \sim RT_0 C. \tag{12.3}$$

In the seawater example, C is approximately 0.035, and the product $RT_0 C$ is the osmotic pressure P_0 of seawater relative to that of pure water. The available-energy-production rate is then represented as a product of force and flux,

$$\Delta A = P_0 n_p. \tag{12.4}$$

The actual flux n_p across the membrane depends on the permeability of the membrane and the pressure difference $\Delta P = P_0 - P$ across the membrane from the freshwater to the seawater side,

$$n_p = K\Delta P = K(P_0 - P). \tag{12.5}$$

When $P = P_0$ equilibrium is reached, and no net flux occurs across the membrane.

The power available is

$$\Delta A = P_0 K(P_0 - P). \tag{12.6}$$

If the pressure across the membrane is used to provide power through a waterwheel device, then the power delivered will be

$$Pn_I = P(n_p + n_r) = Pn_p(1 + 1/Y). \tag{12.7}$$

Since we have already assumed $Y \to 0$, this is not a useful expression. The net power deliverable is given by the flow (equation 12.5) times the pressure P:

$$N_p = K(P_0 - P)P. \tag{12.8}$$

This net power N_p is a maximum when $P = P_0/2$ with a value

$$N_{p\ max} = K(P_0/2)^2. \tag{12.9}$$

The efficiency at this pressure is

$$\eta = \frac{N_{p\,max}}{\Delta A} = 1/2. \qquad (12.10)$$

This system then behaves as an ideal machine, as described in chapter 7. Maximum efficiency, of course, occurs at $P = P_0$; but the net power is then zero.

In the actual case, the concentration of salt in the chamber is found from the salt balance as

$$C_I n_I = C n_r + B n_p. \qquad (12.11)$$

With $n_I = n_r + n_P = n_p(1 + 1/Y)$, then

$$C_I = \frac{C + BY}{Y + 1}.$$

Using the approximation of osmotic pressure,

$$\frac{\Delta A}{n_p} = RT_0 C_I = RT_0 (C + BY)/(Y + 1). \qquad (12.12)$$

This approximation may be used to estimate the effect of the variation of the flow fraction Y on the power and efficiency.

In the actual case Y must be greater than one to minimize the pumping-energy requirement. Optimization then can be attempted by considering the pumping energy as $n_r P/e$, where e is the efficiency of the seawater pump. The net power N_p is then

$$N_p = n_I P - n_r P/e.$$

From continuity,

$$n_I = n_p + n_r \qquad \text{and} \qquad Y = n_p/n_r,$$

the actual net power is

$$N_p = \frac{n_p P}{Y}\left(Y + 1 - \frac{1}{e} \right) \qquad (12.13)$$

Using equation 12.5, the osmotic pressure as approximately $P_0 = RTC_I$, for the case where $B = 0$ gives the relations

$$P_0 = RT_0 C/(Y + 1) \qquad (12.14)$$

and

$$n_p = K\left(\frac{RT_0 C}{Y + 1} - P \right). \qquad (12.15)$$

The net actual power, which is the product of n_p and P, is then given as

$$N_p = n_p\left(\frac{RT_0 C}{Y + 1} - \frac{n_p}{K} \right). \qquad (12.16)$$

The maximum net power occurs when $n_p = KRT_0 C/[2(Y + 1)]$, and the power-flow behavior is shown in figure 12–2. (Compare this with chapter 7.)

The selection of the ratio Y must still be made for maximum power. From equation 12.13, with n_p selected for maximum power, then

$$N_{p\,max} = K\left[\frac{RT_0 C}{2(Y + 1)} \right]^2 \left[\frac{Y + 1 - \dfrac{1}{e}}{Y} \right]. \qquad (12.17)$$

This function is sketched in figure 12–3 as a function of Y for different pump efficiencies e. The maximum power occurs at the value of Y for which

$$2Y^2 + 3Y - 3Y/e + (e - 1)/e = 0. \qquad (12.18)$$

For design purposes one would use equation 12.18 to select Y for a given pump efficiency. The flow rate n_p is then determined at the maximum power point. The efficiency can then be found as

$$\eta = \frac{N_p/n_p}{\Delta A/n_p} = \left[1 - \frac{n_p(Y + 1)}{KRT_0 C} \right] \left[\frac{Y + 1 - \dfrac{1}{e}}{Y} \right]. \qquad (12.19)$$

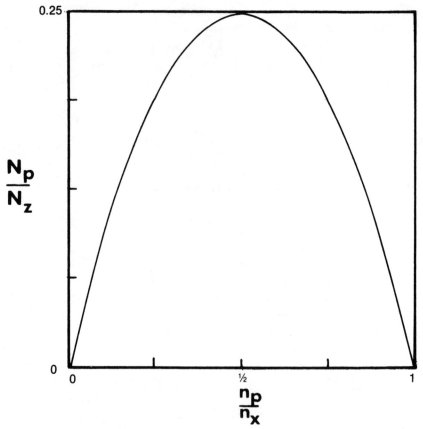

Note: Parameters are made dimensionless by appropriate factors of concentration, pump efficiency, and fresh- to ocean-water flows:

$$N_Z \equiv K \left[\frac{RT_0C}{(Y+1)} \right]^2 \left[\frac{Y+1-\frac{1}{e}}{Y} \right]$$

$$n_x \equiv KRT_0C/(Y+1)$$

Figure 12–2. Net Power Output of an Osmotic Power Cell as a Function of the Freshwater Flow

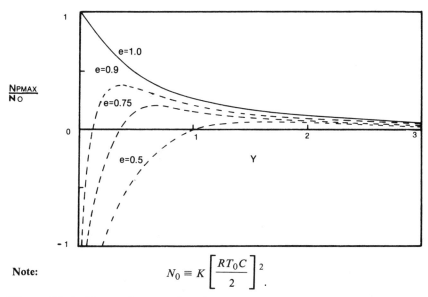

Note:

$$N_0 \equiv K \left[\frac{RT_0 C}{2} \right]^2.$$

Figure 12–3. Power Output of an Osmotic Cell as a Function of the Ratio (Y) of Fresh-to-Ocean-Water Flows for Different Pumping Efficiencies (e)

This is shown as a function of the flow rate n_p in figure 12–4. The efficiency at maximum power as a function of the flow ratio Y is shown in figure 12–5. This can be expressed analytically as

$$\eta_{\text{max power}} = \frac{Y + 1 - \dfrac{1}{e}}{2Y}. \qquad (12.20)$$

Maximum overall efficiencies greater than 50 percent would not be obtained with this type of conversion system, even under ideal conditions.

Actual optimization with consideration of costs would follow from a thermoeconomic analysis, where the cost of the membrane and its conductance are traded off against the available energy required for pumping.

**Freshwater Energy Conversion
with a Dialytic Battery**

Analysis of a dialytic battery for power generation can be made using irreverisble thermodynamics and available energy techniques [1,4]. Figure

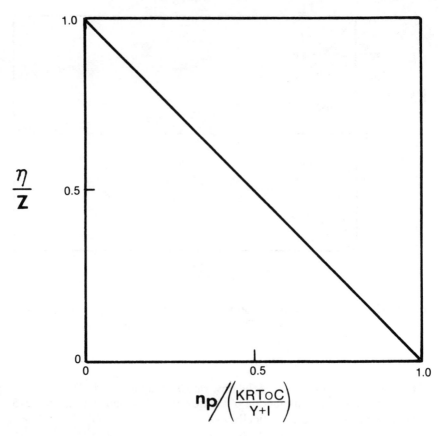

Note: The normalizing factor for the efficiency is $Z \equiv (Y + 1 - 1/e)/[Y(Y + 1)]$.

Figure 12–4. Efficiency Variation of an Osmotic Power System as a Function of the Relative Freshwater Flow

12–6 illustrates the principal elements of a dialytic battery that can be used to produce electrical power directly from the available difference between fresh water and seawater. The available energy of fresh water relative to that of seawater has been calculated in the first section of this chapter. This available-energy loss appears as an entropy-generation term in accordance with the discussion in chapter 8 as

$$T_0 \sigma_c = - n_m \Delta A - i \Delta V. \qquad (12.21)$$

Let us use Odum's formulation discussed earlier to write,

$$n_m = -(n + cf^2)\Delta A + cf\Delta V, \qquad (12.22)$$

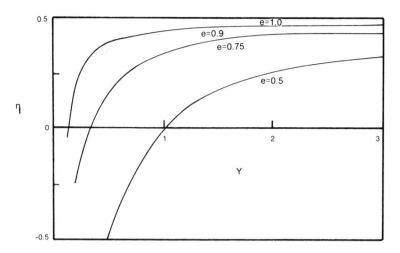

Figure 12–5. Osmotic-Power-System Efficiency at Maximum Power as a Function of Fresh to Ocean Flows (Y) with Different Pumping Efficiencies (e)

$$i = cf\Delta A - c\Delta V. \tag{12.23}$$

Now $f = \Delta V_{oc}/\Delta A$, where ΔV_{oc} is the open-circuit voltage found when $i = 0$. Theoretically this is $f = \alpha/\mathscr{F}$, where α is the permaselectivity and $\mathscr{F}$ is Faraday's constant ($\mathscr{F} = 2.305 \times 10^4$ cal/volt).

With $i = 0$,

$$n_m = -n\Delta A.$$

The available energy is all dissipated, so that

$$T_0 \sigma_c = n_m \Delta A = n(\Delta A)^2. \tag{12.24}$$

n is a loss coefficient obtained by finding the ion flux under open-circuit conditions. The factor c is the reverse internal conductance of the dialytic battery determined by stopping the flow of water and measuring the conductance when $\Delta A = 0$, then $i = -c\Delta V$. The power delivered is $i\Delta V$ or, expressed as net power, N_p,

$$N_p = cf\Delta A \Delta V - c(\Delta V)^2.$$

With $r \equiv \Delta V/f\Delta A$ (the force ratio), this power is

$$N_p = cf^2(\Delta A)^2 r(1 - r).$$

The maximum power is at $r = 1/2$, giving

$$N_{p\ max} = cf^2(\Delta A)^2/4 = c(\Delta V)^2. \qquad (12.25)$$

This occurs when the voltage is

$$\Delta V = f\Delta A/2,\ \text{or one-half the open-circuit voltage (impedence match)}.$$

The efficiency at this point would be

$$\eta = \frac{i\Delta V}{n_m\ \Delta A} = \frac{r(1 - r)}{n/cf^2 + (1 - r)}. \qquad (12.26)$$

With $r = 1/2$, and $n/cf^2 = 2$

$$\eta = 1/4.$$

For maximum efficiency, a high conductance c is required, a high voltage per unit available energy f, and a high input impedence, $1/n$ (a low-permeability membrane).

Order-of-magnitude calculations can now be made to estimate the feasibility of such a battery. The available-energy difference across the membrane is found from chapter 5 as

$$\Delta A \simeq RT_0 \ln\frac{C_o}{C_L}.$$

The voltage at maximum power is then

$$V = \alpha\Delta A/2\mathcal{F} = \frac{\alpha}{2\ \mathcal{F}} RT_0 \ln\frac{C_o}{C_L} \qquad (12.27)$$

where C_o is the average concentration on the seawater side.

C_L is the average concentration on the freshwater side.

Assume a mixed flow n_I with a membrane salt flow n_m. A mass balance gives, from figure 12-6,

$$\text{Total flow: } n_S - n_R = n_I - n_P. \qquad (12.28)$$

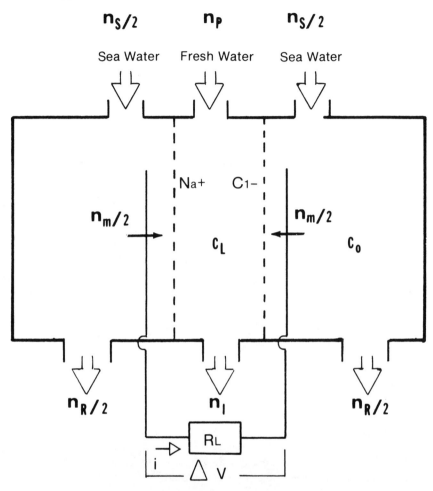

Figure 12–6. Schematic of a Dialytic Cell for Conversion of Available Energy of Fresh Water to Electrical Energy in a Seawater Environment

$$\text{Salt flow: } C_S n_S - n_R C_R = C_I n_I - C_P n_P. \qquad (12.29)$$

Let $Y = n_P/n_S$, $X = C_R/C_I$, and $Z = C_P/C_S$. The average concentrations on each side of the membranes are

$$C_o = \frac{C_S + C_R}{2} \qquad \text{and} \qquad C_L = \frac{C_P + C_I}{2}.$$

If the ocean and freshwater streams are brought to equilibrium on leaving, then their concentrations and available energies will be equal. Then with $X = 1$, the ratio of the concentrations across the membrane can be written as

$$C_o/C_L = \frac{2 + (1 + Z)Y}{2ZY + Z + 1}.$$

If the fresh water has zero salt concentration, $Z = 0$, then

$$C_o/C_L = 2 + Y.$$

The voltage at maximum power for a single membrane is then found from equation 12.27 as

$$\Delta V = \frac{\alpha}{2\,\mathscr{F}}\, RT_0 \ln (2 + Y).$$

For typical cells, $\alpha \simeq 0.6$ (see reference [2]). With $C_S = 1$ molar and $C_P = 0.01$ molar,

$$\Delta V = 0.036 \text{ volts.}$$

An open-circuit voltage of twice this value would be produced in the double cell shown in figure 12–6. Multiple cells are then required for practicable voltage potentials.

The current flow through the cell is determined at the maximum-power point from $i = \Delta V/\bar{R}$. The internal cell resistance $\bar{R}$ is the sum of the electrode resistances at the cathode R_c and anode R_a, plus the solution resistances on the ocean side R_o and river side R_L:

$$\bar{R} = R_C + R_a + R_o + R_L.$$

In a flat membrane cell

$$\frac{1}{c} = \bar{R} = \frac{r\,\ell}{A} = (r_c\ell_c + r_a\ell_a + r_L\ell_L + r_o\ell_o)/A.$$

where A = area of cell.

ℓ = conductance-path length.

r = resistivity.

A good approximation for dilute solutions of ions is that the resistivity is inversely proportional to the concentration of salt in solution, $r \simeq 1/KC$. Then $\bar{R} = (r_c \ell_c + r_a \ell_a + (\ell_o/C_o + \ell_L/C_L)/K)/A$. We will assume the conductance of seawater C_o is much higher than the freshwater conductance C_L. Then with $X = 1$ and $Z = 0$, the concentration $C_L = C_S/2(1 + Y)$. The maximum power per unit area is then expressed as

$$\frac{N_{p\ max}}{A} = \frac{\left(\dfrac{\alpha R T_0}{2 \mathscr{F}} \ \ln\,(2 + Y) \right)^2}{r_c \ell_c + r_a \ell_a + 2 \ell_L (1 + Y)/KC_S}. \qquad (12.30)$$

The optimum freshwater to seawater flow ratio Y can now be found, which maximizes the power output per unit membrane area. This occurs for the value of Y for which

$$\ln\,(2 + Y) = \frac{(r_c \ell_c + r_a \ell_a)\dfrac{KC_S}{\ell_L} + 2(1 + Y)}{2 + Y}.$$

As an illustration, let us take typical values of

$$\alpha = 0.6, \ \ell_L = 0.1, \ T_0 = 300 \text{ K}, \ C_S = 1.0 \text{ mole/liter},$$

$$K = 0.096 \text{ liter/ohm} \cdot \text{mole} \cdot \text{cm}$$

$$r_c \ell_c = r_a \ell_a = 10 \text{ ohm/cm}^2.$$

Then $Y = 17$, and $C_L = 0.028$ moles/liter. The maximum per unit area is then

$$N_{p\ max}/A = 9.3 \times 10^{-2} \text{ W/m}^2.$$

The open-circuit voltage is 0.046 volts.

A check on the overall effectiveness of this system can now be made by comparing the maximum power output with the change in available energy of the freshwater-saltwater dilution. From the first section of this chapter, the available energy of the fresh water relative to that of the seawater that is lost in the dilution is

$$\frac{\Delta A_{ideal}}{n_p} = RT_0 \left[\ln\left(\frac{Y + 1}{1 - C_S + Y} \right) + \frac{1 - C_S}{Y} \right.$$

$$\times \ln \frac{(1 - C_S(Y + 1))}{1 - C_s + Y} + \frac{C_S}{Y} \ln (Y + 1).\Bigg]$$

$$= 11.87 \ \text{W} \cdot \text{s/mole}$$

In order to find the ideal power $\Delta A_{\text{ideal}} n_p / A$ for calculating the effectiveness, one must calculate some other flow ratios. From equation 12.26,

$$\frac{n_m}{A} = i \Delta V / A(\Delta A) = 5.08 \times 10^{-5} \ \text{moles/sec} \cdot \text{m}^2.$$

Then from mass balances, $n_p / n_m = 642$.
Combining these factors gives

$$\text{Effectiveness} = \frac{N_{p \ \text{max}} / A}{\dfrac{\Delta A_{\text{ideal}}}{n_p} \dfrac{n_p}{n_m} \dfrac{N_m}{A}} = 0.25.$$

The overall effectiveness is 25 percent, which is the effectiveness of the maximum-power-conversion system. This is expected since the losses occur across the membrane in the conversion process.

A detailed analysis and experiment are described by Weinstein and Leitz [1]. They consider deviations from ideal chemical solutions and arrive at results quite close to the simple analysis indicated here.

The dialytic-battery membrane operates on a different basis than does the osmotic-membrane power system described earlier. In the dialytic cell the ions of the salt pass through the membrane. In the osmotic cell the water molecules pass through the membrane. Hence, different fluxes are involved, and thus the different approach to a solution. If the pumping energy required to operate the dialytic cell is included, an additional loss is incurred.

In a dialytic-battery electro-osmotic system, flow of the fresh water and seawater must be provided through the cell. This requires pumping energy, which can be estimated as $p = (\Delta P)(Q)$ where ΔP is the pressure drop through the cell and Q is the flow rate. From consideration of momentum conservation in viscous flow through a flat channel [5],

$$\Delta P = \frac{1}{2} \rho v^2 \ 4\text{f}(L/D),$$

where f is the friction factor that is dependent on the Reynolds number and the turbulence in the flow.

v is the average velocity.

L is the channel length.

D is the hydraulic diameter.

ρ is the density.

For laminar flow,

$$f = \frac{64}{vD\rho/\mu}, \quad \Delta P = 128\,\frac{vL\mu}{D^2}.$$

For water the viscosity μ is approximately 10^{-2} gm/cm·sec. For a cell of dimensions 0.1 cm $\times$ 15 cm $\times$ 15 cm the hydraulic diameter $D = 0.1$ cm. The volume rate of flow per unit membrane area is found to be

$$Q/A = 3.77 \times 10^{-4} \text{ m}^3/\text{m}^2 \cdot \text{sec}.$$

The pumping power required per unit membrane area is then PQ/A. The mean flow velocity is found as

$$P_p = (Q/A)(A/A_p) = 5.66 \times 10^{-2} \text{ m/sec}.$$

Here $A = 15$ cm $\times$ 15 cm and A_p (the passage area) $= 15$ cm $\times$ 0.1 cm. The power per unit area for pumping is then

$$P_p/A = 4.1 \times 10^{-2} \text{ W/m}^2.$$

The overall effectiveness is then

$$\frac{\text{Power out} - \text{Pumping power}}{\Delta A_{\text{ideal}} n_p} = 0.14.$$

This overall effectiveness of 14 percent appears feasible for such a device. It is clear that further optimization would require careful consideration of the fluid mechanics and the ion-transport properties of the cell.

References

1. Weinstein, J.N., and Leitz, F.B. "Energy Recovery from Saline Water by Means of Electrochemical Cells." *Science* 191(1976):719.

2. Lakshminarayanaiah, N. *Transport Phenomena in Membranes.* New York: Academic Press, 1966.
3. Norman, R.S. "Water Salination: A Source of Energy." *Science* 186(1975):350.
4. Isshiki, N. "Experiments on a Simple Concentration Difference Energy Engine Are Under Way." *J. Nonequilibrium Thermodynamics* 2(1977):125.
5. Schlicting, H. *Boundary Layer Theory.* New York: McGraw-Hill, 1979.

13

Interaction of Resource Recovery and Environment

I frequently have these meetings with the Environmental Protection Agency. EPA is saying a technology exists, and industry and the Department of Commerce are saying it does not. Ninety-five percent of the time EPA is right. I have concluded, EPA talks to the engineers and the Department of Commerce talks to the board rooms. The problem is the board rooms do not talk to their engineers.

—John Whitaker, White House Domestic Council,
quoted by R.L. Sansom in *The New American Dream Machine* (1976)

Interaction of Energy Systems

Net energy analysis [1–3], as noted in chapter 2, is an attempt to put into input-output format the energy utilized to produce useful products and services. Both direct and indirect energy inputs are determined to indicate the energy effectiveness of different processes. In this chapter available-energy concepts are introduced into the input-output format of chapter 3. The purpose is to add better predictive measures through the combination of resource quality and technological factors. Available-energy accounting of environmental factors is also introduced.

In future resource development it is expected that more concern will be directed at the interaction of the following factors.

1. depletion of the quality of resources;
2. increased energy requirements for processing these lower-quality resources;
3. increased environmental impacts from increased resource development;
4. changing resource-processing technology;
5. increased diversity of materials and energy supplies.

This chapter attempts to combine these five factors systematically to assist in the evaluation of new technological material and energy processing systems. Questions about efficiency (or effectiveness) of energy-and material-conversion systems are explored. The objective is to provide assistance in making rational decisions about the use of natural resources and the environmental impact of technological processing.

First, consider the trade-offs of energy use and environmental impact in resource processing where the objective is the production of electrical energy. The serial nature of fuel processing is an important characteristic of large-scale electrical-generation systems. The analysis illustrates some principles of available-energy accounting in processing chains where decisions can be made to increase efficiency and reduce environmental impacts.

The impacts of changes in a serial process such as oil refining are reflected in new alternatives for other energy processers and users. Changes in technology may indicate a change in the form of the input energy from other sectors. Replacement of diesel- or gas-turbine-driven pumps by electrical-motor drives in the transportation of oil or gas, or in the processing of coal, are typical price-change strategies. For example, in 1979 the Eli Lilly Company replaced 500-horsepower diesel-driven stirrers by electric-driven stirrers because of changes in the price and availability of petroleum.

The nature of the technology of each process chain and the quality of the input energy must be examined before making these decisions. Consideration of the quality of energy inputs increases the importance of the cross-coupling between energy-conversion systems. A method is outlined for dealing with multistage processes in which inputs from other multistage processes are significant.

An examination of economic and engineering literature shows some basic differences in the handling of efficiency. Engineering-systems efficiencies are expressed in terms of the ratio of useful outputs to inputs for each device, machine, or system. Economic efficiencies are expressed in terms of inputs per unit output from each unit or economic sector. In dealing with the available-energy concept in these decisions, we would like to compare the available energy of the process output with the available energy of the inputs. In energy-conversion processes the available energy of the output of a stage may exceed that of the input principal flow of available energy. When electrical energy is used to pump oil in a pipeline, the available energy of the oil leaving the pump is higher than that of the oil entering. However, if the whole system, including pumping losses, is considered, there will be a net loss in available energy. Available energy may also be lost in stages through the disposal or dispersion of process waste to the environment. This may include exhaust products, effluents, thermal discharges, or mine tailings.

With multiple inputs and outputs for a stage, let us arbitrarily select an effectiveness for a stage as the available energy of the principal products out divided by the available energy of the input streams. The available energy of the waste streams relative to the total input of available energy would be a second parameter of use in examining recovery systems. Thermal-energy exchange processes require special discussion in this context.[1]

Suppose we have two alternate processes that require the same total energy, one of which must be maintained at a higher temperature than the other. The higher-temperature process will require greater available-energy input. This must be accounted for in the evaluation of processing systems. The total energy required is not a sufficient measure of the effectiveness of a technology.

Interactions between multistage processes converting oil and coal resources to electrical energy are shown in figure 13–1. In reference to this figure, the following notation is applied. Each chain is indicated by a subscript j and each stage in the chain by a subscript i. The main product output of the ith stage of the jth chain will be denoted by X_{ij}. In this case i goes from 1 to 5 for each stage, and j takes on values 1 for coal and 2 for oil. The final output from j chain is X_{5j}. A single stage of the chain is shown in figure 13–2 to clarify the notation further.

The Leontief notation of input-output analysis as applied to inter-industrial exchanges is employed so that the inputs to each sector are denoted as proportional to the total output of that sector [5,7,10]. The input of oil to the processing of coal, for example, is denoted as $C_{3221}X_{21}$. The coefficient C_{3221} indicates that the oil came from the third element of the second chain and is destined for the second element in the first chain. X_{21} is the total output of the second element of the first chain.

The environmental sector is considered in a manner similar to that used by Converse [6]. The inputs of a sector required *from* the environment are designated $C_{ee11}X_{11}$, for example, and outputs *to* the environment designated as $C_{21ee}X_{21}$. In both cases they are proportional to the output of a stage element. Two subscripts are reserved for environmental factors to allow division of the environmental sector into different elements such as air, water, and land.

For each stage let us write a mass balance and an available-energy balance. This assumes steady flow, which neglects storage, which is typically 2–5 percent in most processes. From figure 13–3 consider the total mass flow into a process as m_1. Of this m_1, C_i is the portion that is potentially convertible to product. The remainder, $m_1(1 - C_i)$, is rock or other unwanted material. m_oC_o represents the quantity of product output from the stage, the remainder being waste. The ratio, $f \equiv m_oC_o/m_iC_i$, represents the *material-recovery factor*—the ratio of total mass of resource recovered to that potentially recoverable.

Now consider the available-energy balance. For the resource, a *recovery ratio* $r \equiv m_oa_o/m_ia_i$ represents the ratio of the available energy in the refined material to that in the natural deposit. a_o and a_i represent the specific available energy of the product and the input resource, respectively.

The ratio f represents an extraction efficiency, which is equal to one if all the resource is extracted and approaches zero at the limit of trying to extract material from a very dilute source. An example of the latter might be the extraction of precious metals from seawater, where large quantities of low-concentration material are processed and large quantities left in the waste,

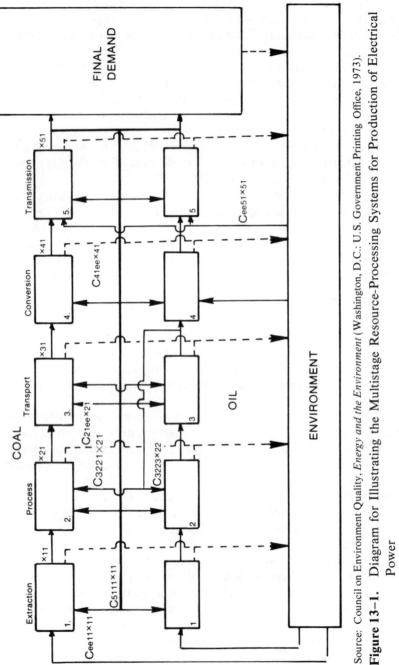

Source: Council on Environment Quality, *Energy and the Environment* (Washington, D.C.: U.S. Government Printing Office, 1973).

Figure 13–1. Diagram for Illustrating the Multistage Resource-Processing Systems for Production of Electrical Power

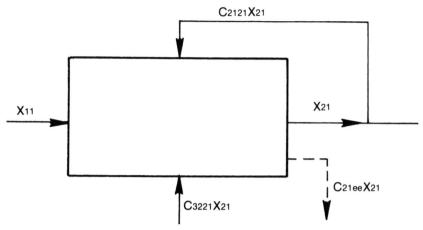

Figure 13–2. Diagram of the Input-Output Nomenclature for a Single Stage
of a Multistage Process

but only a small amount of metal is recovered. This ratio represents a
conventional measure used in mining and in energy-resource-processing
work.

The resource-recovery ratio r improves on this measure by accounting for
the fact that the quality of the resource is also important in an extraction
process. The resource-recovery ratio may be greater than one in cases where
the processing energy is of high quality. This occurs when electrical energy is
used to drive a compressor in a natural-gas pipeline. The exiting gas is at a
higher temperature and pressure than the entering gas and hence has greater
available energy. This also occurs in chemical precipitation processes in
systems with solar-energy inputs.

The dependence of the recovery ratio on concentration for many
extraction processes can be illustrated by considering the available energy[2]

Figure 13–3. Mass-Flow Diagram for the Processing of a Resource of Initial
Concentration C_i to a Product with Concentration C_o

as

$$a_i = -RT \ln \frac{C_i}{C_d}, \qquad (13.1)$$

and

$$a_o = -RT \ln \frac{C_o}{C_d}, \qquad (13.2)$$

where C_d represents a datum concentration.

Substituting this relation in the recovery-ratio expression gives

$$r = \frac{m_o}{m_i} \frac{RT \ln C_o}{RT \ln C_i} = f \frac{C_i}{C_o} \frac{\ln C_o}{\ln C_i}. \qquad (13.3)$$

The recovery ratio as a function of the concentration ratio is shown in figure 13–4 for various output concentrations and $f = 1$. From examination of this figure, it is clear that when high-purity minerals (C_o high) are refined from ores that are low in concentration C_i (as in copper mining), the recovery ratio changes little as the resource concentration decreases with ore depletion. In situations where the concentration of the deposit is high, as in coal fields, small changes in the required processed-coal concentration are reflected in large changes in the recovery ratio. One difficulty with the recovery ratio as a measure is that it is higher for extraction processes that produce lower-concentration products. Coal processing thus appears more efficient than petroleum refining. This difficulty, together with the comment on processing energy, lead one to define a new measure, the effectiveness e of a process. The effectiveness is defined as a ratio of the available energy of the product of a stage to the available energy of the main resource input plus the available energy of all processing inputs:

$$e \equiv \frac{m_o a_o}{m_i a_i + C'(m_o a_o)}. \qquad (13.4)$$

In this notation C' is the fraction of the output total available energy that must be supplied for processing. The effectiveness can be written in terms of the recovery factor as

$$e = \frac{r}{1 + C'r}. \qquad (13.5)$$

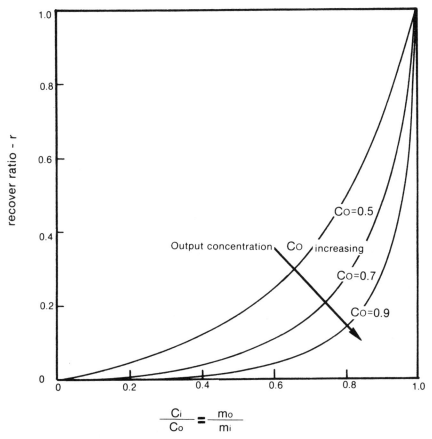

Source: R.H. Edgerton, "Measures of Energy Effectiveness of Interacting Resource Processing Systems," *Energy, Int. J.* 4(1980):1151. Reprinted with permission.

Figure 13–4. Illustration of the Effect of Input Quality of Materials C_i on the Available-Energy Recovery Ratio r for Different Output-Concentration Requirements C_o

C' represents the energy effectiveness of the technology; a high C' indicates an energy-intensive process. The effectiveness then combines the effects of resource depletion and technology in one measure.

Figure 13–5 illustrates those interactions in a convenient fashion. It is expected that as ores are depleted, the recovery ratio will decrease. This is indicated by the direction lines on the effectiveness-recovery ratio plot. The effect of technological improvement on effectiveness is indicated by vertical movement from one line to another. As the resources become more depleted,

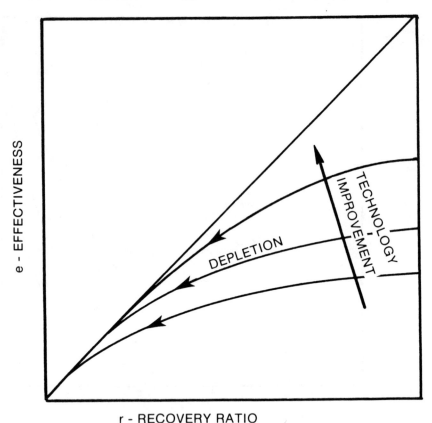

Source: R.H. Edgerton, "Measures of Energy Effectiveness of Interacting Resource Processing Systems," *Energy, Int. J.* 4(1980): 1154. Reprinted with permission.

Figure 13–5. Stage Effectiveness as a Function of the Recovery Ratio for Varying Technology and Resource Depletion

the maximum effectiveness possible continues to decrease. This is indicated by the arrows pointing to the left, representing depletion of resources with time. Discussion of future effectiveness increases must take into account these two factors: technology and depletion. The importance of the application of improved technology as soon as possible in resource recovery is shown by the fact that the maximum effectiveness increase by technology improvement continues to decline as the depletion continues.

In net energy analysis another measure is used: $\zeta \equiv$ energy output divided by the energy input other than the energy of the principal material being processed [8,10,14]. In the notation used here, this efficiency is

$$\zeta = \frac{m_o a_o}{C'(m_o a_o)}, \tag{13.6}$$

in this case, referring to the energy delivered compared with the actual energy required to process the resource. If the constant C' relation is assumed, then $\zeta = 1/C'$. This net energy efficiency then represents only a technology coefficient.

The American Institute of Physics (AIP) [20] measure of effectiveness is also in use. This is defined as

$$E = \frac{\text{minimum change in available energy required}}{\text{actual available energy applied in the process}}.$$

In the notation used in this section, this can be expressed as

$$E = \frac{m_o a_o - m_i a_i}{C'(m_o a_o)}. \tag{13.7}$$

This is very useful from a scientific point of view but not particularly easy to deal with in industrial decisions. It is not considered further because of the following practical deficiencies:

1. It is negative for processes where the available energy of the product stream is less than that of the input stream. This often occurs in the processing of plastics or in the development of composite-type materials.
2. It does not emphasize the output-product importance, which is often of primary concern in industrial decisions. The engineer is more often interested in dealing with efficiency in the sense of what one gets as a useful product compared with the cost of obtaining that product.
3. As initially formulated, it did not attempt to differentiate resource quality and technological status.
4. It is a task measure, not a device measure. For example, in a furnace the largest available-energy loss occurs in the heat exchanger. The effectiveness of the process, however, cannot be increased by improved heat-exchanger design.
5. When there is a solar-energy subsidy, this measure also provides values greater than one.

Note that the AIP effectiveness can include both resource and technology if written as shown in terms of r and C', as

$$E = \frac{(1 - 1/r)}{C'}, \tag{13.8}$$

and also in terms of the effectiveness we have defined,

$$e = \frac{r}{1 + C'r},$$ (13.9)

by eliminating the recovery ratio to give

$$E = \frac{1}{C'(1 - e) + 1}.$$ (13.10)

In terms of both r and C', the effectiveness e is better. E becomes infinite for cases where no processing energy is required and becomes negative for small recovery ratios. We will therefore use the effectiveness e in further discussions. See appendixes 13A and 13B for other comments on effectiveness measures in use.

Pollutant Factors

Consideration of the pollutant effect as proportional to the available energy of the waste output is examined in this section. The waste output really has two parts: that due to dissipation in the system and the available energy in the leaving waste stream. The total can be expressed as the difference between the available energy in the input streams and that in the product streams:

$$P = K(m_i a_i + C'm_o a_o - m_o a_o).$$ (13.11)

The K factor is included as a proportionality constant dependent on local health, sociological, aesthetic, or political considerations. The idea that the pollutant effect of a waste stream is proportional to its available energy emphasizes again that it is both the amount and the quality of the material leaving that are meaningful. In another sense the available energy represents the thermodynamic work required to bring the effluent to the environmental state. In most cases it represents a wasted opportunity. It is a loss that represents positive harm. Note that the measure P does not require measurement of the waste-stream properties. The waste-stream available-energy evaluation is discussed in chapter 11.

Dividing P by $m_o a_o$ to obtain a pollutant effect per unit output of available energy in the product stream, and introducing the recovery ratio, gives

$$\frac{P}{m_o a_o} = K(1/r + C' - 1).$$ (13.12)

The effect of technological efficiency on this pollutant measure is shown in figure 13–6. Again, note the importance of improving technology (reducing C') while r is large. Delaying the implementation of a technological improvement reduces the marginal benefits of pollutant control.

Since the effectiveness in terms of the recovery ratio is $e = r/(1 + C'r)$, the unit pollutant effect can be expressed as

$$\frac{P}{m_o a_o} = K(1/e - 1). \tag{13.13}$$

In this formulation an improved effectiveness is directly related to a reduction of pollutant output per unit product.

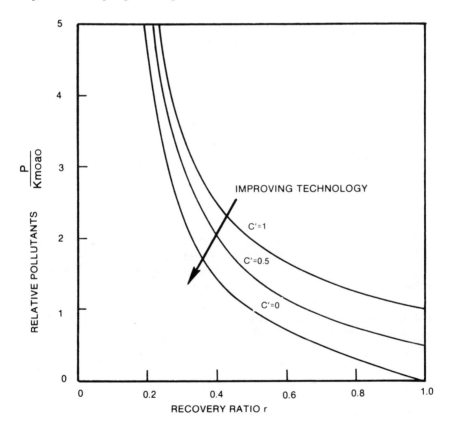

Source: R.H. Edgerton, "Measures of Energy Effectiveness of Interactive Resource Processing Systems," *Energy Int. J.* 4(1980): 1156. Reprinted with permission.

Figure 13–6. Pollutant per Unit Output as a Function of Recovery Ratio for Varying Technnology

This formulation also assigns the pollutant effect to the individual stage or sector where the available energy is dissipated. When the application of greater energy is required to maintain the product flow from a depleting resource, the recovery factor increases, the effectiveness decreases, and the pollutant output is expanded. This is characteristic of what in the past has often been called the application of improved technology to improve a process. It is now clear that if the environmental effect is included, many of these technological changes that rely on only additional energy input to maintain or improve output are pseudosolutions. Solutions that increase C' are no longer broadly acceptable. Future engineering technology should be charged with providing smaller values of C' to reduce both energy use and environmental impact. Pollution-control technology has been shown by others [10–13] to increase the total mass flow of pollutants to the environment (see chapter 3). Pollution-control technology is concerned with improving the quality of the material outputs of a process, not only the total quantity of discharges. From an available-energy viewpoint, the process of dilution of effluents used in the past is a waste process. It often also defers the problem or transfers it to a different environmental sector.

Multiple-Chain Processing Systems

This section is directed at available-energy accounting in resource-processing systems that are characterized by multiple interconnected serial stages (chains). Although the input-output format of chapter 3 is applicable to these systems, that format is not specifically designed to evaluate serial interconnected systems. The trade-offs and priorities for technological decisions about resource utilization and environmental impacts can be examined in detail with the multiple-chain analysis system.

The application of effectiveness concepts to the interrelated energy chains shown in figure 13–1 provides further insight into the analysis of complicated energy and materials problems. If the X_{ij}s represent the available-energy output of each sector, then the effectiveness of the first element of the coal extraction process gives

$$e_{11} = X_{11}/(C_{ee11}X_{11} + C_{s111}X_{11}). \qquad (13.14)$$

Similarly, the recovery ratio is

$$r_{11} = \frac{X_{11}}{C_{ee11}X_{11}} = \frac{1}{C_{ee11}}. \qquad (13.15)$$

Since the stage recovery ratio is dependent only on the major inputs and outputs, it may be designated for each stage as

$$r_{nj} = \frac{X_{nj}}{C_{n-1jnj}X_{nj}} = \frac{1}{C_{n-1jnj}} . \tag{13.16}$$

Looking at the second element in the chain,

$$e_{21} = \frac{\dfrac{X_{21}}{X_{11}}}{C_{1121}\dfrac{X_{21}}{X_{11}} + C_{5121}\dfrac{X_{21}}{X_{11}} + C_{3221}\dfrac{X_{21}}{X_{11}}} . \tag{13.17}$$

Since $C_{11121}X_{21} = X_{11}$, this may be rewritten as

$$e_{21} = \frac{r_{21}}{1 + \sum_{kl} C_{kl21}r_{21}} . \tag{13.18}$$

For each stage this may be written as

$$e_{nj} = \frac{r_{nj}}{1 + \sum_{kl} C_{lknj}r_{nj}} . \tag{13.19}$$

Often we are interested not only in identifying where losses occur or where effectiveness is low in a multiple-step process, but also in determining the overall effectiveness of a whole chain of processes. This may, of course, be determined by looking at the whole chain as one element, considering only the inputs and outputs. Sometimes we are interested in combining different elements, new technologies, and so forth, which include changes in individual elements. This requires expressing the overall effectiveness in terms of separate element effectiveness. For an n-element chain this can be expressed as

$$\bar{e}_{nj} = \frac{\displaystyle\prod_{s=1}^{n} r_{sj}}{1 + \displaystyle\sum_{s=1}^{n} \left(\frac{r_{sj}}{e_{sj}} - 1\right) \prod_{t=1}^{s} r_{t-1j}} . \tag{13.20}$$

Usual chain calculations use the product of each stage efficiency to calculate

320   Available Energy and Environmental Economics

overall efficiencies. In coupled-chain systems this overestimates the effectiveness of each chain.

An important observation from equation 13.20 is that the same relative increase in effectiveness of a stage increases the overall effectiveness more if it occurs at the beginning of a chain. This means that if one can secure, at equal cost, the same percentage improvement in effectiveness from two stages in a chain, the benefits are greater if the improvement is attained in the earlier element in the process chain. In process terms, greater overall improved effectiveness is to be expected from efforts to develop better extraction technology than from improvements in conversion technology.

Environmental Factors

The environmental effects of a stage can be expressed in terms of the available energy of the waste stream m_o and the dissipation defined in equation 13.13:

$$P = m_o a_o K\left(\frac{1}{e} - 1\right) = KX\left(\frac{1}{e} - 1\right).$$

The pollutant effect generated for the rj stage can be written with the proportionality constant K_{rj} as

$$P_{rj} = K_{rj} X_{rj}\left(\frac{1}{e_{rj}} - 1\right). \tag{13.21}$$

The total pollutants generated in a chain are then

$$P_j = \sum_{r=1}^{n} P_{rj} \sum_{i=1}^{n} K_{ij} X_{ij}\left(\frac{1}{e_{ij}} - 1\right). \tag{13.22}$$

This can be expressed as:

$$P_j = \sum_{i=1}^{n} \frac{K_{ij}\left(\frac{1}{e_{ij}} - 1\right)}{\prod_{s=1}^{n} r_{sj}} X_j \tag{13.23}$$

or in terms of r_{ij} only as:

$$P_j = \sum_{i=1}^{n} \frac{K_{ij}\left[\dfrac{1 + C'r_{ij}}{r_{ij}} - 1\right]}{\displaystyle\prod_{s=1}^{n} r_{sj}} X_j. \qquad (13.24)$$

From this formulation, the result of increasing the recovery ratio or effectiveness of a later stage in a chain can be shown to decrease the environmental impact proportionately more than the same improved effectiveness in an earlier stage. This can be used as an argument for placing priority on improving end-of-process conversion efficiency and the extra benefits environmentally from consumer conservation measures. The pollutant effect of the kth chain is, similarly,

$$P_k = \sum_{i=1}^{n} \frac{K_{ik}\left(\dfrac{1}{e_{ik}} - 1\right) X_k}{\displaystyle\prod_{s=i}^{n} r_{sk}} .$$

Often it is important to determine the effect of the change in the effectiveness of a stage in another chain on the effectiveness or pollutant output of a given chain. An example might be the switch from a petroleum-based solvent to an electrolytic process in coal liquefaction. How are the total pollutant effects of both the j and k chains affected by a decrease in the effectiveness of the mth stage of the jth chain? The direct effect will be seen in the jth chain as:

$$P'_j = \sum_{i=1}^{n} \frac{K_{ij}\left(\dfrac{1}{e'_{ij}} - 1\right) X'_j}{\displaystyle\prod_{s=1}^{n} r'_{sj}}$$

where the primes refer to the new e_{ij}. (Note that r'_{ij} may still equal r_{ij} if the recovery ratios remain the same, except that it requires more energy input for the processing.) This will, of course, be reflected in reduced e_{ij} and have a smaller X' total output. Similarly, with the same X_j, an increase in one of the C_{k1ij} coefficients reduces the output X_k.

Another quality of interest is the increased extraction of resources from the environment if additional energy is required to diminish a pollutant effect. In this case one would need to evaluate the change in $\Sigma E_{ij} \equiv \Sigma C_{eeij} X_{ij}$ due to this increase in outputs X_{ij} to meet environmental requirements.

Application Example: Aluminum Refining

The production of aluminum metal from bauxite (aluminum ore) provides an illustration of the materials-processing available-energy analysis. For simplicity let us break down the process into a two-element chain. The first element is the processing of the raw ore to alumina (Al_2O_3) (Beyer process). The second is the electrolytic refining to aluminum metal (Hall process). Each process is indicated in figure 13–7 together with the input flows of materials [19,22]. The input flows of materials that require processing before use are also indicated, together with their energy requirements. Where thermal inputs are required, the available energy of the fuel is assumed. Where electrical-energy inputs are used, the available energy of the fuel used to generate the electricity is assumed [4,9,15,16,18,21]. The processing energy for the materials required for the aluminum extraction are averages without regard to the source of the energy. This assumption is based on the fact that the difference between the heating value of fuels and their available energy is small compared with the variation that appears in the referenced energy-accounting estimates.

The first decision with respect to the accounting is the determination of the available energy of aluminum metal. This available energy is taken as the minimum energy required to extract the metal in pure form from the earth crust (8 percent aluminum in the form of Al_2O_3). This includes the unmixing available energy and the chemical available energy. The available energy of the fuels is taken as the available energy relative to that of the complete-combustion products.

The available energy of materials is based on the available energy of the ideal products of the reaction required. This involves a mixing factor when dilution occurs as well. One of these factors is the dilutrion of reactants in forming a water solution. This occurs in the forming of a saturated salt solution in the preparation of the caustic solution for the alumina extraction.

This method of accounting for the available-energy use of each input is consistent with the fuel accounting but does not properly account for the pollutant effect of the effluents, since they still have available energy relative to the environment. This will overestimate the effectiveness of the system and underestimate the pollutant effect. This can, in part, be corrected in the pollutant-effect formulation through the addition of the available energy of

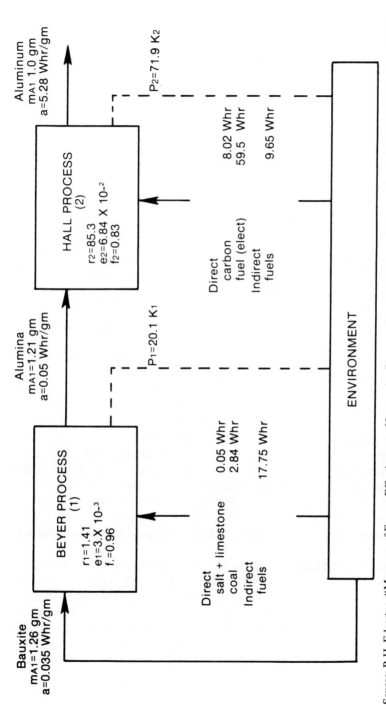

Source: R.H. Edgerton, "Measures of Energy Effectiveness of Interacting Resource Processing Systems," *Energy, Int. J.* 4(1980):1158. Reprinted with permission.

Note: Overall recovery ratio $\bar{r} = 119.5$. Overall effectiveness $\bar{e} = 6.77 \times 10^{-2}$. Overall mass-flow ratio $\bar{f} = 0.79$.

Figure 13-7. Aluminum-Production Available-Energy Schematic

the effluent products. In a more complete analysis this factor would be included to indicate better the effectiveness and pollutant effects. It is, however, useful for the evaluation of single-stage performance, since it allocates to each stage the available energy required by that stage.

Aluminum has potential as a fuel once it is refined, in the same way that lead has potential as a fuel in a lead-acid battery. In fact, research has been done to examine the chemical use in a fuel-cell-type arrangement. The feasibility has been claimed [23] to be due to the idea that much aluminum is produced from the ore using electrical energy generated hydroelectrically. The data in figure 13–7 indicate that just the available energy in the carbon elecrodes used exceeds the available energy of the metal aluminum produced. Research efforts at development of an aluminum fuel cell can therefore easily be demonstrated to be interesting but thermodynamically and economically misdirected.

The analysis in terms of the effectiveness is surprising. The analysis shows that the effectiveness of the process for producing alumina from bauxite is lower than the effectiveness of the electrolytic processing of alumina to aluminum. Although the available-energy requirements of the alumina-from-ore stage are smaller than those of the electrolytic refining stage, there is a very large energy expended relative to the change in available energy of the aluminum in this stage.

This first stage is the one for which more energy will be required in the future as ores become more depleted. The changes in available energy required for this stage are small for the aluminum-refining case, indicating the poor thermodynamic performance of this technology. It could reasonably be expected that the technology improvement of this stage will take place faster than the ore depletion. The effectiveness of this stage (and of the whole aluminum-refining system) will be expected to increase in the future.

Research on improved effectiveness of the process probably should be based on chemical processing of the ore to forms other than alumina. The chemical process of this first stage is not effective since it does not change the energy of the aluminum bond to the oxygen. Breaking these energy bonds is the principle energy cost of the extraction process, not the concentration process.

Similar conclusions can be reached with regard to the pollutant effect of each stage. This is reflected in the values of P_1 and P_2 on the diagram. The pollutant effect of the first stage is less than that of the second stage because, although the effectiveness of the second stage is greater, the total available energy lost to the environment is greater than the first stage. A final point must be made here about the very low values of effectiveness indicated. Metal processes are characteristically inefficient in terms of available-energy use because of the availability in the past of cheap energy relative to the cost of materials. This is certain to change in the future as the relative cost of energy approaches that of materials.

An interesting example of the cross-chain effects is the Michigan Air Pollution Control Law, which in 1975 limited the sulfur content of fuels used outside Wayne County (Detroit metropolitan area) to 2 percent (1.5 percent for the large power plants) and imposed a limit of 0.5 percent within Wayne County. The Marathon Oil Company refinery located in Wayne Country must now reduce the sulfur content of its final-stage product (heating fuel) from 2.7 percent to 0.5 percent. The final stage for sulfur removal is an electrochemical process that reduces the sulfur from 1.5 percent to 0.5 percent, using electrical energy supplied by the Detroit Edison Company. A simple calculation of the minimum theoretical available energy required to reduce the sulfur content from 1.5 percent to 0.5 percent indicates 47.5 kW-hr/lb of sulfur removed.

Currently Detroit Edison uses approximately 0.9 lb of coal per kW-hr of electrical energy generated. The sulfur content of the coal is 3–5 percent but is being reduced to the 2 percent required with the use of sulfur-removal equipment and the burning of low-sulfur western coal. Using the 3-percent sulfur-emission rate, the removal of 1 lb of sulfur by Marathon Oil Company introduces a minimum of 1.2 lb of sulfur to the atmosphere by the electrical generation process. In reality, the actual sulfur-removal process requires about 500 kW-hr/lb of sulfur. Even with equal requirements for sulfur emissions, the sulfur-removal process introduces more net sulfur into the atmosphere.

Conclusions

The use of available energy in an accounting system has been helpful in many cases [14,17,18,21]. The introduction of available-energy concepts into net-energy analysis with input-output techniques provides a new basis for evaluating resource-processing systems. It allows for a better measure of the efficiency of a process that includes both resource depletion and techno-logical factors. Chemical and thermal environmental impacts can be included in this formulation.

Interacting elements in chain processes can be analyzed using these techniques to arrive at both stage performance and overall management decisions. The advantage of this formulation is its direct applicability to technical processes to identify where efforts can best be applied to improve system effectiveness and reduce environmental impact.

Notes

1. Often the available energy actually used in a low-temperature process is the same as that in a high-temperature process. This discrepancy is due either to the lack of proper technology or to its high cost [1,2].

2. This available-energy relationship to concentrations is applicable to low concentrations. Corrections can be made in a more complete analysis, but the logarithmic dependence is basic.

References

1. Ballard, C., and Herendeen, R. "Energy Costs of Goods and Services." Document 140, Center for Advanced Computation. Urbana: University of Illinois, 1975.
2. Makhijani, A., and Lichtenberg, A. "Energy and Well Being." *Environment* 14, no. 5(1972):10.
3. Hirst, E. "How Much Overall Energy Does the Automobile Require?" *Auto. Engr.*, 80, no. 7(1972):35.
4. Council on Environmental Quality. *Energy and the Environment.* Washington, D.C.: U.S. Government Printing Office, 1973.
5. Leontief, W. "The Structure of the U.S. Economy." *Sci. Amer.* 212, no. 4(1965):3.
6. Converse, A.O. "On the Extension of Input-Output Analysis to Account for Environmental Externalities." *Amer. Econ. Rev.* 61(1971):197.
7. Herendeen, R. "Use of Input-Output Analysis to Determine the Energy Cost of Goods and Services." In *Energy: Demand, Conservation, and Institutional Problems*, ed. M.S. Macrakis. Cambridge, Mass.: MIT Press, 1973.
8. Berry, R.S., and Fels, M. "The Energy Cost of Automobiles." *Sci. Public Affairs*, (*Bull. At. Scient.*) 29(December 1973):11.
9. Institute for Energy Analysis. "Net Energy from Nuclear Power." Oak Ridge Associated Universities, IEA-75-3, November 1975.
10. Kneese, A.V. "Analysis of Environmental Pollution." In *Economics of the Environment*, eds. R. Dorfman and N.S. Dorfman, New York: Norton, 1971, p. 39.
11. Maxim, L.D., and Braize, C.L. "A Multistage Input-Output Model for Evaluation of the Environmental Impact of Energy Systems." *IEEE Trans., Systems, Man and Cybernetics SMC-3* (1973):583.
12. Leontief, W. "Environmental Repercussions and the Economic Structure: An Input-Output Approach." *Rev. Econ. Stat.* 52(1970):262.

13. Ayers, R.V., and Kneese, A.V. "Production, Consumption and Externalities." *Amer. Econ. Rev.* 59(1969):282.
14. Gyftopoulos, E.P.; Lazaridis, L.J.; and Widmer, T.F. *Potential Fuel Effectiveness in Industry.* Report to the Energy Policy Project, Ford Foundation. Cambridge, Mass.: Ballinger, 1974.
15. Hittman Associates. *Environmental Impacts, Efficiency, and Cost of Energy Supply and End Use,* vol. I. Washington, D.C.: Council on Environmental Quality, 1974.
16. Franklin, W.E.; Bendersky, D.; Park, W.R.; and Hunt, R.G. "Potential Energy Conservation from Recycling Metals in Urban Solid Waste." In *The Energy Conservation Papers,* ed. R.H. Williams. Cambridge, Mass.: Ballinger, 1975, p. 171.
17. Ross, M.H., and Williams, R.H. *Our Energy: Regaining Control.* New York: McGraw-Hill, 1981.
18. Berg, C.A. *Energy Conservation in Industry: The Present Approach, the Future Opportunities.* Report for the Council on Environmental Quality, May 1979.
19. Just, J. "Impacts of New Energy Technology Using Generalized Input-Output Analysis." *In Energy: Demand, Conservation, and Institutional Problems,* ed. M.S. Macrakis. Cambridge, Mass.: MIT Press, 1973, p. 113.
20. Ford, K.W.; Rochlin, G.I.; and Scolow, R.H. (eds.) "Efficient Use of Energy." *APS Studies on the Technical Aspects of the More Efficient Use of Energy.* New York: American Institute of Physics, 1975.
21. Berg, C.A. "Process Inovation and Changes in Industrial Energy Use." *Science* 199(1978):200.
22. Saxton, J.C., and Ayres, R.U. "The Materials Process Product Model: Theory and Applications." In *Mineral Materials Modeling—A State of the Art Review,* ed. W.A. Vogely. Baltimore, Md.: Resources for the Future. Johns Hopkins University Press, 1975, p. 178.
23. Cooper, J., and Littaner, E., *Proceedings,* Thirteenth Intersociety Energy Conversion Engineering Conference. San Diego, Calif., 1978.
24. Edgerton, R.H. "Measures of Energy Effectiveness of Interacting Resource Processing Systems." *Energy, Int. J.* 4(1980):1151.

Appendix 13A:
Energy Effectiveness

Effectiveness is a measure of technical efficiency or second-law efficiency that was first introduced by Darrieus [1] to describe the performance of steam turbines. It has since been extended by others [2–5] to many other applications. Riestad [6] summarizes general measures of effectiveness and the advantages of each. Most effectiveness measures are based on the principle that the best technical performance is obtainable when the available energy required to produce a desired result is a minimum.

In keeping with the concept of efficiency, it is useful to have a measure that is normalized to values of one for best performance to zero for poorest performance. The most commonly used measures are:

1. the available-energy increase in the desired output divided by the available-energy decrease in the resource input;
2. the minimum available-energy change required in a process divided by the actual available energy of the resource input;
3. available energy out of the system divided by available energy into the system.

The unfortunate differences in definitions for the same name designation arises because of the different points of view about the value of the inputs and outputs. Riestad [6] discusses several examples, particularly the effectivenesses of heat exchangers. The basic difference in the measures lies in whether the measure is applied to a single component or to a total system of many components. These differences are discussed in chapter 13 in regard to resource-recovery systems. A simple illustration for a turbine would be

$$\text{effectiveness (I)} = \frac{\text{Workout of the turbine}}{\text{Actual available energy supplied}}.$$

$$\text{effectiveness (II)} = \frac{\text{Work out of an isentropic ideal turbine}}{\text{Actual available energy supplied}}.$$

$$\text{effectiveness (III)} = \frac{\text{Work of the turbine plus available energy of exhaust}}{\text{Actual available energy supplied}}.$$

The effectiveness (III) measure is based on the idea that there is available energy in the exhaust that could be utilized at a later stage. The differences are based in part on the value of the "waste" stream.

In some energy analyses of fuel-production systems, a measure termed the *amplification* is used. Amplification refers to the energy value of the fuel produced divided by the energy required for the processing.

References

1. Darrieus, G. "the Rational Definition of Steam Turbine Efficiencies." *Engineering* 130(1930):382.
2. Keenan, J.H. *Thermodynamics*. New York: Wiley, 1941.
3. Obert, E.F. *Thermodynamics*. New York: McGraw-Hill, 1948.
4. Bosnjakovic, F. *Technische Thermodynamik*, vol. 1. Leipzig: Teil, Dresden University, 1948.
5. Ford, K.W., et al. *Efficient Use of Energy: APS Study on Technical Aspects of the More Efficient Use of Energy*. New York: American Institute of Physics, 1975.
6. Riestad, G.M. "Availability: Concepts and Applications." Ph.D. diss., University of Wisconsin, 1970.

Appendix 13B:
Second-Law Efficiency
and Task-Oriented
Effectiveness

$$E \equiv \frac{\text{Heat or work usually transferred by a given device or system}}{\substack{\text{Maximum possible heat or work usefully transferable for the} \\ \text{same function by any device or system using the same energy} \\ \text{input as the given device or system}}}$$

This definition of a second-law efficiency proposed by the AIPS study [1] is qualified by a note that " . . . this denominator is a *task* maximum, not a *device* maximum." This qualification is important in the context of energy-system analysis, where we often wish to choose among alternative ways to perform a task. This measure theoretically gives a comparable measure of thermodynamic performance. The comparability is not as direct as one might suppose from their approach because of the difficulty in determining the minimum energy required for a particular task. As long as the effectiveness measure is used to evaluate a single device, the problem is a technical one; the device is tested and the inputs and outputs measured and computed. In the task effectiveness, a minimum-energy-use scheme is hypothesized to use as a standard. Two examples of this difficulty are as follows.

First, conventional home heating involves the combustion of a hydrocarbon fuel. The high-temperature products of combustion supply thermal energy to the house by way of a heat exchanger. An energy balance indicates that approximately 40 percent of the fuel energy is exhausted up the chimney, only 60 percent being useful for heating the house.

From a second-law point of view, however, this situation gives an effectiveness of approximately 0.08. This low value represents the low available energy required to heat a house with only 316 K air. In heating applications the effectiveness is related to efficiency approximately by the relation

$$\varepsilon = \left(1 - \frac{T_0}{T} \right) \frac{Q}{W}$$

(see chapter 15). Q/W is the energy efficiency, T_0 is the environment temperature, and T is the temperature required for the task. For $\eta = 0.6$, then $T_0 = 273$ K and $T = 316$ K; this gives $\varepsilon = 0.082$. If the air being heated is at 294 K (21°C) then the effectiveness is only $\varepsilon = 0.043$.

Figure 13–B1 indicates the approximate variation of the effectiveness

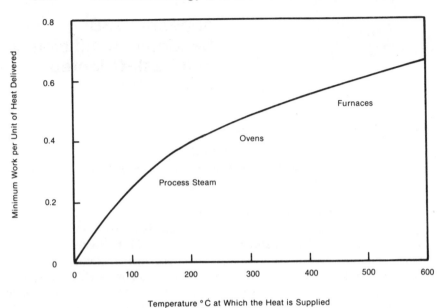

Figure 13B-1. Minimum Thermodynamic Work per Unit of Heat Energy
Deliverable with an Ideal Heat Pump

with the temperature that the application requires. This is crucial to the
evaluation of the effectiveness of heating applications. If the temperature
required for an application such as a chemical process is reduced by the use
of an improved catalyst, the effectiveness will be decreased. If the environmental temperature increases, the effectiveness will decrease. For example,
the same furnace operating in Florida, where $T_0 = 290$ K, would have a
lower effectiveness than one operating in Minnesota, where $T_0 = 275$ K. The
effectiveness is certainly a task, not a device, measure in this example. In the
furnace application the principal available-energy loss occurs in the heat
exchanger; yet it cannot be decreased by an improved heat-exchanger design.
It is a task-*limited* measure.

Now let us hypothesize a way in which the minimum available energy for
heating could be applied. In thermodynamic science the concept of an ideal
heat pump is introduced. This ideal heat pump has a relationship between
heat supplied Q to a constant temperature sink at T from a reservoir at T_0 to
the minimum work input W_{min}

$$\frac{Q}{W_{min}} = \frac{1}{1 - T_0/T}$$

(see chapter 15). This is an ideal heat pump in the same sense in which a

Carnot engine is an ideal heat engine. It represents an ideal thermodynamic device that is unattainable in practice.

In theory, the fuel available energy could be applied through the vehicle of an ideal heat pump to supply heat to a system. However, the closest one can come to such a system is probably a fuel cell coupled to a vapor-compression heat-pump system. A system that involves the conversion of the fuel available energy to thermal energy will usually involve such a large available-energy loss that the use of the compression heat pump and engine does not provide more heat than would a direct-burning heat-exchange device. This should be emphasized, since the alternative to house heating that would involve a heat pump as opposed to direct combustion heating must be connected to a conversion device that will produce work from the available energy of fuel reversibly. A heat-pump technology for heating purposes utilizing central-station electrical energy usually does not prove more effective than direct-combustion heating. The principle advantage is relative to electrical heating, which is clearly indicated by the multiplier factor

$$Q = \left(\frac{1}{1 - T_0/T} \right) W.$$

Technological improvements are possible for both direct-heating and heat-pump systems.[1] A critical technical need is the design of improved heat-exchanger systems. Most heat pump and heating systems currently have the heat exchangers designed to deliver maximum energy at minimum cost. In some cases the system is designed for lowest initial cost rather than lowest total cost. It is this class of economic-energy-sales-design situation that must be clarified in terms of consumer and investor behavior to improve efficiency of energy use. Any competent engineer can design a heat pump system with a coefficient of performance (COP) of 4 by increasing the heat-exchanger size of present systems. Although the total cost of such a system distributed over ten years would be attractive, the initial cost of such a system does not make it competitive with a system with a COP of 2. Note that the effectiveness of any heat-pump heating system requires specification of the temperature T_0 of the environment and the temperature T desired. Performance for particular devices must include this information. The effectiveness of a heat pump used at an outside temperature of 10°C in Texas will be much lower at 0°C in Maine.

A particularly promising future heating system is the combination of a heat pump and a solar collector. In such a system the solar collector can act as a source of direct heating or indirectly as the evaporator for a heat-pump system. Alternatively, in a solar air-conditioning system the solar collector can act as the heat source in a gas-absorption system. General heat-pump theory is discussed more extensively in chapter 15.

The concept of task-oriented effectiveness is an important one from the point of view of comparing alternative energy systems. Care must be taken to avoid wandering from the available-energy basis, particularly in dealing with biological systems and solar-energy systems. A particular example of the problems involved in taking the *task* idea too literally in these cases is outlined by Commoner [2]. Commoner compares the energy required to produce a pound of fertilizer commercially (19,700 But/lb) with a "task minimum" of 2,700 Btu/lb that could be accomplished by planting a legume cover crop. The basic fallacy is that the cover crop has a solar-energy input. On this basis, any task that could be done with solar energy alone would have a thermodynamic efficiency of zero if done any other way. A house in Texas would automatically have a lower effectiveness than one in Alberta, since there is a greater sunlight subsidy.

The basic argument, of course, revolves around whether renewable energy sources are to be included as energy inputs in determining effectiveness. If they are not included, then one cannot talk about the effectiveness of a wind machine, a hydroelectric-or tidal-power system, a solar heating system, or a bioconversion system. But it is precisely these systems that will be more important in future energy applications. If these systems are excluded, then a separate "device" effectiveness is required. The thermodynamic minimum should be established for the task in the effectiveness measure. In this way an account can be kept of the "benefit" of environmental energy inputs. This will also be important if such factors as weather modification and geological alteration are practiced more extensively in the future.

Nuclear energy requires a special comment about its available-energy measure. It is currently assumed that the available energy of nuclear energy is the available energy of the thermal energy generated in a nuclear electric-power station. This underestimates the available energy by many times. The assumption is based on our present technology, which is material and device limited. Other possibilities exist for utilizing the high-energy particles produced in a nuclear reactor. If the real available energy of these particles is included, the low effectiveness of nuclear reactors will be apparent. This is a case where an intermediate technology is wasteful of available energy. A long-range technical plan for uranium utilization would be directed at reducing inefficient fission-reactor burning and emphasize the conservation of fuel until a system is devised to utilize the high available energy of a fission nuclear reaction. The Breeder reactor is a step in this direction since it utilizes the higher-energy particles for production of high-available-energy fuel (plutonium). Decisions to develop the breeder reactor before further depleting uranium supplies inefficiently are supported by thermodynamic analysis.

Note

1. Note that a hydrogen power cycle utilizing a hydride absorption system to reduce compressor work-input requirements, and using low-temperature solar or geothermal energy, has been proposed by scientists at Brookhaven National Laboratory [3].

References

1. Ford, K.W.; Rochlin, G.I.; and Socolow, R.H. (eds.) *Efficient Use of Energy: APS Study on Technical Aspects of the More Efficient Use of Energy*. New York: American Institute of Physics, 1975.
2. Commoner, B. *The Poverty of Power*. New York: Knopf, 1976.
3. Powell, J.R., et al. "A High Efficiency Power Cycle in Which Hydrogen is Compressed by Absorption in Metal Hydrides." *Science* 193(1976):314.

14 Available Energy of Solar Radiation

We conclude—and future scientists will have to shed further light on the subject—that the heat radiating in space must necessarily be able to convert itself into some other form of motion: and that in that new form it can immediately be reconcentrated and reactivated. In this way we are free of the one main obstacle to dead suns being reconverted into incandescent nebulae.

—F. Engels, *Dialectics of Nature* (1892)

Introduction

Renewable energy resources are all basically solar-energy derived. The electromagnetic radiation from the sun is converted by the earth's atmosphere and biosphere to other energy forms. In this conversion to other energy forms, the available energy is changed. This chapter evaluates the available energy of radiation. The dependence of the available energy on the spectral and spatial distributions is described.

In the determination of second-law efficiencies of energy-conversion devices, the available energy of the input is required. If the input is a solar-energy or other radiation flux, the available energy can be calculated from the results of this chapter.

The general impression one gets from the evaluation of solar-energy conversion systems is that solar energy is of a lower quality than conventional fuels. This is not true. It will be shown that solar energy is one of the highest-quality energy sources available.

Emphasis will be placed on the available energy of radiation. This will be followed by a discussion of the limitations of currently utilized systems for converting this energy to other forms. These systems include thermal, photochemical, and photovoltaic converters.

The nature of radiation and the relationship of electromagnetic theory and thermodynamics was formulated by Planck in his classic treatise *The Theory of Heat Radiation* [1]. This chapter extends Planck's work to the consideration of solar energy. We will begin by outlining and defining radiation processes in geometrical and thermodynamic terms. The effects of spectral and spatial distributions on the available energy of radiation will then be developed. This will include specification of temperature, entropy, and available energy.

337

Black-Body Radiation

Black-body radiation has been examined thermodynamically by many researchers [1–5]. They begin with the assumption of quantized photon energies $\varepsilon_i = h\nu_i$. h is Planck's constant, and ν_i is the frequency of the photon's electromagnetic wave. The energy and number of photons in a given volume are then calculated at equilibrium by statistical thermodynamic techniques. This volume is adiabatically isolated from the environment, and the radiation is uniformly distributed in the space. The expected number of photons in this equilibrium volume V at a temperature T can then be shown to be

$$\langle n \rangle = 0.488\left(\frac{2\pi kT}{hc}\right)^3 V, \tag{14.1}$$

where k is Boltzmann's constant.

   c is the velocity of light.

The expected energy (or internal energy) is

$$\langle E \rangle = 0.658\left(\frac{2\pi kT}{hc}\right)^3 kTV. \tag{14.2}$$

The thermodynamic pressure is

$$P = \frac{\langle E \rangle}{3}. \tag{14.3}$$

From these the following thermodynamic properties are then found.

$$\text{Entropy: } \bar{S} = \frac{4\langle E \rangle}{3T}. \tag{14.4}$$

$$\text{Pressure-volume: } PV = \frac{\langle E \rangle}{3}. \tag{14.5}$$

$$\text{Helmholtz free energy: } F \equiv \langle E \rangle - T\bar{S} = -\frac{\langle E \rangle}{3}. \tag{14.6}$$

$$\text{Gibbs free energy: } G \equiv H + T\bar{S} = \langle E \rangle + PV - T\bar{S}. \tag{14.7}$$

Available energy: $B \equiv H - T_0 \bar{S} = \dfrac{4}{3}\left(1 - \dfrac{T_0}{T}\right) \langle E \rangle$

$$= \left(1 - \frac{T_0}{T}\right) H. \qquad (14.8)$$

Note that T_0 is the datum-state temperature, and H is the enthalpy. The Gibbs free energy is proposed by some as a measure of the quality of energy. Thermal radiation in this context has no value for conversion to work forms. We follow the more acceptable procedure of using available energy as the measure of the value of energy. In this case, radiation, as shown in equation 14.8, has the expected available-energy–enthalpy Carnot factor relationship.

Thermodynamic quantities require an interpretation in terms other than expected energy because radiation processes are usually described in terms of energy fluxes through or on surfaces. The connection between energy density and energy flux in an equilibrium volume can be determined simply. If each particle has an average energy $\bar{e}$, then the energy impinging on a surface per unit area is:

$$e = \frac{c \langle n \rangle \bar{e}}{4V} = \frac{c \langle E \rangle}{4V}. \qquad (14.9)$$

The radiant flux per unit area can then be written as

$$e = \frac{2\pi^5(kT)}{15h^3c^2} = \sigma T^4. \qquad (14.10)$$

where σ is the Stefan-Boltzmann constant, $(5.67 \times 10^{-12}\ \text{W/cm}^2\text{K}^4)$.

The standard relationships between the radiation parameters of intensity, energy density, and energy flux per unit area can be derived from geometrical relations [1]. They may be summarized for axisymmetric geometries as

$$\text{Energy density, } u = \langle E \rangle / V = \frac{2e}{c(1 - \Omega/4\pi)} \qquad (14.11)$$

$$\text{or } u = \frac{I\Omega}{c}. \qquad (14.12)$$

$$\text{Energy flux, } e = I\Omega(1 - \Omega/4\pi). \qquad (14.13)$$

where I is the intensity or energy flux per unit area per unit solid angle, and Ω is the solid angle of the radiation cone.

Equations 14.10 and 14.12 can be combined to relate energy density to temperature for any black-body radiation in a solid angle Ω as

$$u = \frac{\sigma \Omega T^4}{c\pi}. \tag{14.14}$$

This is sufficient for cases where the angular distribution of the radiation is known, as well as the temperature of the radiation. If the processing of the radiation is done reversibly (as on perfect reflection), the temperature remains constant. In observable terms, the color of the radiation is determined by the frequency distribution that is specified by the temperature.

The energy-flux relationship to the angular distribution and temperature can be found using equation 14.13 as

$$e = I\Omega(1 - \Omega/4\pi) = \frac{\sigma T^4}{\pi} \Omega(1 - \Omega/4\pi). \tag{14.15}$$

For this special case of uniform radiation in all directions, the energy flux on a surface is found by taking $\Omega = 2\pi$, to give

$$e = \sigma T^4.$$

The net flux across a transparent surface is, of course, zero for $\Omega = 4\pi$. In these expressions e is the net flux per unit area in one direction across an area.

Inverting equation 14.15 to solve for T^4 gives

$$T^4 = \frac{\pi e}{\sigma \Omega(1 - \Omega/4\pi)},$$

or, from equation 14.12,

$$T^4 = \pi I/\sigma. \tag{14.16}$$

Note that in optical problems, a constant temperature of radiation is indicated if the intensity remains the same in the process.

The temperature may be determined by measurement of the energy flux per unit area in the solid angle of the radiation. The "temperature" of solar radiation may then be calculated from measurement of the radiation flux

$e = 1{,}353$ W/m^2 and $\Omega = 6.78 \times 10^{-5}$, the solid angle of direct radiation from the solar disk at the mean earth distance.

$$T^4 = \frac{\pi(1353)}{6.78 \times 10^{-5}\, \sigma\left(1 - \dfrac{6.78 \times 10^{-5}}{4\pi}\right)}.$$

$$T \simeq 5{,}900 \text{ K}$$

This temperature closely represents the energy distribution of radiation from the sun, as shown in figure 14–1. This temperature is the temperature that would be observed if the sun emitted radiation as a black body. Another way to consider it is as the temperature of radiation if the space were filled with radiation moving in all directions with the same intensity as sunlight.

Another interesting "temperature" is the temperature that would be obtained if this radiation is absorbed in a black cavity. The cavity will come to equilibrium with the flux leaving equal to the flux entering. The difference will be that the flux leaving will be distributed over the angle $\Omega = 2\pi$. Equations 14.13 and 14.14 give

$$e = \frac{\sigma T^4}{\pi}\, \Omega\left(1 - \frac{\Omega}{4\pi}\right).$$

Solving for T with $e = 1{,}353$ W/m^2 gives $T = 393$ K. This is a significant temperature, as it represents the maximum temperature attainable by a black surface outside the earth's atmosphere at the earth-sun distance.

If the earth and its atmosphere acted as a black body, the flux leaving would be distributed over a solid sphere with $\Omega = 4\pi$. To calculate the temperature of the earth in an equilibrium situation, assume the energy flux from the sun is uniformly distributed over the hemisphere toward the sun. The radiation from the earth would be uniformly distributed over the full sphere. Then the equilibrium temperature of the earth, for $\Omega = 2\pi$ and $e = 1.353/2$, would be $T_e = 330$ K. This temperature is significantly higher that an average earth temperature of approximately 300 K. The difference indicates that the earth does not behave as a black body. The atmosphere reflects solar radiation and selectively absorbs and reemits radiation.

The next thermodynamic variable of concern is the entropy. Equation 14.4 indicates that the entropy of the radiation is determined by the energy

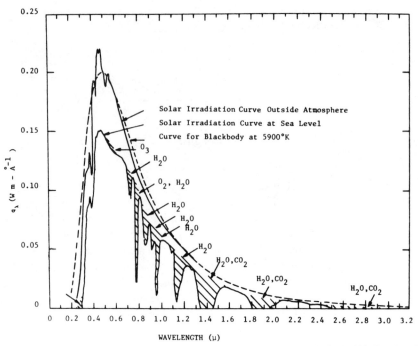

WAVELENGTH (µ)

Source: R.H. Edgerton, "Second Law and Radiation," *Energy, Int. J.* 5(1980):693. Reprinted with permission.

Figure 14–1. Representative Spectral Distributions of Solar Energy at the Earth

and the temperature. The entropy per unit volume for uniform radiation at equilibrium is found using equation 14.10 as

$$S = \frac{\bar{S}}{V} = \frac{16\sigma T^3}{3c}.$$ (14.17)

The entropy intensity is then

$$S_I = Sc/\Omega = \frac{4\sigma T^3}{3\pi}.$$ (14.18)

The entropy flux S_e follows from equation 14.15 as

$$S_e = \frac{4\sigma T^3}{3\pi} \Omega(1 - \Omega/4\pi).$$ (14.19)

In uniform flux in all directions, the entropy flux incident on one side of a unit area is

$$S_e = \frac{4\sigma T^3}{3}.$$

The entropy flux may be written in terms of the energy flux by combining this with equation 14.15 to give

$$S_e = \frac{4}{3}(e)^{3/4}\left[\frac{\sigma}{\pi}(1 - \Omega/4\pi)\right]^{1/4}. \qquad (14.20)$$

The entropy flux for solar radiation at the earth-sun distance is then found for $\Omega = 6.78 \times 10^{-5}$ and $e = 1.353$ kW/m^2 as

$$S_e = 0.313 \text{ W/m}^2 \text{ K}.$$

The entropy flux reradiated from a black surface for this incident solar-energy flux is

$$S_{eb} = \frac{4}{3} e^{3/4} \sigma^{1/4} = 4.59 \text{ W/m}^2 \text{ K}.$$

The rate of entropy generation at the black surface is the difference between the entropy leaving and that incident, or

$$\dot{\sigma} = S_{eb} - S_e = 4.28 \text{ W/m}^2 \text{ K}.$$

The available-energy flux can similarly be found as

$$b = \frac{4}{3}\left(1 - \frac{T_0}{T}\right)\frac{\sigma T^4}{\pi}(1 - \Omega/4\pi). \qquad (14.21)$$

The Carnot factor $(1 - T_0/T)$ is applicable, and the factor 4/3 accounts for the radiation pressure, which is also available for work processes. This is analogous in steady flow systems to applying the Carnot factor to the enthalpy, not the internal energy. For radiation problems in classical heat-transfer theory, the momentum flux is assumed to be zero, thus neglecting the radiation-pressure effects. In fact, this energy is assumed not available because of the technological difficulty of utilizing it. Boltzmann [6] originally hypothesized a radiation-expansion engine in his consideration of the thermodynamics of black-body radiation. Utilizing this effect in a through-flow machine that includes the flow of gases would not be feasible because of the low radiation pressure relative to practical gas pressures. In space applications, where the density of particles is extremely low, the expansion-energy utilization might prove useful. Ashkin [18] has proposed that solar radiation might be utilized to drive a particle accelerator in a fusion device.

The available-energy flux can be found in terms of the energy flux by substitution for the temperature from equation 14.20 to give

$$ b = \frac{4}{3} e \left\{ 1 - \frac{T_0}{\left[\dfrac{\pi e}{\sigma \Omega (1 - \Omega / 4\pi)} \right]^{1/4}} \right\} . \quad (14.22) $$

The rate of available energy loss when sunlight is absorbed or diffusely scattered at a surface can then be calculated if one knows the initial solid angle of the light. The available-energy loss when direct solar energy is diffusely scattered or absorbed by black surface is

$$ b_{\text{loss}} = b_{\Omega = 6.78 \times 10^{-5}} - b_{\Omega = 2\pi} . $$

The ratio of the available-energy ratio at the surface to that incident for $e = 1.353 \text{ kW/m}^2$ is

$$ b_{\Omega = 2\pi} / b_{\Omega = 6.78 \times 10^{-5}} \simeq 0.24 . $$

Energy-conversion systems that rely on black absorber surfaces have a maximum second-law effectiveness of approximately 24 percent. The available energy is low because the solid-angle available energy of the incident radiation has not been utilized. In the process of distributing the radiation over the angle of 2π, the temperature of the radiation has been decreased. A focusing system used to concentrate the radiation utilizes the small angle of the incident radiation to advantage. If the focusing system is ideal, the energy flux per unit area can be increased in the ratio of $e_{\text{receiver}} / e_{\text{solar}} \simeq 4.6 \times 10^4$. This will give a receiver temperature equal to the sun temperature, and the available energy will be that of solar radiation [17].

Equation 14.15 expresses a relationship between a measured radiation flux and the temperature of equilibrium radiation that is basic to an understanding of radiation in space. Suppose we measure the energy flux per unit area on an area pointed at the sun. The temperature of this radiation, which represents the distribution of energy over the energy states or wavelengths, is fixed by the radiating object. From this information we can calculate the solid angle the sun subtends and hence the distance r from the sun that we are in space. If we know the radius of the sun, this distance r can be found from the solid-angle relation $\Omega = (r_s^2)/r^2$. As the radiation propagates in free space, the distribution of energy over the energy states, temperature, does not change. The energy flux per unit area decreases depending on the solid-angle relationship of equation 14.15. The energy flux per unit area at the earth surface may be small, but the temperature will remain that at the sun surface. The available energy flux b per unit energy flux e in the propagation through free space has not changed, as equation 14.22 shows. The propagation in free space is reversible, and the energy distribution over the energy states (temperature) is constant.

Nonequilibrium Spectral Distributions of Radiation

Solar radiation in the atmosphere, as shown in figure 14–1, is not spectrally distributed in the black-body solar-energy spectrum of 5,900 K. The atmosphere alters the distribution through absorption and scattering processes. The major absorption is by the gas constituents of the atmosphere; the scattering is by these gases as well as dust particles. These two processes reduce not only the energy flux, but also the available energy of the solar radiation. The absorption processes principally change the spectral distribution and the scattering processes the spatial distribution of the radiation. To examine these effects, we will consider spectral changes in this section and, the next section, the spatial influences.

Consider the radiant energy between frequency of v and $v + dv$ as

$$d\langle E \rangle = \frac{\partial \langle E \rangle}{\partial v} \, dv \qquad (14.23)$$

Now the expected energy per unit frequency can be written in statistical mechanics [1] as

$$\frac{\partial \langle E \rangle}{\partial v} = \frac{8\pi V}{c^3} \, hv^3 \, \frac{e^{-\beta hv}}{1 - e^{-\beta hv}} \qquad (14.24)$$

Note that $\beta = 1/kT$.

In the equilibrium situation the distribution of the energy over the spectrum is determined by the temperature. If the radiation is in a state that is not in equilibrium, the specification of the energy distribution over the spectrum requires measurement of the energy at each wavelength or frequency. Equation 14.24 indicates that measurement of the energy at a given frequency $u_\nu \equiv \partial E/,N$ can be used to specify a β or temperature for that specific wavelength. Inverting this expression to solve for β gives

$$\beta = \frac{1}{h\nu} \ln\left(1 + \frac{8\pi V h \nu^3}{c^3 u_\nu} \right) . \qquad (14.25)$$

The energy of a photon with frequency ν is $h\nu$; therefore, the number N_ν of photons at a given frequency is given by

$$N_\nu = \frac{u_\nu}{h\nu} .$$

In flux terms, the flux of photons across a surface is

$$n_\nu = \frac{e_\nu}{h\nu} .$$

From equation 14.24, the number of photons per unit volume at a given frequency ν is

$$N_\nu = \frac{8\pi}{c^3} \nu^2 \frac{e^{-\beta h\nu}}{1 - e^{-\beta h\nu}} . \qquad (14.26)$$

The number of photons at a given energy level is specified by the temperature. In terms of the discussion in chapter 4, the energy in the volume can be changed by altering the allowed energy states $h\nu$ or the number of particles with this energy N_ν. Changing the number of particles of each frequency changes the entropy. From the expression for N_ν, the temperature determines this distribution among the energy states.

If the radiation is in thermodynamic equilibrium, then the distribution is given by N_ν, with β constant. If the radiation is not in equilibrium, the number of photons at each energy $h\nu$ will be specified by a different β. Equilibrium is then specified as temperature equilibrium.

Note, however, that energy exchange between photons does not occur as it would in a gas. If the photons were in a perfectly reflecting container the nonequilibrium distribution would be maintained indefinitely. If the walls of the container are not perfectly reflecting, then interaction of the photons with the energy states of the container will bring about an equilibrium distribution. In this process available energy would be lost, as the equilibrium distribution has a lower available energy.

Let us now find the entropy and available-energy functions in terms of the energy flux e_ν is uniform in all directions, then

$$e_\nu = \frac{u_\nu c}{4V} . \tag{14.27}$$

β can then be written from equation 14.25 in terms of measurable e_ν as

$$\beta = \frac{1}{h\nu} \left[\ln \left(1 + \frac{e_\nu c^2}{2\pi h\nu^3} \right) - \ln \left(\frac{e_\nu c^2}{2\pi h\nu^3} \right) \right] \tag{14.28}$$

The following relationships then follow, in which the temperature or β do not appear: (1) the pV function:

$$\frac{\partial (pV)}{\partial \nu} = \frac{8\pi V h \nu^3}{c^3} \left\{ \frac{\ln \left(1 + \dfrac{e_\nu c^2}{2\pi h\nu^3} \right)}{\ln \left(1 + \dfrac{2\pi h\nu^3}{e_\nu c^2} \right)} \right\} ; \tag{14.29}$$

(2) the entropy function:

$$S_\nu = \frac{k 8\pi V \nu^2}{c^3} \left[\ln \left(1 + \frac{e_\nu c^2}{2\pi h\nu^3} \right) \right] + \frac{k 4V}{h\nu c} e_\nu \left[\ln \left(1 + \frac{e_\nu c^2}{2\pi h\nu^3} \right) \right.$$

$$\left. - \ln \left(\frac{e_\nu c^2}{2\pi h\nu^3} \right) \right] . \tag{14.30}$$

The available-energy function is then

$$A_\nu = \frac{\partial A}{\partial \nu} = \frac{4 V e_\nu}{c} + \frac{8\pi V h \nu^3}{c^3} \times \frac{\ln \left(1 + \dfrac{e_\nu c^2}{2\pi h\nu^3} \right)}{\ln \left(1 + \dfrac{2\pi h\nu^3}{e_\nu c^2} \right)}$$

$$- T_0 k \left[\frac{8\pi V \nu^2}{c^3} \ln \left(1 + \frac{e_\nu c^2}{2\pi h\nu^3} \right) \right.$$

$$\left. + \frac{4 V e_\nu}{ch\nu} \ln \left(1 + \frac{2\pi h\nu^3}{e_\nu c^2} \right) \right] . \tag{14.31}$$

The available-energy flux $b_\nu = (\partial A/\partial \nu)(c/4V)$ is then

$$b_\nu = e_\nu + \frac{2\pi h \nu^3}{c^2} \cdot \frac{\ln\left(1 + \dfrac{e_\nu c^2}{2\pi h \nu^3}\right)}{\ln\left(1 + \dfrac{2\pi h \nu^3}{e_\nu c^2}\right)}$$

$$- T_0 k \left[\frac{2\pi \nu^2}{c^2} \ln\left(1 + \frac{e_\nu c^2}{2\pi h \nu^3}\right) \right.$$

$$\left. + \frac{e_\nu}{h\nu} \ln\left(1 + \frac{2\pi h \nu^3}{e_\nu c^2}\right) \right]. \qquad (14.32)$$

The available-energy in terms of wavelength can also be obtained by the derivative relation

$$\frac{\partial}{\partial \lambda} = - \frac{c}{\lambda^2} \frac{\partial}{\partial \nu} .$$

Then $e_\lambda = \partial e/\partial \lambda$, and the available-energy flux per unit area for uniform radiation becomes

$$b_\lambda = e_\lambda - \frac{2\pi h c^2}{\lambda^5} \cdot \frac{\ln\left(1 - \dfrac{e_\lambda \lambda^5}{2\pi h c^2}\right)}{\ln\left(1 - \dfrac{2\pi h c^2}{e_\lambda \lambda^5}\right)}$$

$$+ kT_0 \left[\frac{2\pi c}{\lambda^4} \ln\left(1 - \frac{e_\lambda \lambda^5}{2\pi h c^2}\right) - \frac{e_\lambda \lambda}{hc} \ln\left(1 - \frac{2\pi h c^2}{e_\lambda \lambda^5}\right) \right].$$

$$(14.33)$$

The total available-energy flux per unit area in any direction will then be

$$b = \int_0^\infty b_\lambda \, d\lambda.$$

Solar energy-conversion devices are designed to utilize the particular spectral distribution of solar radiation. These conversion devices are usually characterized as either thermal or photo devices. *Thermal devices* convert the solar radiation to energy with an equilibrium energy distribution described by a single temperature. This thermal energy is then used to generate electrical or chemical energy. The most efficient devices selectively absorb the short-wavelength radiation to produce high temperatures. Focusing systems are often utilized to improve the efficiency of these converters by increasing the surface temperature. These converters are Carnot-efficiency limited [19].

Photo devices utilize the energy of photons to produce selective electronic transitions in the receiver [10,14]. These are limited by the minimum energy required to provide an electronic transition. Photons with energy less than this minimum energy are not useful. Photons with energy greater than the electron-transition energy only contribute the electron-transition energy. Excess energy is not utilizable directly. Quantum mechanics allows only one electronic transition per photon. These devices have limits that are difficult to specify thermodynamically [7–9].

This chapter outlines the method of determining the thermodynamic available energy of radiation. This gives an upper bound on energy conversion. Any device will have constraints that limit its conversion potential to a lower value.

Figures 14–2 and 14–3 show the effects of atmospheric absorption and scattering on the available energy of solar radiation for the case where the radiation is assumed uniformly distributed over the sky. The air mass m is a measure of the thickness of the atmospheric layer through which the radiation passes. The coefficients α and β are atmospheric absorption or turbidity coefficients. Data on the atmosphere were taken from the work of Thekaekara [11–13] and computations from Edgerton [17].

The results show that the turbidity has significantly more effect at high air mass than at low air mass. This would imply that turbid atmospheres will affect photoconversion devices much less when the sun is near the zenith than when it is at large zenith angles, as in the early morning or late afternoon.

The available-energy versus cutoff-wavelength plot (figure 14–2) shows an interesting characteristic. The available energy reaches a maximum between 2.5 and 3.5μ. This indicates the trade-off between narrow bandwidth and total energy in the determination of the temperature of the radiation. Beyond the peak wavelength, low-energy photons reduce the available energy of the high-energy photons and thus reduce the energy convertible to work. This would indicate the importance of a cutoff filter in this range ($2.5–3.5\mu$) for photoconversion processes. Photons beyond this range are detrimental in that they provide heating with little quality energy. In thermal devices a cutoff filter can be used to increase effectiveness. This effect is very small, however, and may not be of practical significance.

The ratio of available energy to total energy flux is shown across the spectrum in figure 14–3. For the spectral distributions examined, the energy ratio ranges from 0.5 to 0.6 over the full spectrum. This is considerably

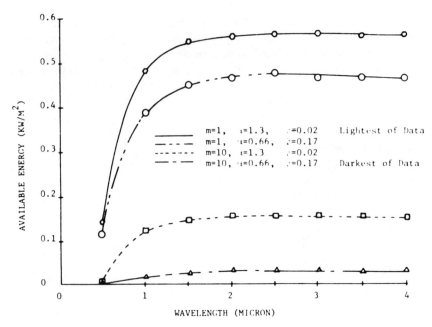

Source: R.H. Edgerton, "Second Law and Radiation," *Energy, Int. J.* 5(1980): 703.

Figure 14–2. Available-Energy Variation with Cutoff Wavelength for Sunlight through Different Atmospheres

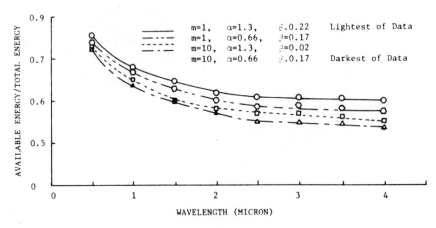

Source: R.H. Edgerton, "Second Law and Radiation," *Energy, Int. J.* 5(1980): 704. Reprinted with permission.

Figure 14–3. Ratio of Available Energy to Total Energy at the Earth's Surface as a Function of Cutoff Wavelength for Different Atmospheres

higher than the ratio of available energy to total energy for radiation at equilibrium, as shown in table 14–1. This difference is the most important consideration in the development of better solar energy-conversion systems. The use of spectrally selective surfaces for thermal converters takes advantage of this difference to increase the temperature of a collector. In the future, optimization of solar thermal converters should be based on the variation of the selective cutoff wavelength with atmospheric conditions.

**Combined Spatial and Spectral Contributions
to Radiation Availability**

We next examine the available energy of a radiation flux that is distributed both spatially and spectrally. The determination will be made in terms of spherical coordinates, as shown in figure 14–4. The availability calculation will assume that an intensity or energy flux per unit area per solid angle $e_{\omega\nu} \equiv \partial^2 e/\partial\omega\partial\nu$ across a surface can be found experimentally, or theoretically from radiation absorption and scattering models. The energy flux per unit area across a surface can the be found as

$$ e = \int \int \int e_{\omega\nu} \sin\phi \, d\phi \, d\theta \, d\nu. \qquad (14.34) $$

The specific energy flux $e_{\omega\nu} = I_A \cos\phi$, where I_A is the usual intensity and ϕ is the angle from the normal to the surface. The angle ϕ will be taken as the angle from the normal to a surface that is in the zenith direction in this discussion. A measurement by a spectral pyrheliometer pointed in the direction (r, ϕ, θ) will indicate a value $e_{\Omega p\nu}$ at a given frequency with an aperture of solid angle Ω. The intensity is then

$$ I_A = e_{\Omega_{p\nu}}/\Omega. $$

The energy flux per unit area across the surface A is then

$$ e = \int \int \int e_{\Omega_{p\nu}} \cos\phi \, \sin\phi \, d\phi \, d\theta \, d\nu. \qquad (14.35) $$

The energy-density relationship as a function of frequency ν and solid angle ω is

$$ u_\nu \equiv \frac{\partial u}{\partial \nu} = \frac{\omega}{4\pi} \frac{8\pi h \nu^3}{c^3} \frac{e^{-\beta h \nu}}{1 - e^{-\beta h \nu}}. \qquad (14.36) $$

Table 14–1
Available Energy of Solar Spectral Energy with Different Atmospheres and Loss in Available Energy When Converted to Thermal Energy

Air Mass	α	β	Available Energy Flux kW/m^2	Total Energy Flux kW/m^2	Available Energy Flux if Brought to Thermal Equilibrium	Thermal Equilibrium Temperature	Available Energy Lost in Conversion to Thermal Energy kW/m^2
1	1.3	0.02	0.56	0.94	0.23	396	0.33
1	1.3	0.04	0.54	0.91	0.22	394	0.32
1	0.66	0.085	0.52	0.88	0.21	392	0.31
1	0.66	0.17	0.46	0.79	0.17	385	0.29
4	1.3	0.02	0.34	0.58	0.10	368	0.24
4	1.3	0.04	0.29	0.52	0.09	362	0.20
4	0.66	0.085	0.24	0.44	0.07	355	0.17
4	0.66	0.17	0.16	0.30	0.04	340	0.12
7	1.3	0.02	0.22	0.40	0.06	350	0.16
7	1.3	0.04	0.18	0.33	0.04	343	0.14
7	0.66	0.085	0.13	0.25	0.03	334	0.10
7	0.66	0.17	0.06	0.13	0.01	319	0.05
10	1.3	0.02	0.16	0.29	0.03	339	0.13
10	1.3	0.04	0.12	0.23	0.02	332	0.10
10	0.66	0.085	0.08	0.15	0.01	322	0.065
10	0.66	0.17	0.03	0.06	0.002	309	0.025

Source: R.H. Edgerton, "Second Law and Radiation," *Energy, Int. J.* 5(1980): 693. Reprinted with permission.

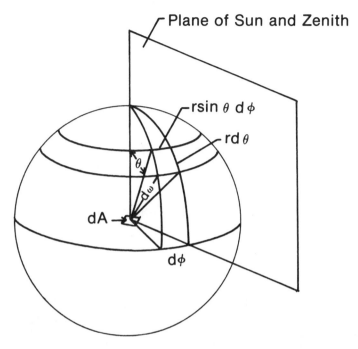

Plane of Sun and Zenith

$r\sin\theta\ d\phi$

$rd\theta$

dA

$d\phi$

Note: θ is the zenith angle and ϕ is the azimuth angle from the plane of the sun and the zenith.
Figure 14-4. Spatial Geometry of Solar-Radiation Descriptions in Spherical Coordinates

The energy density per solid angle per unit frequency is then

$$u_{\omega v} \equiv \frac{\partial u_v}{\partial \omega} = \frac{2hv^3}{c^3}\ \frac{e^{-\beta hv}}{1 - e^{-\beta hv}}. \qquad (14.37)$$

Then, using the flux to density relation,

$$e_{wv} = cu_{wv}\cos\phi = \frac{2hv^3}{c^2}\left(\frac{e^{-\beta hv}}{1 - e^{-\beta hv}}\right)\cos\phi. \qquad (14.38)$$

Using the measured variable obtained with a spectral pyrheliometer,

$$e_{\Omega_{pv}} = \frac{e_{wv}\ \Omega}{\cos\phi} = I_A\Omega = 2\ \frac{\Omega hv^3}{c^2}\left(\frac{e^{-\beta hv}}{1 - e^{-\beta hv}}\right). \qquad (14.39)$$

Note that

$$e_{\Omega_{pv}} \equiv \frac{\partial e_{\Omega p}}{\partial v} \qquad \text{and} \qquad e_{wv} \equiv \frac{e_{\Omega pv}}{\Omega} \cos \phi.$$

Then β or $1/kT$ can be evaluated from inversion of this relationship:

$$\beta = \frac{1}{hv} \ln\left(1 + \frac{2hv^3}{c^2 I_A} \right) \qquad \text{or} \qquad T = \frac{hv}{k \ln\left(1 + \frac{2hv^3\Omega}{c^2 e_{\Omega_{pv}}} \right)}.$$

The entropy flux per unit frequency per unit solid angle is then found as

$$S_{wv} = \frac{2hkv^2}{c^2} \cos \phi \ln\left(1 + \frac{e_{\Omega_{pv}}c^2}{2hv^3\Omega} \right).$$

$$+ \frac{k}{hv\pi} \cos \phi \left[\ln\left(1 + \frac{e_{\Omega_{pv}}c^2}{2hv^3\Omega} \right) - \ln\left(\frac{c^2 e_{\Omega_{pv}}}{2hv^3\Omega} \right) \right].$$

$$(14.40)$$

The total entropy flux through the area is then

$$S = \int \int \int S_{wv} \sin \phi \, d\phi \, d\theta \, dv.$$

The available-energy flux $b_{v,w}$ can then be found as

$$b_{wv} = e_{wv} + \frac{2hv^3}{c^2} \frac{\ln\left(1 + \frac{e_{\Omega_{pv}}c^2}{2hv^3\Omega} \right)}{\ln\left(1 + \frac{2hv^3\Omega}{e_{\Omega_{pv}}c^2} \right)} \cos \phi$$

$$- T_0 k \left[\frac{2hv^3}{c^2} \cos \phi \left[\ln\left(1 + \frac{e_{\Omega_{pv}}c^2}{2hv^3\Omega} \right) \right] \right.$$

$$\left. + \frac{\cos \phi}{hv\pi} \left\{ \ln\left(1 + \frac{e_{\Omega_{pv}}c^2}{2hv^3\Omega} \right) - \ln\left(\frac{e_{\Omega_{pv}}c^2}{2hv^3\Omega} \right) \right\} \right]. \qquad (14.41)$$

Using $\nu = c/\lambda$, $e_{\Omega p\lambda} = -e_{\Omega p\nu} c/\lambda^2$ and $e_{\omega\lambda} = -e_{\omega\nu} c/\lambda^2$, the available energy in terms of wavelength is

$$
b_{\omega\lambda} = -e_{\omega\lambda} \frac{2hc^2}{\lambda^5} \; \frac{\ln\left(1 - \dfrac{\lambda^5 e_{\Omega p\lambda}}{2hc^2}\right)}{\ln\left(1 - \dfrac{2hc^2}{\lambda^5 e_{\Omega p\lambda}}\right)} \; \cos\phi
$$

$$
+ T_0 k \left[\frac{2hc^2}{\lambda^5} \cos\phi \left[\ln\left(1 - \frac{\lambda^5 e_{\Omega p\lambda}}{2hc^2\Omega}\right) \right] + \frac{\cos\phi}{\pi h\lambda} \right.
$$

$$
\left. \times \left\{ \ln\left(1 - \frac{\lambda^5 e_{\Omega p\lambda}}{2hc^2\Omega}\right) - \ln\left(-\frac{\lambda^5 e_{\Omega p\lambda}}{2hc^2\Omega}\right) \right\} \right]. \qquad (14.42)
$$

The available energy is then

$$
b = \int \int \int b_{\omega\lambda} \sin\phi \, d\phi \, d\theta \, d\lambda.
$$

No computation of the available energy of solar energy from both spatial and spectral measurements has been completed to date. This leaves as uncertain for the present the available-energy input to solar energy-conversion systems. Such computations would assist in the prediction of the performance of solar converters at different geographical locations. This would require consideration of the variable optical characteristics of the atmosphere, including altitude, clouds, and atmospheric constituent gases and particles. Loferski [15] notes that the efficiency of a solar cell can be increased by water-vapor absorption of solar energy. He illustrates a situation in which the electrical output of a silicon solar cell increased by 4 percent with water absorption.

At present, the comparison of solar-heating collectors is based on performance under standard condittions at one location. The estimation of performance at other locations requires more detailed spectral and spatial information than is currently available.

An instructive way to look at the solar radiation is that the total energy is nearly all available energy outside the atmosphere, since it is a stream of particles from a source at about 5,900 K. The atmosphere reduces this availability by dispersion and absorption processes. The solar energy-conversion processes do not utilize all the available energy because of their limited ability to convert photons over the full spectrum or the available radiation space. Conversion to thermal energy at a black surface is the least efficient thermodynamic process, as it does not utilize either the non-equilibrium energy-distribution effect or the directional characteristics. The temperature calculated across the spectrum and the resultant available energy are measures of the usefulness of the radiation for devices that are designed to utilize the whole solar spectrum within the atmosphere. If spectral selective devices or directional collectors are proposed, then the available energy and temperature of that portion of the spectrum can be calculated. This will provide a measure of the potential of spectral selective conversion devices. A solar cell with sensitivity out to a cutoff wavelength can be evaluated by this procedure.

Mallinson and Landsberg [14] have attempted this to a limited extent in trying to optimize photodevices for use in Great Britain, where the solar-energy character is significantly different from that in the bright-sky areas where most photodevice research has been conducted. Available-energy comparison of photodevices based on effectiveness measures would help to clarify the many discussions of different advantages of devices with different spectral sensitivities. Initial attempts have been made to assign available-energy losses to the components of optical photosystems [20].

Another factor that must be included in the discussion of the output of photo conversion devices is the effect of the energy outside the spectral-response wavelengths or frequencies. This is typically dealt with in terms of the heating effects. With spectral sharing systems, the systematic evaluation of this remaining energy becomes more important.

Presently, experiments with photothermovoltaic cells for solar-to-electric conversion are being carried out [16]. A potentially useful optimization of these converters involves the exchange of *spatial* availability, in terms of a small-angle cone of radiation, for *spectral* availability, a high temperature distribution.

References

1. Planck, M. *The Theory of Heat Radiation*, trans. of the 2nd ed. of Waermestrahlung, 1913. New York: Dover, 1957.
2. Bose, S.N. "Ueber das Gasetz der Energieverteilung im Normal-spectrum." *Ann. Physik* 4(1901):553.
3. Einstein, A. "Quantentheirie des Einatomigen Idealen Gases." *Berl. Ber.* 2(1924):261.
4. Petela, R. "Exergy of Heat Radiation." *J. Heat Transfer* 86(1964):187.
5. Landsberg, P.T., and Tonge, G. "Thermodynamics of the Conversion of Dilute Radiation." *J. Phys. A: Math. Gen.* 12(1979):551.
6. Boltzmann, L. "Ableitung der Stephan'schen Gesetzes betreffen der abhängigkeit den Warmestrahl Von der Temperturen aus der electro-

magnetische Lichttheorie." *Wiedmannsche, Annalen der Physik* 22(1884):291.

7. Leontovich, N.A. "Maximum Efficiency of Direct Utilization of Radiation." *Sov. Phys. Usp.* 17(1975):963.

8. Press, W.H. "Theoretical Maximum for Energy from Direct and Diffuse Sunlight." *Nature* 264(1976):734.

9. Thoma, J.U. *Entropy Radiation and Negentropy Accumulation with Photocells, Chemical Reactions, and Plant Growth*, RM-78-14, International Institute of Applied Systems Analysis, Laxenburg, 1978.

10. Schwerzel, R.E. "Methods of Photochemical Utilization of Solar Energy." In *Progress in Astronautics and Aeronautics*, ed. K.W. Billman. vol. 61, New York: AIAA, 1978, p 626.

11. Thekaekara, M.P. "Data on Incident Solar Energy." *Proc. Inst. Envir. Sci.* 2(1974):21.

12. American Society for Testing Materials. "Standard Specification for Solar Constant and Air Mass Zero Solar Spectral Irradiance." ASTM Standard E490-73a, *1974 Annual Book of ASTM*, Philadelphia, 1974.

13. National Aeronautics and Space Administration. *Solar Electromagnetic Radiation*. NASA SP 8005, Washington, D.C., 1971.

14. Mallinson, J.R., and Landsberg, P.T. "Meterological Effects on Solar Cells." *Proc. Roy. Soc. (London)*, ser. A 355(1977):115.

15. Loferski, J.J. "Theoretical Considerations Governing the Choice of the Optimum Semiconductor for Photovoltaic Solar Energy Conversion." *J. Appl. Phys.* 27(1956):777.

16. Swanson, R.M. "A Proposed Thermophotovoltaic Solar Energy Conversion System." *Proc. IEEE* 67(1979):446.

17. Edgerton, R.H. "Second Law and Radiation." *Energy, Int. J.* 5(1980):693.

18. Ashkin, A. "Acceleration and Trapping of Particles by Radiation." *Phys. Rev. Lett.* 24(1970):156.

19. Shaffer, L.H. "Wavelength-Dependent (Selective) Processes for the Utilization of Solar Energy." *J. Solar Energy Sci. Engr.* (July–August 1958):21.

20. Kreider, J.F. "Second Law Analysis of Solar Thermal Processes." *Energy Res.* 3(1979):325.

15 Useful Dissipative Energy Systems

At the conclusion of our survey of the ways in which human intelligence calls art to its aid in counterfeiting nature, we cannot but marvel at the fact that fire is necessary for almost every operation.... Fire is the immeasurable, uncontrollable element, concerning which it is hard to say whether it consumes more or produces more.
— Plinius Secundus (Pliny), *Historia Naturalis* (ca A.D. 50)

Heat as Energy

The useful form of energy is not always work. Many applications of available energy have the purpose of heating. Typical examples are the heating of buildings and of domestic hot water, and the acceleration of chemical-processing reactions. Other applications include air-conditioning and refrigeration processes (cooling).

From a thermodynamic point of view, the lower the temperature required in the heating application, the smaller the available energy required. The loss in available energy of fuels on combustion is a major available-energy loss in many applications. This chemical- to thermal-energy conversion is very inefficient. Once the energy is in thermal form (for example, 1,000°C combustion temperature), another large loss in available energy occurs if this energy is used to heat a building to 25°C.

From a first-law point of view (conservation of energy), the efficiency of a heating device is measured by the amount of energy utilized compared with the total energy supplied. This is commonly the useful heating effect compared with the heating value of the fuel. The principal loss of fuel energy in heating applications is the exhaust-gas ventilation. Typically, 30–40 percent of the fuel energy is wasted in this way. Electrical heating improves on this efficiency because all the electrical energy is converted to useful heat without any exhaust requirements. However, if the fuel-to-electrical conversion loss at a central station is taken into account, the efficiency is generally lower than that of combustion heating.

In a simple application such as the heating of a house, a great deal of effort can be expended in just defining the energy required. Conventionally, the heating requirement is taken as the energy required to maintain the living space at a given temperature while meeting certain ventilation and humidity constraints. The energy efficiency of a house depends on these constraints and also on the insulation of the house from the outside environment. The efficiency can generally be improved in many ways. Two specific means are clear in this context: (1) reducing the energy requirement and (2) increasing the efficiency of the heating system. The engineer must deal with both of these—the first when he defines the problem and the second when he evaluates his solution.

Energy conservation generally confuses these two functions. Energy can be saved by reducing the temperature of the living space, and it can also be saved by increasing the insulation. The thermodynamic efficiency involves the way in which the necessary energy is in fact supplied. This section will examine the latter in the context of heating. What is the most efficient way to meet a given heating requirement? From a first-law point of view, it makes no difference what the source of energy is for heating. Fuel, sunlight, electricity, and so forth have equal energy value when converted to heat.

Let us next ask about the available energy required for a heating process. If a quantity of heat Q is required at a temperature T, the available-energy requirement A is found (from chapter 6):

$$A = Q(1 - T_0/T). \qquad (15.1)$$

Therefore, a heating process that requires a higher temperature T will require greater available energy than will a low-temperature heating process.

If electrical energy is used to provide heat to a process that operates at a temperature T, then the available energy lost is

$$A_{\text{lost}} = A_{\text{initial}} - A_{\text{final}}$$

$$= Q - Q(1 - T_0/T) = Q\left(\frac{T_0}{T}\right). \qquad (15.2)$$

At first this loss in available energy may appear unavoidable and a consequence of using primary energy for heating. Certainly, if low-temperature waste heat is available from some source, this—rather than electrical energy—should be used for heating.

The supply of waste heat usually has two characteristics that limit its use. First, it is often produced at a different place from where it is needed. Second, its temperature is often lower than the temperature needed for heating.

One might assume that the temperature of the waste heat must be higher than the temperature required if the heating application is to be useful, but this is not the case. In many applications the waste heat may be used for preheating. In other cases the waste heat can be used to provide heat to a heat-pump system. The heat-pump alternative provides the mechanisms whereby the quantity of heat delivered can exceed the quantity of high-quality (available) energy used.

Heat Pumps

A heat pump is a device that allows a quantity of high-quality energy to supply more heat at a given temperature than can be supplied by simply converting the energy to heat. A common device for doing this is a refrigerator. A refrigerator absorbs thermal energy at a low temperature and releases it at a higher temperature as a result of the addition of high-quality energy. Two devices to do this are represented schematically in figures 15–1 and 15–2. The first figure shows a vapor-compression heat pump in which mechanical work is applied through a compression to provide heat through a condenser. A simple energy balance will show that the heat provided by the

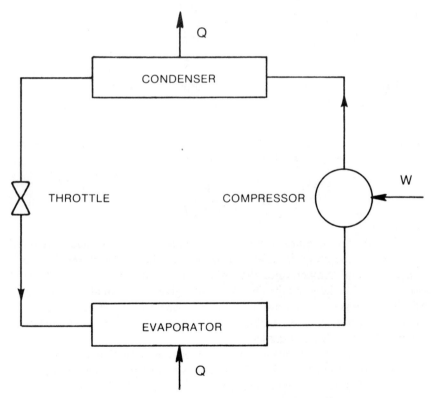

Figure 15–1. Schematic of a Vapor-Compression Heat-Pump–
         Refrigeration System

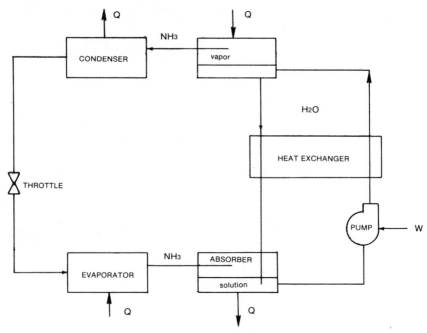

Figure 15–2. Schematic of an Absorption Heat-Pump—Refrigeration System

condenser is equal to the heat picked up in the evaporation plus the work done by the compressor. For an ideal heat pump,

$$Q = \frac{W}{1 - T_0/T} . \qquad (15.3)$$

T_0 is the temperature of the environment from which the heat is removed. T is the temperature at which the heat is applied. The closer T is to the environmental temperature, the greater the quantity of heat that can be supplied by the heat pump. The expression given is for an ideal heat pump in the same sense that a Carnot engine represents an ideal heat engine. The equation represents a reversible heat pump, an ideal thermodynamic device that is unattainable in practice.

Figure 15–2 represents an absorption heat pump with a principal input of (high-available-energy) thermal energy instead of work. The limitations of this type of system will be discussed in the next section. First we will explore the possibilities of heat pumps that have a work input.

The first thing to note from the ideal-performance heat pump is the limitation for most practical applications to low-temperature heating. To see this, consider the relationship

$$\frac{W}{Q} = \left(1 - \frac{T_0}{T}\right).$$

This is plotted in figure 15–3. It represents the minimum work required per unit of heat delivered at a temperature T. If this ideal heat pump were used to replace process steam, approximately 1 workunit would be required to supply 3 units of heat. For industrial-furnace heat of 500°C, approximately 0.6 units of work would be required for each unit of heat delivered. Therefore, the use of heat-pump systems for high-temperature heating applications is very limited. If the input energy to an electric heat pump is supplied by a central power station without exhaust heat use, the heat delivered will be less than that from direct heating if the temperature required is high. For example, assume an environmental temperature of $T_0 = 300$ K and an electrical-generation efficiency of 33 percent. Even under ideal

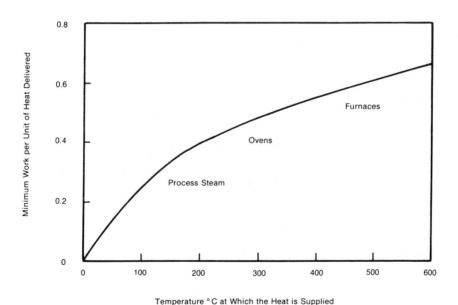

Temperature °C at Which the Heat is Supplied

Note: Reference temperature 0°C.

Figure 15–3. Minimum Thermodynamic Work Required per Unit of Heat Energy Delivered by an Ideal Heat Pump

conditions, more heat can be supplied by direct heating if the temperature required is greater than about 460 K.

With an ideal heat pump, the heat supplied is

$$Q_s = \frac{1}{1 - \dfrac{T_0}{T}} W.$$

Note that the ratio Q_s/W is defined as the *coefficient of performance for a heat pump*.

If the efficiency of the conversion of fuel available energy A_f to work is $\eta = W/A_f$, then the overall effectiveness may be expressed as

$$E = Q / \left[\left(1 - \frac{T_0}{T} \right) / \eta \, A_f \right].$$

If the electrical energy is generated with an efficiency of 0.33, then $W = 0.33 A_f$. Then for an ideal system $E = 1$ and

$$Q_s = \left(\frac{1}{1 - \dfrac{T_0}{T}} \right) 0.33 \, A_f.$$

With $T_0 = 300$ K, Q_S is less than A_f if T is greater than 460 K. As a rough estimate: in the ideal case, if the temperature rise required is greater than 160 K, then more heat is supplied by direct heating. In actual systems the temperature increase must be less than 60 K if there is to be a heat-pump advantage, because actual heat-pump performance is much lower than what is theoretically possible, as shown in figure 15–4 [1]. This actual performance severely limits the use of heat pumps to small-temperature difference applications.

Heating and Electrical-Energy Substitutions

In any preliminary survey of energy use in an industry, home, or utility, a prime target for change is the electrical-energy use. Electrical energy is often substituted for chemical fuels in heating applications. In most such cases the cost is higher, and the net energy use is greater. Convenience and control factors are highly valued in these applications. If proper maintenance and

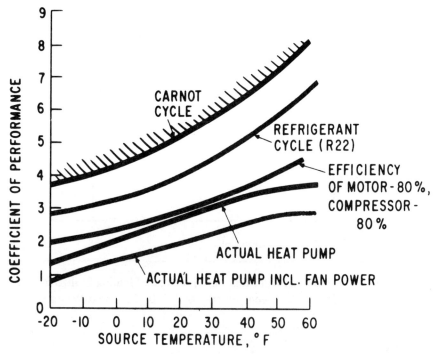

Note: Supply temperature is 120°F.

Figure 15–4. Heat-Pump Heating Efficiency as a Function of the Source Temperature for Real and Ideal Systems

design are used, the fuel heating system is more efficient, except where a heat-pump principle is applicable or where the high-temperature combustion products are used to produce work as well as heat effects.

Berg [2] has described an ideal heat-pump use for domestic hot-water heating, shown in figure 15–5.

For an ideal Carnot process the work delivered is

$$W_k = Q_H \left(1 + \frac{T_0}{T_H} \right).$$

For an ideal Carnot-process heat pump, the heat delivered is

$$Q_p = \frac{W_k}{1 - T_0/T_p}.$$

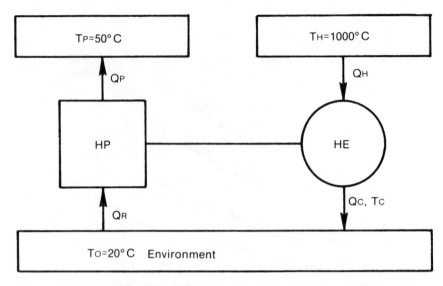

Figure 15–5. Idealized Heat Pump for Hot-Water Heating Utilizing a Fuel-Combustion Temperature of 1,000°C to Heat Water to 50°C

$$\frac{Q_p}{Q_H} = \frac{1 - T_0/T_H}{1 - T_0/T_p}.$$

For the values of temperature indicated, the ratio of the heat delivered to that produced by the combustion process is about 8.3. This means that through an ideal process, more than eight times the heating value of the fuel is made available for heating water to 50°C.

If commerically available components are used to construct this system, then the ratio

$$\frac{W_k}{Q_H} = 0.4$$

is reasonable for small steam engines or turbines.

For commercial heat pumps, typical coefficients of performance Q_p/W_k for better systems at these temperatures are 1.2–1.5. In actual use $Q_p/Q_H \simeq 0.6$; actual performance is very much reduced from theoretical reversible-system estimates.

Particular irreversibilities can be reduced by the use of larger heat exchangers and larger more efficient compressors. For hot-water appli-cations, the practicality of this has not been demonstrated on a commercial scale. One difficulty is the high temperature required for the hot-water-heating system.

In house heating applications, the picture is altered by the presence of a lower temperature requirement T_p. In a heat pump required for a house, $T_p \simeq 25°C$. The environmental temperature for heating systems depends on geographical and climatological conditions. If northern latitudes above $42°$ are assumed, then $T_0 \simeq 0°C$ is a reasonable assumption for estimating performance. Coefficients of performance of 10 percent higher than estimates for the water-heating system are expected, giving

$$\frac{Q_p}{Q_H} \simeq 0.66.$$

Again, the overall performance is of the same magnitude as for household-furnace systems, thus demonstrating no advantage of the larger capital investment in engine, heat exchangers, and compressors.

A better technological alternative for heating is the use of an absorption heat-pump system. The performance of commercial equipment for air conditioning indicates that a coefficient of performance for heating a house to $25°C$ from a $0°C$ environment is approximately 0.85–0.90. Such an alternative, which does not require an engine or a compressor, is competitive in certain geographical areas of the United States. Conventional heat-pump systems also compete effectively with furnace heating in Florida and other areas where the outside temperature is not as low as in the illustration. This environmental temperature is what determines the feasibility.

An alternative environmental heat source, the ground, is an attractive possibility. The ground temperature below about 5 meters remains constant throughout the year. In latitudes of $40°–50°$ N, this ground temperature remains about $10°C$. If this source can be used as the environmental temperature for heat-pump applications, the overall performance of the turbine-heat-pump system can be increased to about 1.1. Similarly, the gas-absorption heat-pump-system performance can be increased to approximately

$$\frac{Q_p}{Q_H} = 1.5.$$

This makes sense but presents some technical and economic difficulties:

1. The heat-transfer characteristics of soil, sand, and rock are very poor, reducing the system's performance or necessitating a very large buried heat exchanger.

2. Actual temperatures of the underground coil must be lower than 10°C to provide heat flow into the coil.
3. Installation of systems in rock or hard-packed soils is difficult.

Techniques now being developed for hydraulic fracturing of rock could prove useful in reducing the last difficulty. Groundwater poses both special advantages and disadvantages. It supplies good heat-transfer characteristics for buried coils; however, its temperature needs to be monitored carefully, to avoid ice formation around the coils, which reduces the performance. Future systems might include the production of porous rock beds that are sealed so that the heat-pump working fluid can be pumped directly through the beds without internal piping.

With the introduction of heat-pump systems, the use of solar collectors as the heat source can also improve performance (see figures 15–6 and 15–7). In most locations solar collectors alone are not expected to provide sufficient heating. In middle latitudes the combination of a heat-pump system with a solar collector can offer a better alternative than auxiliary-furnace heating. In these areas, peak heating requirements are not extreme, and cold-weather protection of the solar collectors is not expensive.

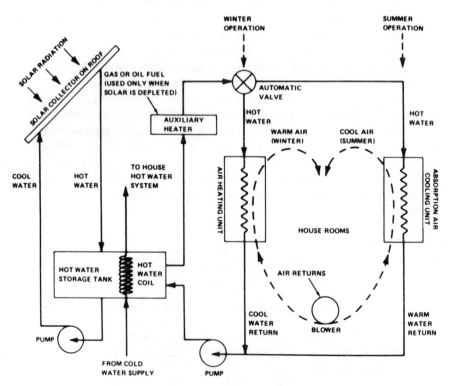

Figure 15–6. Simple Solar-Assisted Heat-Pump Schematic

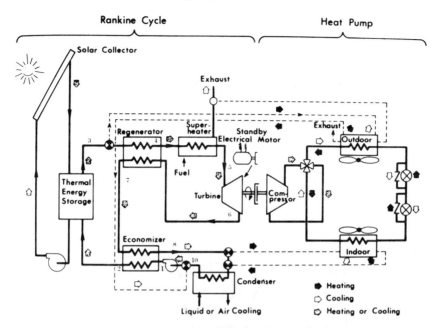

Source: From W. R. Martini, "An Efficient Heat Engine-Heat Pump," in *Solar Energy Heat Pump Systems for Heating and Cooling Buildings*, ed. S. Gilman, Workshop proceedings (University Park, Pa.: Pennsylvania State Univ., 1975), p. 201.

Figure 15-7. Sophisticated Solar–Steam-Turbine-Driven Heat-Pump Schematic

In chemical-processing applications the use of heat-pump systems can provide increased effectiveness and better product control. An example is the use of a vapor-compression system to extract solutes from solution, as shown in figure 15–8. This system requires approximately one-quarter the amount of fuel needed in older open-boiling evaporation systems. This includes the efficiency loss in the electrical-energy-production process. Most modern maple-syrup production systems use these vapor-compression techniques for concentration of the sugar solution.

A total energy system used for electrical-energy generation as exhaust thermal heating can easily provide more thermal energy than direct-combustion heating. This is apparent from consideration of the combined engine–heat pump illustrated in figure 15–5, as noted earlier.

An additional possibility of this combination is to use the heat rejected by the engine Q_c. This can be included by considering the effect of the heat-sink temperature T_c on the performance. For the ideal heat engine,

$$W_k = Q_H(1 - T_c/T_H).$$

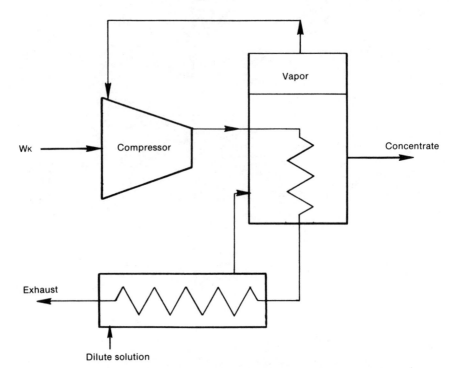

Note: In maple-syrup processing, the fuel energy required is reduced to approximately 1 MJ/kg of water from 4 MJ/kg of water in open boiling.

Figure 15–8. A Vapor-Compression Technique for Concentrating Liquid Solutions

The combination shown then gives the heat delivered as

$$Q = Q_H \frac{\left(1 - \dfrac{T_c}{T_H}\right)}{(1 - T_0/T_p)}$$

The numerator is largest when $T_c = T_0$. T_H is again the combustion temperature.
Then

$$Q = Q_H \frac{\left(1 - \dfrac{T_0}{T_H}\right)}{\left(1 - \dfrac{T_0}{T_p}\right)}.$$

If the exhaust of the engine were used for heating, then $T_c = T_p$, and the total heat delivered is

$$Q_T = Q + Q_c = Q_H \frac{\left(1 - \dfrac{T_p}{T_H}\right)}{\left(1 - \dfrac{T_0}{T_p}\right)} + Q_H \left(\frac{T_p}{T_H}\right) = \frac{Q_H}{\left(1 - \dfrac{T_0}{T_p}\right)}.$$

This would then provide a greater total heating effect.

Since real engines have lower efficiencies than the hypothesized engine, let us next examine the effect of the efficiency η of the engine on the total heat delivered when the exhaust is used for heating.

$$W_k = \eta Q_H.$$

Then

$$Q_c = Q_H - W_k = Q_H(1 - \eta).$$

Now

$$Q_T = Q + Q_c.$$

With

$$Q = W_k \frac{1}{1 - T_0/T}$$

from equation 15.3, then

$$Q_T = \frac{\eta Q_H}{1 - \dfrac{T_0}{T}} + Q_H(1 - \eta).$$

After some algebraic manipulation, this can be reduced to

$$Q_T = Q_H \left[1 + \eta \frac{1}{\left(\dfrac{T}{T_0} - 1\right)} \right]. \qquad (15.4)$$

The greater the efficiency η, the greater the total heat delivered. This indicates the importance of the engine in a total energy system if the exhaust is to be used for heating.

In terms of real components, engine and heat pump, the heat delivered is without exhaust heat use,

$$Q = Q_H \eta (COP).$$

COP is the conventional coefficient of performance of a heat-pump system. With exhaust use, the relationship is

$$Q = Q_H (1 + \eta \{(COP) - 1\}).$$

Commercially available small-capacity systems provide values of engine efficiency of $\eta = 0.25$ and COP of 2. This would give a heat-delivered/heat-supplied ratio, without exhaust recovery, of

$$\frac{Q}{Q_H} = (0.25)(2) = (0.5).$$

With exhaust recovery this increases to

$$\frac{Q}{Q_H} = 1.25.$$

With central-station electrical generation, efficiency $\eta = \frac{1}{3}$; and the use of this electrical energy with a household heat pump would give

$$\frac{Q}{Q_H} = (\tfrac{1}{3})(2) = \tfrac{2}{3}.$$

Again it is demonstrated that in current systems for utilizing heat pumps for heating, there is little increased efficiency over direct-combustion heating.

Performance of Heat Pumps and Cooling Devices Operating with Thermal Energy

The conventional measure of the performance of heating and cooling units that operate on a heat-pump principle is the coefficient of performance

(*COP*). In a heat pump it is usually defined as the thermal energy delivered per unit of mechanical work supplied:

$$COP_H = \frac{Q_p}{W_k}.$$ (15.5)

In a cooling device the *COP* is defined as the heat removed per unit of mechanical work supplied:

$$COP_R = \frac{Q_R}{W}.$$ (15.6)

An ideal cooling or heating cycle is often useful for comparison with the actual performance. This ideal-cycle *COP* can be readily examined by hypothesizing a reversed Carnot cycle, with the result that the maximum possible coefficient of performance for a heat pump operating to supply heat at T_p from an environment at T_0 is

$$COP_H = \frac{1}{1 - \dfrac{T_0}{T_p}}.$$

Similarly, for an ideal cooling cycle,

$$COP_R = \frac{1}{\dfrac{T_0}{T_R} - 1}.$$

A better way to define the efficiency in each case would be the useful effect per unit of available energy supplied:

$$COP = \frac{Q}{A_H}.$$ (15.7)

This definition allows the use of *COP* for absorption devices where the principal energy supplied is heat rather than work. If the thermal energy supplied is from a source at temperature T_H in an environment at a temperature T_0,

$$A_H = Q_H \left(1 - \frac{T_0}{T_H} \right).$$ (15.8)

The *COP* is then, from equations 15.7 and 15.8,

$$COP = \frac{Q}{Q_H \left(1 - \dfrac{T_0}{T_H}\right)}.$$
(15.9)

For a heat pump with a thermal-energy supply,

$$COP_H = \frac{Q_p}{Q_H} \left(\frac{1}{1 - \dfrac{T_0}{T_H}} \right).$$
(15.10)

This may be visualized as a first-law efficiency Q_p/Q_H divided by a Carnot factor.

It is remarkable that the Carnot factor appears in this formulation, since the complete cycle need not involve any conversion to work at all. To examine this further let us ask: What is the maximum heating effect obtainable from a heat source using a heat-pump principle? The answer will have a direct bearing on the evaluation of the potential uses of waste heat and solar energy.

Figure 15–9 shows a schematic of an ideal heat pump with only thermal inputs and outputs. Let us tentatively measure the amplification a as the thermal energy delivered Q_p divided by the thermal energy supplied Q_H:

$$a = \frac{Q_p}{Q_H} = \frac{Q_p}{A_H} \left(1 - \frac{T_0}{T_H} \right).$$
(15.11)

The first law tells us that

$$Q_p = Q_H + Q_R.$$

The second law tells us that in a reversible process (the best we can do),

$$\Delta S = \frac{Q_p}{T_p} - \frac{Q_H}{T_H} - \frac{Q_R}{T_R} = 0.$$

Substituting these into the amplification equation and eliminating the Qs gives

$$A = \left(\frac{T_p}{T_H}\right) \frac{\left(\frac{T_H}{T_R} - 1\right)}{\left(\frac{T_p}{T_R} - 1\right)}. \tag{15.12}$$

For a heat pump to operate as assumed, $T_H > T_p > T_R$.
Special cases can now be examined:

1. In the case where the supply temperature T_H equals the hot temperature T_p, $a = 1$. This means that direct heating may as well be used, as the heat-pump amplification is 1 and $Q_p = Q_H$.
2. In the case where the supply temperature T_H equals the cold-reservoir temperature T_R, $a = 0$. This means that no heat can be delivered to a system at a higher temperature by a source at a lower temperature alone.
3. In the case of $T_p \to T_R$ the amplification approaches infinity, and a very small heat supply Q_H can provide large amplification of heating.
4. The case where $T_p \gg T_R$ shows that $a \to 1 + (T_R/T_p) - (T_R/T_H)$. This indicates the limitation of the heat pump for high-temperature applications.

The last case requires further clarification, which is indicated in figure 15–10. In this figure the amplification is shown as a function of the temperature factors T_p/T_R and T_p/T_H. The expression for this is

$$a = 1 + \frac{1 - (T_p/T_H)}{(T_p/T_R) - 1}. \tag{15.13}$$

The amplification increases significantly for low-temperature heating applications when a high-temperature source is available. In home-heating applications, T_p/T_R is of the order of magnitude of 7/6 for an outside temperature of 300 K and hot-water-heating system designed for 350 K. In solar-heating applications, a 400 K source temperature is feasible. Such a system would ideally give an application of about 1.5. A focusing collector that increased the temperature to 700 K would, however, give an amplification of 4 times. A significant gain is obtained by using a focusing collector.

In practical heating systems, amplification is typically about 20–30 percent of the ideal. Systems with theoretical amplifications of 3 are usually required for feasible designs. A large loss in potential amplification is lost in

Qp, Tp

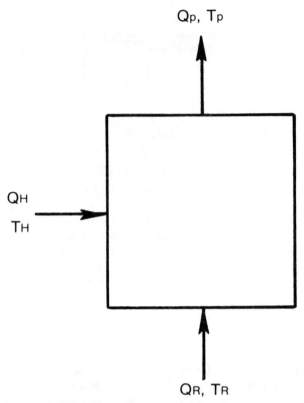

QH
TH

QR, TR

Figure 15.9. Block Schematic of a Device That Produces a Heat-Amplifi-
cation Effect

conversion from combustion temperatures to heat-pump condenser temperatures.

The relationship of the amplification to the coefficient of performance is
readily found from equations 15.10 and 15.11 as

$$COP_H = a \left(\frac{1}{1 - \dfrac{T_0}{T_H}} \right) = \frac{T_P(T_H/T_R - 1)/T_H}{\left(\dfrac{T_p}{T_R} - 1 \right) \left(1 - \dfrac{T_0}{T_H} \right)}. \qquad (15.14)$$

If a low-temperature heat source such as a waste heat system or ground
heat is not feasible, then $T_0 = T_R$. This reduces to

$$COP_H = \frac{1}{1 - (T_R/T_p)}.$$

a

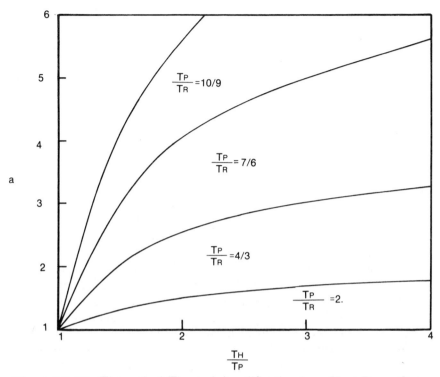

Figure 15–10. Theoretical Thermal Amplification by a Heat Pump Operating with a Heat-Source Temperature T_H Delivering Heat at a Temperature T_P From a Thermal Reservoir at Temperature T_R

Although the COP_H in this form is simple, it is not as useful, since it does not indicate the possible design conditions with respect to the source-heat temperature T_H and the cold-reservoir temperature T_R.

Process Steam

The energy use of process steam by industry is estimated to account for about 17 percent of the total energy used in the United States [3]. This energy consumption by industry represents 40 percent of the industrial energy use. This process steam, which is used for heating and temperature control, is predominant in the chemical industry. It is generated by the burning of fossil fuel with an efficiency of approximately 65–70 percent. Dow Chemical Company in Midland, Michigan, is cooperating with Consumers Power, an electrical utility, in the construction of a nuclear plant to generate steam for chemical processing. This will be the first such system in the United States. Figure 15–11 is a schematic diagram of this combined electrical-process steam facility.

The advantage of steam in process heating is the control of the heating temperature by the condensing temperature of the steam, which in turn is controlled through pressure regulation. The condensing temperature remains constant in a heat exchanger through the latent-heat region. This allows for very small temperature variation in the heat exchanger, which is extremely important for chemical-reactor processes. Where the temperature of a chemical reaction or a biological process, such as the fermentation of beer, must be uniform for optimal quality, steam provides the best heating mechanism through the condensing process. Table 15–1 illustrates the distribution of the uses of process steam over different temperature and pressure ranges. It also illustrates the substantial available-energy losses in the condensation processes.

The combustion temperatures of coal, natural gas, and fuel oil shown in table 15–2 are several times higher than these condensation temperatures used in chemical processing. A large loss in available energy from the flame to the steam used for processing is then incurred, in addition to the loss in available energy in combustion. The available-energy loss of combustion is primarily associated with the heating of the air in the combustion process. With the use of oxygen instead of air, this available-energy loss of combustion can be substantially reduced. However, if account is taken of the available energy required to separate oxygen from air, there is a net loss. Other losses in available energy occur as a result of heat loss to the environment and incomplete combustion. Available-energy-loss diagrams for typical power systems are indicated in figure 15–12.

Process-steam systems are prime candidates for either "topping" cycles or total energy systems. The available energy of the fuel is poorly utilized and could be extracted with very little loss in energy to the processing function.

Most steam systems return condensed steam to the boiler for recycling, so that little condensed steam is wasted directly. This recycling is important from two points of view. First, the heating value of the condensed steam is still appreciable. Second, the recycle reduces the requirement for conditioning intake water to remove minerals or microorganisms that corrode or precipitate on heat-exchanger surfaces. (Note that Watt's contribution to steam-engine development was primarily the closing of the cycle, which reduced this need to condition the water.) As discussed in appendix 1A, the efficiency improvement was not the most important consideration in the coal fields of England.

Although industrial people often point out that the clouds of billowing white vapor from industrial sites are not environmentally harmful, these

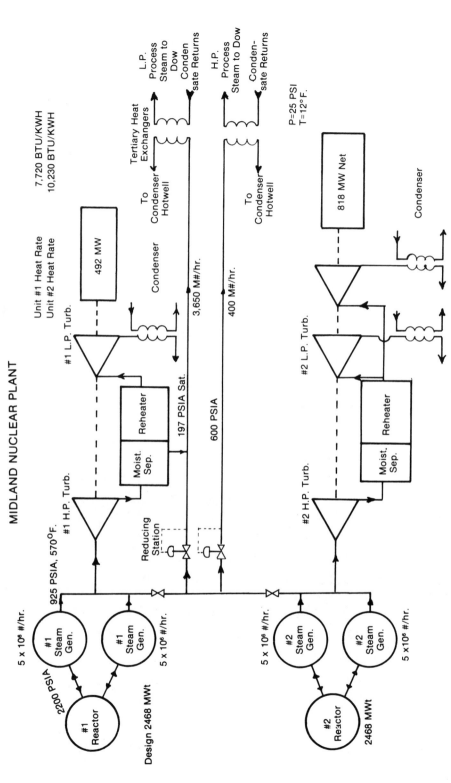

Figure 15-11. Schematic Diagram of the Midland, Michigan, Combined Electrical-Power and Process-Steam Nuclear Facility

Table 15–1
Industrial-Process Steam Available-Energy Loss on Condensation

Industrial sector	Pressure Range kPa	Temperature (K)	ΔH_{fg} KJ/Kg	ΔA_{fg} KJ/Kg
Chemicals and allied products	3100–6900	510–560	1780–1570	710–700
	1450–3100	470–510	1950–1780	705–710
	780–1450	440–470	2050–1950	655–705
	< 780	< 440	< 2050	<655
Petroleum refining and associated industries	1100–4200	400–525	2000–1700	690–740
	< 1100	< 400	< 2000	< 690
Paper and allied products	780–1450	440–470	2050–1950	655–705
	< 780	< 440	<2050	< 655
Food and kindred products	440–780	410–440	2140–2050	625–655
	< 440	< 410	< 2140	< 625

Note: Adapted from J. Burroughs, "Technical Aspects of the Conservation of Energy for Industrial Processes" (Midland, Mich.: Dow Chemical Company, 1974).

Available Energy and   Diagrams for Primemovers

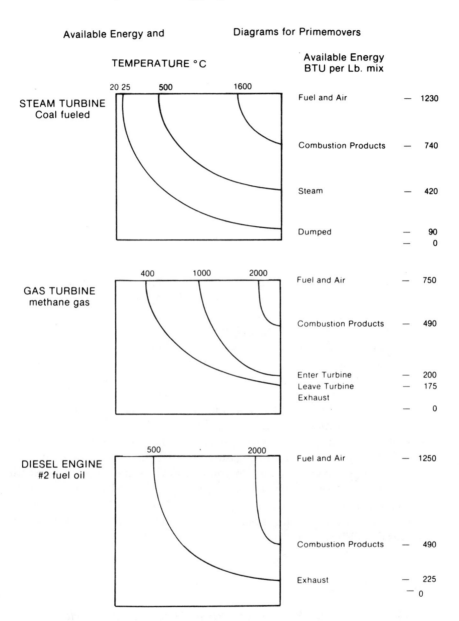

Figure 15–12. Available-Energy–Temperature Diagrams for Typical Fossil-Fuel Power Systems

Table 15–2

Adiabatic-Flame Temperatures for Combustion of Fuels at Stoichiometric Air/Fuel Ratios

Fuel	Pressure (Atm.)	Adiabatic Flame Temperature (K)		
		1	*10*	*100*
Methane (CH_4)		2,223	2,266	2,293
Propane (C_3H_8)		3,633	4,069	
n-octane (C_8H_{18})		3,668	4,113	
Hydrogen (H_2)		2,400		
Acetylene		2,600		
Coal		2,250		
Carbon monoxide (CO)		2,350		

releases usually represent an unnecessary waste of thermal energy. Topping-cycle energy systems can be used to improve the available-energy output of energy-conversion systems as well as that of heating systems. Conventional energy-conversion systems are limited in efficiency by metallurgical or cost considerations rather than thermodynamically by the combustion temperature. The development of devices to convert this available-energy loss from the 2,000°C combustion temperature to the 700°C working-fluid temperature is the objective of topping-cycle research. Figure 15–12 indicates that the available-energy loss in the combustion process far exceeds that in the exhaust stream. Thus energy-conversion research on this topping end is much more promising than research on conversion devices based on waste-heat recovery conversion. There is a lot of waste heat in the exhaust of an engine, but proportionately little available energy. In addition, conversion of present process steam systems to topping or total energy systems will usually require the replacement of low-pressure boilers by high-pressure boilers.

Magnetohydrodynamic and thermionic devices are primary research devices under development as topping systems. They both require high temperatures for operation and can be designed to reach these temperatures without some of the metallurgical problems associated with conventional devices.

References

1. Gilman, S.F., ed. *Solar Energy Heat Pump Systems for Heating and Cooling Buildings*. Pennsylvania State University, ERDA Doc. C00-2560-1, 1975.
2. Berg, C.A. "A Technical Basis for Energy Conservation." *Mech. Eng.* 86, no. 5 (1974): 30.

3. Stanford Research Institute. *Patterns of Energy Consumption in the U.S.*, Office of Science and Technology, Office of the President, Washington, D.C., January 1972.
4. Burroughs, J. "Technical Aspects of the Conservation of Energy for Industrial Processes." Midland, Mich.: Dow Chemical Company, 1974.

16 Energy Storage

The creative mind sees the world as one great workshop and the present as no more than material for the future, for the thing that is yet destined to be. —H.G. Wells, *The Discovery of the Future* (1902)

Energy-Storage Thermodynamics

Fossil fuels represent solar energy stored in the form of chemical energy. This stored chemical energy is readily converted to thermal energy by the addition of oxygen. It is also directly convertible to electrical energy in a fuel cell. The energy stored in fossil fuels is indicated in table 16–1. Typical values are 20–60 MJ/kg for fossil fuels. These high energy densities are used to advantage in mobile applications. They are also of great importance in stationary power systems such as electrical-power plants located on navigable waters that are frozen over in winter. Large coal piles are characteristic features of power plants located on the Great Lakes. Oil-storage facilities are also an important requirement in refinery installations and in any industry or home-heating systems.

The energy-storage utility of fossil fuels has been underestimated in the past. When alternative energy sources are examined, this utility is especially evident. Most renewable energy alternatives require energy storage as part of their system in order to prove feasible.

Associated with flows and forces in most physical systems are static energy-storage elements. These may be springs that store elastic energy, capacitors that store electrical energy, compressible systems that store internal energy, gravitational systems where potential energy is stored, or molecular bonds where chemical energy is stored. In these the stored energy is potentially available for work processes and can be stored indefinitely without losses. In addition to these static systems, there are kinetic systems such as linear or rotational kinetic storage, electrodynamic-field storage, thermal storage, and so forth. In these, restrictions associated with the irreversibilities of dissipative processes occur. Available energy must be supplied continually to replace rate losses. Storage of kinetic energy necessitates techniques to minimize losses. This may take the form of vacuum-storage, low-temperature, or isolation systems.

Static energy-storage systems are characterized by physical components in which the generalized displacement of the system from its equilibrium state

Table 16–1
Energy-Storage Densities

	Density, MJ/kg
Beryllium-oxide	65
Lead-acid battery	0.1
Fused silica flywheel	3.0
Carbon-fiber flywheel	1.0
Linear kinetic, 160 Km/hr	0.001
Rotational kinetic, 20,000 rpm disk	0.85
Gravity, 100 m	0.01
Magnetic field, 2 weber/m^2 (air)	0.0016 MJ/liter
Electric field, 6.5×10^8 V/m	0.006 MJ/liter
Combustion-fuel: Natural gas	0.04 MJ/liter
Coal	28
Petroleum	43
LNG	27 MJ/liter
Hydrogen recombination	216
Municipal waste	14
Compressed air, 3 M Pa	0.18
Thermal-latent heat: Glauber salt	0.25
lithium hydride	4.2
water fusion	0.34
water, 100°C	0.43
Nuclear: Fission, U^{238}	83,000
Breeder, U^{235}	6,000,000
Fusion, $D + T \rightarrow He_4 + 17.6$ MeV	340,000

Sources: R.F. Post and S.F. Post, "Flywheels," *Scientific American* 229(1973): 17–23; A.R. Foster and R.L. Wright, *Basic Nuclear Energy* (Boston: Allyn and Bacon, 1968); J. O'M. Bockris, *Energy, The Solar Hydrogen Alternative* (New York: Wiley, 1975); L.C. Lichty, *Combustion Engine Processes*, 6th ed. (New York: McGraw-Hill, 1967).

is a function of the force applied to the system. Alternatively, it may be viewed as a system in which the potential developed depends on the displacement from equilibrium. In an electrical capacitor the voltage (force) developed is proportional to the charge displaced as

$$V = \frac{q}{C},$$

where C is the capacitance. In a mechanical spring the force developed in the spring is proportional to the displacement from equilibrium

$$\langle F \rangle = \frac{X}{C}.$$

The energy stored is equal to the work done on the system

$$W_k = \int \langle F \rangle \, dX = \int \frac{X}{C} \, dX = \frac{X^2}{C2},$$

$$W_k = \frac{C}{2} \langle F \rangle^2.$$

This represents the potential energy stored in a field. For an electrical capacitor,

$$W_k = \frac{CV^2}{2} = \frac{q^2}{2C}.$$

In these cases the force field is affected by the displacement. In some cases, as in hydroelectric storage, the force field can be approximated as constant independent of the displacement. In this case the force of gravity can be assumed constant as g, so the force on a mass m is

$$\bar{F} = m\bar{g} = \text{grad } \phi.$$

For displacement of a mass in the direction of the gravitational field force, the potential energy stored is

$$E = W_k = \int mg \, dX = mg \, X.$$

This represents the static potential energy stored, and the force can be expressed as the gradient of the potential, as shown. These conditions may also apply to electrostatic fields, and represent fields in which energy is not required to maintain the energy in storage.

Solar energy-conversion systems, including wind, biomass, hydro, and so forth, are characterized and often limited by the time variability. Most require energy storage to provide matching of the energy use to the energy supply. The type of energy storage is determined in part by the application. If the energy requirement is for low-temperature heat, the potential for thermal storage is improved. Typical examples include solar heating of buildings and hot water, where sensible-heat storage is often both technically and economically feasible. The temperature required in these applications are usually less than $T = 70°C$ (343 K) in an environment at $T_0 = 20°C$ (293 K). The thermodynamic potential (available energy) required is then only

$$1 - \frac{T_0}{T} = 0.15$$

of the total energy value. Sensible-heat-storage capacity for selected materials is shown in table 16–2. The high specific heat of water, combined with its low cost, make it one of the best sensible-heat-storage systems.

A second thermal-storage method uses the latent heat of a substance. Latent-heat storage has an advantage over sensible-heat storage in that the thermal energy stored in the phase change can be attained with a smaller

Table 16-2
Sensible-Heat Storage Capacity of Selected Materials

Material	Phase	kJ/kg K
H_2O	Liquid	4.17
Isobutanol	Liquid	3.00
Caloria HT43	Liquid	2.80
Therminol 55	Liquid	2.35
Hitec	Liquid	1.43
NaK [56-44]	Liquid	0.91
M_g	Solid	1.04
Granite	Solid	0.78
Iron	Solid	0.45
Sodium Carbonate	Solid	1.09
Brick	Solid	0.84
Aluminum	Solid	0.88

Sources: F. Kreith, "An Overview of Energy Technology for Intermediate Temperature Applications," *Eng. Sci. Perspectives* 3(1978):7–18; R.H. Turner, *High Temperature Thermal Energy Storage* (Philadelphia: Franklin Institute Press, 1978); T.S. Dean, *Thermal Storage* (Philadelphia: Franklin Institute Press, 1978).

temperature difference between the storage and the environment. This advantage is discussed with reference to heat-exchanger performance in chapter 15. Latent-heat systems include many different phase-transition materials. In order to reduce the volume required for latent-heat storage, only solid-solid and liquid-solid phase transitions are usually considered. Table 16–3 indicates typical storage capacities for phase-transition latent heats. Critical to the selection of these materials is the need to match the phase-transition temperature to the load-temperature requirement. If the phase-transition temperature is much higher than the load temperature, the thermal-

Table 16-3
Phase Transition Latent Heats and Temperatures

Material	Phase Change	Latent Heat kJ/kg	Transition Temperature K
Water	Solid-liquid	335	273
Sulfur	Solid-Liquid	56	383
Paraffin	Solid-liquid	230	347
Boron hydride	Solid-liquid	267	372
Ammonium thiocynate	Solid-liquid	260	419
Sodium hyroxide	Solid-liquid	167	591

Sources: F. Kreith, "An Overview of Energy Technology for Intermediate Temperature Applications," *Eng. Sci. Perspectives* 3(1978): 7–18; R.H. Turner, *High Temperature Thermal Energy Storage* (Philadelphia: Franklin Institute Press, 1978); P.G. Grodzka, "Phase-Change Storage Systems," *Solar Energy Technology Handbook*, part A, ed. W.C. Dickinson and P.N. Cheremisinoff, pp. 795–809 (New York: M. Dekker, 1980).

energy losses to the environment are high. Phase-transition-material research has been directed at eutectic salts because of their phase-transition temperatures, which are close to storage temperatures obtainable with non-focusing solar collector systems and near the temperature needed for water and space heating.

Chemical-reaction thermal storage is a third type of storage mechanism that allows storage of energy in chemical bonds. Chemical thermal-storage systems allow for longer-term energy storage than do sensible or phase-transition systems because the principal storage is chemical rather than thermal. Heat-transfer losses are therefore minimal.

The simplest thermochemical system is the easily observed heating of water as a result of the addition of sulfuric acid. This experiment, done in many introductory chemistry laboratories, provides heat as a result of the hydrogen H^+ and SO_4^{--} ions reacting with water molecules. This exothermic process is reversed by the addition of heat to evaporate the water molecules from the dilute sulfuric acid solution.

If the water is evaporated from the sulfuric acid solution by solar heating, the water vapor can be circulated to a heat exchanger and recondensed, supplying the latent heat of water as well. A schematic of such a system is shown in figure 16–1 [5,9].

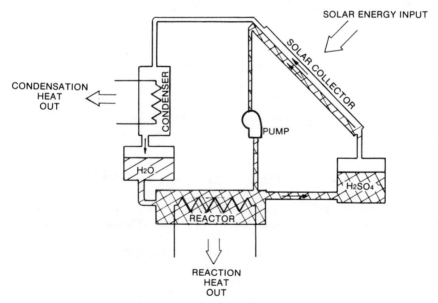

Note: The solar-energy input evaporates water from the dilute sulfuric acid solution. The water is then reacted with the concentrated sulfuric acid solution to produce heat as needed. Condensation is also available for low-temperature applications.

Figure 16–1. Heat-of-Reaction Chemical-Energy Storage System

Concentration-dilution systems require special consideration from a thermodynamic viewpoint. In chapter 5, the Gibbs free energy change for the mixing of an ideal solute a and solvent b was shown to be

$$\Delta G_{\text{tot}} = RT \left(n_a \cdot \ln \frac{n_a}{n} + n_b \ln \frac{n_b}{n} \right).$$

At constant temperature this change in Gibbs free energy is due to the entropy associated with mixing. No energy change is involved in this process.

The differential of the Gibbs free energy,

$$dG = dH - Tds - sdT,$$

is reduced to

$$dG = -Tds$$

for the ideal solutions where dH and dT are zero. If the solutions are not ideal, there will be a change in enthalpy (the heat of solution); and the concentration terms must involve activities a_i instead of concentrations C_i. The mixing entropy contribution at constant temperature is then

$$-Tds = RT\{n_a \ln a_a + n_b \ln a_b\}.$$

The actual enthalpy change can be found with the aid of chemical-thermodynamic tables, using the chemical balance

$$H_2SO_4 \ (\text{liquid}) \rightleftharpoons H_2SO_{4(aq)},$$

or

$$H_2SO_4 + nH_2O \rightleftharpoons 2H^+ + SO_4^{--}{}_{(aq)} + nH_2O.$$

If the sulfuric acid is completely dissolved in an infinite amount of water $n \to \infty$, the tabulated heat of solution at infinite dilution is

$$\Delta H = \Delta H_f(H_2SO_4)_{aq} - \Delta H_f(H_2SO_4)_{liq}$$

$$= -216.90 - (-193.91)$$

$$\Delta H = -22.99 \ \text{Kcal/mole} \ H_2SO_4.$$

The entropy change ΔS is also found as

$$\Delta S = S\,(H_2SO_4)_{aq} - S\,(H_2SO_4)_{liq}$$

$$\Delta S = 4.1 - 37.5 \text{ cal/mole } H_2SO_4/K$$

$$= -33.4 \text{ cal/mole } H_2SO_4/K.$$

Then

$$\Delta G = \Delta H - T\Delta S \qquad \text{at } T = 298 \text{ K is}$$

$$\Delta G = -22.99 + (298)(33.4) \times 10^{-3} \text{ Kcal/mole } H_2SO_4,$$

$$\Delta G = -22.99 + 9.95 = -13.04 \text{ Kcal/mole } H_2SO_4.$$

The reaction goes nearly to all ionic components in solution. The ions in solution are not free, however, and react to form molecular complexes with the polar water molecules. The number and kinds of molecular complexes depend on the ratio of the ionic components to the water molecules present. There is an enthalpy change associated with the reaction that depends on the concentration of the solute (H_2SO_4) in the solvent (H_2O).

The change in enthalpy tabulated as -22.99 Kcal/mole H_2SO_4 is for infinite dilution. For finite dilution the reaction is not complete, and the enthalpy of the reaction per mole of H_2SO_4 is less. The dependence is shown in figure 16–2. This figure indicates that the heating effect is highest at high dilution. The heat of reaction decreases as the quantity of water available for reaction decreases. This decrease in heating effect limits the temperature attainable in the recombining of acid and water. If the Gibbs free energy for recombination ΔG is to remain negative, then $\Delta H - T\Delta S < 0$. Equating $\Delta H = T\Delta S$ and solving for this equilibrium temperature $T = \Delta H/\Delta S$ gives $T = 688$ K. This is approximately the limit of heating for this dilution reaction.

Conventional absorption refrigeration systems can also be used for energy storage. The familiar ammonia-water and lithium bromide-water absorption systems can be used as long-term thermal-storage systems. (Chapter 15 discusses these absorption systems as used in heat-pump applications.)

Decomposition and recombination of many chemicals, with associated heat absorption and release, can also be obtained by pressure control. An example is NH_4HSO_4, a liquid that decomposes to ammonia, water, and SO_3

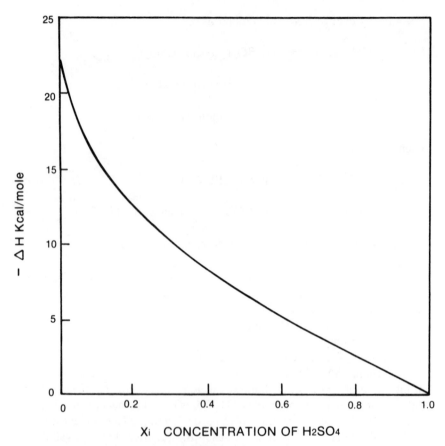

Figure 16–2. Heat of Solution per Mole of H_2SO_4 When Reacted with H_2O

gas at 740 K with absorption of 337 kJ/mole. Other low-temperature decomposition reactions are shown in table 16–4. The reactions listed have one phase that is either a solid or a liquid. This provides a higher storage density than do gas reactions. These chemical storage systems can be very low-loss systems if the storage-reaction temperature is near the environmental temperature.

The use of chemicals that react reversibly at elevated temperature to store and transport thermal energy has recently been explored in what is known as a chemical heat pipe. As originally envisioned, a heat pipe was a system that absorbed heat at one location and transported it to a second location, where it was released. The most common systems are refrigerant fluids, which evaporate at the heat source. The vapor is either pumped or convected to another location. At this second location it condenses, releasing the heat of vaporization. The liquid is then returned to be reevaporated.

Table 16–4
Typical Decomposition-Recombination Chemical Reactions

			$\Delta H\,^{\circ}$ KJ/mole
$Ca(OH)_2$	$\rightleftharpoons$	$C_aO + H_2O$	109
$M_g(OH)_2$	$\rightleftharpoons$	$M_gO + H_2O$	81
$NH_4\,Cl$	$\rightleftharpoons$	$NH_3 + HCl$	176
$2H_2O$	$\rightleftharpoons$	$2H_2 + O_2$	242
$ZnSO_4\,H_2O$	$\rightleftharpoons$	$ZnSO_4 + H_2O$	19
$NH_4\,(HSO_4)$	$\rightleftharpoons$	$NH_3 + H_2O + SO_3$	337
H_3PO_4	$\rightleftharpoons$	$PH_3 + 2O_2$	1283
H_2SO_4	$\rightleftharpoons$	$SO_3 + H_2$	176
$2NO_2$	$\rightleftharpoons$	$2NO + O_2$	24
CH_3OH	$\rightleftharpoons$	$2H_2 + CO$	92
$2H_2 + 2Cl_2$	$\rightleftharpoons$	$4\,HCl + O_2$	114

Sources: R.W. Mar and T.T. Bramlette, *Thermochemical Energy Storage Systems*, Sandia Laboratories Report SAND 77–8051 (1978); C.G. Miller, *Temperature Transformer Energy Conversion System*, NASA Technical Brief (Pasadena, Calif.: JPL, 1977); M.D. Silverman and J.R. Engel, *Survey of Technology for Storage of Thermal Energy in Heat Transfer Salt*, ORNL TM–5682 (1977).

A chemical heat pipe provides the same function, but the heat absorption and release are accomplished by chemical reactions. Figure 16–3 indicates a proposed chemical-heat-pipe system using solar energy. Methane and water are heated by solar energy to 960 K. In the presence of a catalyst, they will react at a sufficient rate to produce CO and H_2, both good reducing agents. The CO and H_2 can then be pumped to another location, where they react to release heat and recover the methane and water. An advantage of this system is the long-term storage potential and low cost of the methane and water reactants.

Energy storage in many of these systems can be optimized using standard thermodynamic analysis. Crucial to most systems are the kinetics of the reactions. The development of new catalysts will be necessary to provide better control of reactions. Hydrocarbon combustion has usually not required careful attention to catalyst chemistry.

Future energy-storage systems will probably rely less on combustion heating and will emphasize reversible chemical heat-absorption–release processes.

Solar Work Storage (Available Energy)

The systems discussed with respect to solar heating can be used to provide work storage as well. For work storage the production of thermal energy is to be avoided because of the available energy lost in the conversion from thermal energy to work, associated with the Carnot limitations.

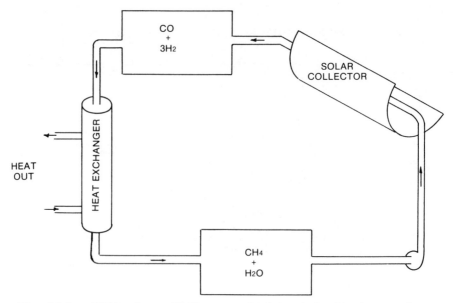

Note: Methane (CH_4) and water (H_2O) are reacted endothermically with solar-energy input. The product gases H_2 and CO are then reacted in a heat-exchanger–reactor to provide heat [10, 11]

Figure 16–3. Chemical Heat-Pipe Illustration

Chapter 14 discusses the available energy of solar energy and the important fact that solar energy is not thermal energy. Any conversion system that converts solar energy to thermal energy is an inefficient work or available-energy storage. Table 16–1 illustrates the available-energy storage except for the specific thermal-energy storage systems, which are, of course, very poor available-energy storage systems.

Chemical-energy storage systems generally have a loss associated with the heat required to carry out the reaction at a useful rate.

Studies of the production and use of hydrogen as a storage system have resulted in the development of hydrides. A typical one is lithium hydride, in which the hydrodgen is absorbed at a low temperature and given off when the hydride is heated. These systems provide safe storage of hydrogen at much higher densities than pressurized tank storage permits.

The objectives of energy storage in large power-utility systems are directed at load leveling. Load leveling allows an electrical utility to operate its largest generating units at a uniform generating rate. In theory this is supposed to allow the use of newer, larger, more efficient units, rather than smaller, inefficient units. The smaller, inefficient units are conventionally brought on line to meet peak-load demands. In many applications the peak load is supplied by oil- or gas-fired turbine generators, and the baseload is supplied by coal or nuclear systems. Thus the use of energy storage on a large system usually allows a shift from oil and gas to nuclear or coal fuels.

The storage in electrical-power systems occurs in many forms. For example, pumped hydroelectric systems are in use at Ludington, Michigan. In this system water is pumped to a reservoir and returned through turbines to meet peak power demands.

More conventionally, electrical water heaters in homes provide energy storage for electrical utilities. The electrical hot-water tank in a house reduces the house's peak electrical demand on the electrical system. The homeowner, however, pays for the energy-storage system and the energy lost as heat from the storage tank. To compensate for this extra cost, he usually receives a special low electrical-energy rate for the water heating from the utility.

Most large industrial users of electricity are similarly charged on a peak-use rate. The use of storage by an industry can often be used to reduce the total cost of electrical energy.

This energy storage–rate of use game is one that is continually being played to reduce the cost of electrical energy by both producer and user. The net cost and net energy used both be minimized concurrently, although this is not inherent in the system. A shift from an efficient gas turbine to an inefficient nuclear power plant may be made on the basis of the availability of nuclear fuel rather than natural gas.

Photochemical Energy Conversion: Storage

A photochemical energy-storage system may be characterized by reference to an energy-state diagram shown in figure 16–4 [13–15]. The photon absorbed must have an energy greater than that required to provide an electronic transition, moving an electron to an excited state. The excited state must have enough energy to provide charge separation sufficient to change the chemical bonding. If this does not occur, the excited state will fluoresce with light emitted as the electron returns to its initial state. The chemical-bond energy change may be small in cases like the transition from a trans to a cis configuration in an organic molecule, which is simply a bond rotation [16]. In this process the energy to move the atoms is lost. In addition, if the new configuration is to be temporarily stable to provide storage, then the energy state must be lower than the excited electronic state. If this is not the case, spontaneous transition to the electronic state is more probable with return to the original state with photon emission. This emission after molecular change is termed *phosphorescence*.[1]

In the diagram, the higher the energy barrier between the energy-storage state B and the original state A, the more stable the energy storage. The

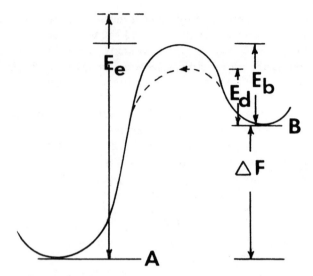

Figure 16–4. Photochemical-Energy Diagram Illustrating the Molecular-
Energy States Required for Energy Storage

higher this barrier, however, the less efficient the system, since the energy
storage is represented by the difference in energy between A and B shown as
ΔF. The efficiency is given by $\Delta F/E_e$. The solar energy-conversion efficiency is lower than this, since only a fraction of the photons in the solar
spectral distribution have energy in excess of this electronic-transition energy
$E_e = h\nu$. Photons with energy less than this value are usually detrimental, as
they contribute to the decay from state B to state A. The same applies to the
excess energy of a photon over the electronic transition energy.

Several other processes of electronic transitions are of secondary interest:

1. photon-caused thermal transitions to higher energy states;
2. photon-excited states emitting light energy in the process of going to state
   B;
3. radiative jumps from the excited state to an intermediate state that decays
   to state B.

The most promising photochemical systems for energy storage appear to
be the isomer molecules [16]. Typical are indigo and thioindigo dyes, which
are cis-trans configuration-change chemicals, or nitrones, that have valence
isomers where chemical bonds change without any change in chemical
composition [17]. Both of these have very small allowable electronic-
transition energy-level changes. Dissociation reactions are also useful, but
they have higher electronic-transition energy requirements, are less stable,

and return to the undissociated state very rapidly. Energy-storage efficiencies in both are in the range of 10–25 percent, with electronic-transition energy in the range of 5–100 kJ/mole.

Although these mechanisms are of interest for energy storage, their value is principally for releasing the energy stored as heat. The more important photochemical reactions are those that produce bulk electrical-charge separation or reaction products that may be used as a more versatile fuel.

The allowed molecular changes are limited depending on the energy of the photon absorbed. In order to provide changes in the electron orbital motion, the energy must be greater than about 40 kCal/mole or have a wavelength shorter than 0.7 micron (frequency $4.2E + 14$ Sec^{-1}). This means the photon must be in the visible or ultraviolet spectral range. If the photon is in the infrared spectrum between 0.7 and 10 microns (2.8–40 kCal/mole), nuclei vibrational-energy transfer can occur. If the energy is in the microwave region from 10 to 10^{-5} microns, electron-spin precessional motion is possible. If the energy is lower than this, only nuclear-spin precessional motion is possible. These limitations restrict the electrical-energy and fuel useful spectrum to the visible and ultraviolet range. Schemes to increase the useful spectral region will need to rely on processes to convert infrared radiation to higher specific energies. An alternative would be to utilize these long-wavelength components for heating purposes. These combined systems may have application where both short-term and long-term storage is needed [29].

Photo Thermal Chemical Processes

The systems just discussed are principally ones where thermal processes are detrimental. Basic thermochemical-reaction shift mechanisms are also useful for energy storage [13, 16]. These are discussed in more detail in chapter 5. They take advantage of the chemical-equilibrium shift between reactants and products with changes in temperature and pressure. Catalysts are again crucial in controlling reaction rates in these shift reaction systems [18].

The photochemical systems discussed so far are oriented toward the production of a fuel or reactive chemical. Since these sysems involve electronic transitions, any photon process involving charge spearation can potentially be used to produce separate chemical species through dissociation while simultaneously producing electrical potentials and a current source.

Simple dissociation reactions have so far not proved useful because of the rapid recombination rates. The practicality of these systems will require the development of chemical reactions in which the dissociation products can be separated before recombination, and which do not absorb the infrared spectral components [17,19,20]. The more complicated multistep photo-chemical dissociation processes have concentrated on separating water molecules into hydrogen and oxygen (direct photolysis) [18].

Newer developments have concentrated on the simultaneous generation of hydrogen and oxygen from water molecules, with charge separation producing a photovoltaic effect [26]. Figure 16–5 illustrates a typical arrangement for this photoelectrolysis. In this arrangement, the photon absorption by the n-type TiO_2 electrode produces a photovoltaic potential sufficient for electrolysis of water. The n-type TiO_2 electrode acts as the anode [27]. The photon interacts with this anode to raise an electron to the conduction band. The electron then flows through the external circuit to the cathode. When the electron is raised to the conduction band, it leaves a hole that has sufficient electropotential to separate the OH^- ion from the hydrogen bond. This leaves an H^+ radical that is attracted to the cathode, where it reacts with the electron excess to produce hydrogen gas. The proton H^+ is separated in the

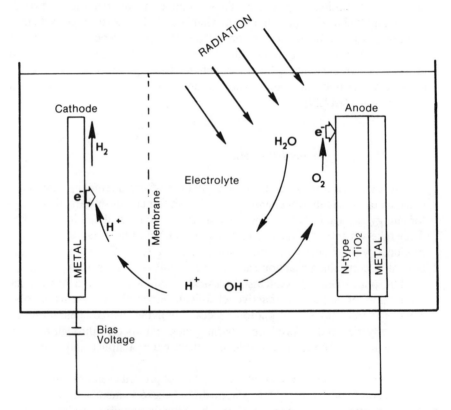

Figure 16–5. Photoelectrolysis Diagram for Production of Hydrogen and Oxygen with a Semiconductor Electrode

solution from the OH⁻ by the semipermeable membrane, permeable to the proton. The OH⁻ is not stable and reacts at the anode to give oxygen. An external bias is required to provide an external OH or pH gradient in these systems.

A further development is the photodiode, which does not require an external bias potential. A photodiode consists of *p*-type and *n*-type electrodes in contact. No external circuit is used. The active element may be in the form of microspheres suspended in a liquid solution. The cell produces hydrogen and oxygen. The oxygen is absorbed by an electrolyte, like acetate, supplied to the cell (see figure 16–6) [26,27]. This cell acts as a chemical reactor, producing hydrogen and various oxides [14,25].

An ideal electrical-storage system might consist of a photodiode that produces hydrogen and oxygen that are stored for use in a fuel cell. An alternative might be a reversible system in which either electricity or fuel is produced, depending on the demand.

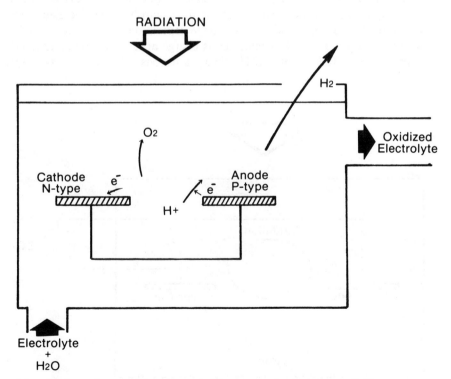

Note: Hydrogen is generated and separated from oxygen by electrolyte oxidation.

Figure 16–6. Schematic of a Photodiode with Semiconductor Electrodes

Electrical Energy from Electrochemical Photosystems

Although photochemical fuel-formation systems are important for energy storage, there are other classes of direct photoenergy conversion where electrical energy is produced. The electrical energy in these systems may then be stored electrochemically in "batteries." In the past, research has concentrated on solid-state photovoltaic devices. Liquid-junction photovoltaic devices have also been explored, as illustrated in figure 16–7. One advantage of these liquid-junction devices appears to be the reduced requirement for high purity of components. The principal difference between solid-state and liquid-junction devices is that the junction layer is a liquid rather than a solid. These cells may also have advantages with respect to control of detrimental charge-concentration effects, since the electrolyte is controllable through flow changes. They may also be more adaptable to cooling and waste-heat utilization. No efficiency advantages relative to solid-state devices have been demonstrated so far.

Photogalvanic devices represent another type of photoelectric system under development. These devices differ from photovoltaic devices in that they rely on a photochemical reaction in the electrolyte near the electrodes, rather than in the electrodes. The charge separation is achieved by redox reactions in which the electrodes act as charge acceptors. The original cell constructed by Rabinowitch [21] utilized a ferrous salt and a thiazine dye.

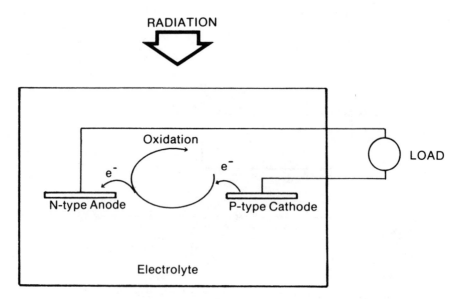

Figure 16–7. Photovoltaic Device with a Liquid-Electrolyte Junction

The reaction occurs near the illuminated electrode. Various other cells have been constructed in an attempt to slow down the reverse reaction. These have generally proved unsuccessful in that the efficiencies reported are less than 2 percent [22,23].

Cells with a membrane separating the electrodes and allowing only one type of charged particle to pass have been proposed. These would act in a similar manner to the membrane cells in the photosynthesis process. Figure 16–8 shows a hypothetical cell proposed by Calvin [24]. The electron from the donor atom tunnels through the membrane to the acceptor molecule and is prevented from returning by reacting on the acceptor side with a molecule that is too large to pass back through the membrane [19,28]. This system, if successful, would provide an important alternative to the photovoltaic cell, as energy storage in chemical form could be accomplished in the cell itself.

Note

1. The difference between fluorescence and phosphorescence is experimentally observed by the lifetime of the excited state. Fluorescence lifetimes range from micro- to nanoseconds, compared with a range from seconds to milliseconds for phosphorescence lifetimes.

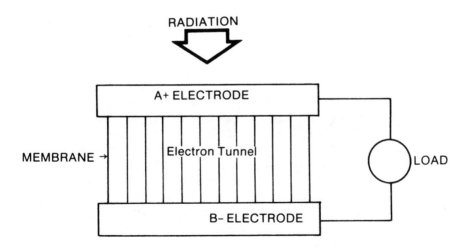

Source: Adapted from M. Calvin, "Photosynthesis as a Resource for Energy and Materials," *Amer. Sci.* 64(1976): 270.

Figure 16–8. Photogalvanic Cell Diagram of a System Proposed by Calvin with Electron Tunneling Through a Membrane from an Excited A Complex

References

1. Post, R.F., and Post, S.F. "Flywheels." *Scientific American* 229 (1973):17–23.
2. Foster, A.R., and Wright, R.L. *Basic Nuclear Energy.* Boston: Allyn and Bacon, 1968.
3. Bockris, J.O'M. *Energy, The Solar Hydrogen Alternative.* New York: Wiley, 1975.
4. Lichty, L.C. *Combustion Engine Processes,* 6th ed. New York: McGraw-Hill, 1967.
5. Kreith, F. "An Overview of Energy Technology for Intermediate Temperature Applications." *Eng. Sci. Perspectives* 3(1978):7–18.
6. Turner, R.H. *High Temperature Thermal Energy Storage.* Philadelphia: Franklin Institute Press, 1978.
7. Dean, T.S. *Thermal Storage.* Philadelphia: Franklin Institute Press, 1978.
8. Grodzka, P.G. "Phase-Change Storage Systems." In *Solar Energy Technology Handbook,* part A, ed. W.C. Dickinson and P.N. Cheremisinoff, pp. 795–809. New York: M. Dekker, 1980.
9. Green, R.M.; Ottesen, D.K.; Bartel, J.J.; and Bramlette, T.T. "Chemical Storage." In *Sharing the Sun: Solar Technology in the Seventies,* vol. 8, ed. K.W. Böer, p. 4. American Section, International Solar Energy Society. Newark: University of Delaware, 1976.
10. Mar, R.W., and Bramlette, T.T. *Thermochemical Energy Storage Systems.* Sandia Laboratories Report SAND 77–8051, 1978.
11. Miller, C.G. *Temperature Transformer Energy Conversion System.* NASA Technical Brief. Pasadena, Calif.: Jet Propulsion Laboratory, 1977.
12. Silverman, M.D., and Engel, J.R. *Survey of Technology for Storage of Thermal Energy in Heat Transfer Salt.* ORNL TM-5682, 1977.
13. Calvert, J.G., "Photochemical Processes for Utilization of Solar Energy." In *Introduction to the Utilization of Solar Energy,* ed. A.M. Zarem and D.D. Erway, p. 140. New York: McGraw-Hill, 1977.
14. Schwerzel, R.E. "Methods of Photochemical Utilization of Solar Energy." In *Radiation Energy Conversion in Space,* ed. K.W. Billman, p. 626. Progress in Astronautics and Aeronautics, vol. 61. New York: Ameican Institute of Aeronautics and Astronautics, 1978.
15. Turro, N.J. *Modern Molecular Photochemistry.* Menlo Park, Calif.: Benjamin Cummings, 1978.
16. Kirsch, A.D., and Wynan, C.M. "Excited State Chemistry of Indigoid Dyes." *J. Phys. Chem.* 81(1977):413.
17. Chang, C.K.; Heller, A.; Schwartz, S.; Menezes, S.; and Miller, B.

"Stable Semiconductor Liquid Junction Cell with 9 Percent Solar to Electrical Conversion." *Science* 196(June 1977):1097.

18. Archer, M.D. "Electrochemical Aspects of Solar Energy Conversion." *J. Appl. Electrochem*. 5(1975):17.

19. "The Current State of Knowledge of the Photochemical Formation of Fuel." (Paper presented by N.N. Lichton,) NSF-RANN Workshop at Osgood Hall, September 1974.

20. Forrester, A.R.; Hay, J.M.; and Thomson, R.H. *Organic Chemistry of Stable Free Radicals*. New York: Academic Press, 1968.

21. Rabinowitch, E. "The Photogalvanic Effect I and II." *J. Chem. Phys.* 8(1940):551, 560.

22. Clark, W.D.K., and Eckert, J.A. "Photogalvanic Cells." *Solar Energy* 17(1975):147.

23. Zaromb, S.; Lasser, M.E.; and Kalhammer, F. "Cyclic Photogalvanic Silver Halide Cells." *J. Electrochem. Soc.* 108(1951):42.

24. Calvin, M. "Photosynthesis as a Resource for Energy and Materials." *Amer. Sci.* 64(1976):270.

25. Balzani, V.; Moggi, L.; Manfrin, M.; Bulleta, F.; and Gleria, M. "Solar Energy Conversion by Photodissociation." *Science* 189(1975):852.

26. Gerischer, J. "Electrochemical Photo and Solar Cells: Principles and Some Experiments." *J. Electroanal. Chem.* 58(1975):263.

27. Manassen, J.; Hodes, G.; and Cahen, D. "Photoelectrochemical Energy Conversion and Storage." *J. Electrochem. Soc.* 124(1977):532.

28. Bolton, J.R. *Solar Power Fuels*. New York: Academic Press, 1977.

17 Bioenergetics and Thermodynamics

Ecology is really an extension of economics to the whole world of life.
—H.G. Wells, J. Huxley, and G.P. Wells, "The Science of Life" (1933)

Thermodynamic Considerations of Energy Flow in Biological Systems

The principal characteristics of biological systems, from a thermodynamic perspective, is their energy-flow behavior. The application of available-energy analysis to these systems requires an understanding of their non-equilibrium nature. This chapter will emphasize the dependence of these systems on energy flow to maintain nonequilibrium states. In appendix 17A, a summary of the concept of diversity, developed extensively in ecological science, is outlined in the perspective of its application to engineering.

We can calculate empirically the energy used in the production of a quantity of wheat, corn, or chickens. This accounting process, however, does not get at the ideal in the sense of what minimum energy is required to produce a biochemical product. A *biological free energy* analogous to the free energy for extraction of pure water from seawater would be useful if an available-energy function is to be used as a standard basis.

The efficiency of primary production systems is often determined from the input sunlight energy and the heat of combustion of the resultant plant. This energy efficiency neglects the entropy change associated with the informational or structural available energy. It neglects the importance of structure in biology. It is the informational structure that is important, not the total energy stored; otherwise we could live on hot air. Gorski[1], in a careful analysis of the entropy change in the production of biomass, has calculated the entropy associated with growth. He showed that if the structural entropy is neglected, there is an increase in entropy of the biochemical constituents. This means that the entropy of the biochemical products of growth is higher than the entropy of the growth media used.

In an analysis of a biological process, the Gibbs free energy, $G \equiv H - TS$ is utilized extensively, since the process is often a steady-flow process at constant temperature. The TS factor involves structural factors that arise from the configurational entropy of organic molecules, biomass, and so forth. The Gibbs free energy is not as useful a measure for biological processes that require radiation energy, as in plant growth. One reason is that in the approximation of black-body radiation, the Gibbs free energy of radiation is zero. The consideration of a flux of free energy from the sun must be reconstructed in terms of available energy. The available-energy flux $H - T_0 S$ provides a better measure of the input-energy quality. (See chapter 14 for a more complete discussion of the available energy of solar radiation.)

Living systems are sustained by available-energy fluxes and can be considered as open systems that use these fluxes to build and maintain structure. Systems far from equilibrium have high information and available-energy content. Macroscopic measurements on such systems are expected to provide more information about the microscopic state than do measurements on equilibrium systems. This follows from Jaynes's analysis [2] that equilibrium states are ones of maximum uncertainty. Callen [3] expressed this by stating that equilibrium states are characterized by very few measurables such as internal energy, volume, and mole numbers of chemical species.

Departure from equilibrium is sometimes usefully described by the rate of approach of the system to equilibrium. Observations from rate experiments have shown that:

1. The further from equilibrium a system is, the faster is its rate of approach to equilibrium.
2. The higher the temperature, the faster the approach to equilibrium.
3. The equilibrium state depends on the temperature.
4. At high temperatures it is difficult to maintain nonequilibrium states.

In terms of the thermodynamic discussion of exergy in chapter 6, the available energy of the nonequilibrium state with probabilities P_i compared with the equilibrium state with probabilities P_{io}, is

$$A = kT_0 \Sigma P_i \ln \frac{P_i}{P_{i0}}.$$

In nonequilibrium biological states, certain energy states are highly populated compared with what is expected in an equilibrium distribution. Their energy distributions are thus similar to what in solid-state science are called negative temperature systems.

The measure of the distance of such a nonequilibrium state from an equilibrium state is difficult to specify. Several operational means have been proposed [4,21].

1. Take a nonequilibrium state and isolate it adiabatically. Then allow it to decay into some equilibrium state. Measure the characteristics of this state compared with those of the original nonequilibrium state.

2. Allow the nonequilibrium system to contact a constant-temperature

reservoir at the kinetic temperature of the nonequilibrium system. Use this resultant state as the reference condition.

The available energy is a convenient measure because it is easily calculated from the partition function and because it is applicable to open systems. The difference in this function between nonequilibrium and equilibrium states would be a useful measure of nonequilibrium. However, it does not give an indication of the energy flow required to maintain the nonequilibrium state. The measure

$$I \equiv A(T')/kT'$$

can be taken as a starting point, where T' is the kinetic temperature of the nonequilibrium state, and $A(T')$ is then the available energy of the nonequilibrium state. The available energy appears as the energy of chemical bonds compared with kinetic energy at equilibrium. The temperature T' is assumed as the temperature attained if this chemical energy were converted to kinetic energy. The measure is dimensionless, and the energy difference is primarily represented by the internal energy modes of potential interaction energy, mechanical and electric potentials, and molecular bondings. These nonequilibrium modes of energy are characterized by slower approach to equilibrium than kinetic-energy states like rotation, vibration, and translation. They also require a flow of energy through the system to maintain the nonequilibrium states.

It is important to remember that flow of energy into a system requires kinetic-energy flow. It is not the ordered states of a source that induce flow. Solar-radiation absorption by a plant is a kinetic-energy form. This kinetic-energy form is then converted to ordered forms, which then decay to lower-temperature kinetic energy. A question that arises, however, is whether the flow into the system is governed by the temperature or the probability-distribution difference between the source and the system. In the case of photosynthesis, energy is transferred into a plant through a selective spectral absorption. Thermal energy is also absorbed in a nonselective way. How this total energy absorption is apportioned is examined later in this chapter.

In many cases the biological structure is maintained and energy stored in a manner that varies with the input-energy flow. At low energy input, little energy storage or ordering occurs. At high flows, the energy storage or ordered flow output (work) declines as the capacity of the system is exceeded. This implies a performance characteristic, as shown in figure 17-1.

Consideration of a hydraulic analogy following Morowitz [4] is helpful in describing the nonequilibrium state maintained by a throughflow, (see figure 17-2). Water flows into a cylinder that has holes that increase in size from

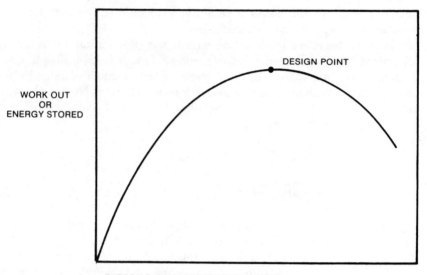

Note: Shows a design-point maximum at an intermediate-energy flow-through the system.

Figure 17–1. Energy-Conversion Characteristic Curve

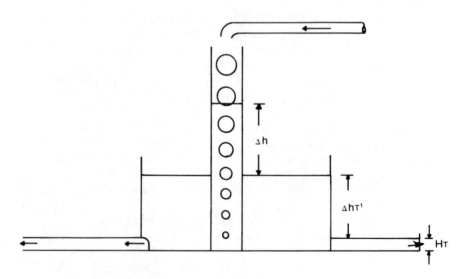

Figure 17–2. Water-Flow Analogy of a Thermodynamic-Bioenergetic System Maintained in a Nonequilibrium State by Through-flow of Energy

the bottom of the tank. The middle tank has one hole through which water flows into the reference-level tank. At low input flows the levels in all tanks are equal. In energy terms, not enough energy flow is available to build order. In biological systems this low energy flow goes into maintaining respiration losses. As the flow is increased, differences in the liquid levels in the tanks occur. The multiholed tank, representing structure in biological systems, has an increasing level relative to that of the middle tank, representing the kinetic temperature state. The increasing size of the holes allows flow between the multiholed tank and the middle tank to increase as the middle-tank height increases. This simulates the increase in the energy exchange between structured states and kinetic states at higher kinetic temperatures. A maximum height difference or nonequilibrium is attained at some intermediate flow rate by this exchange. At higher flow rates the kinetic and structured states interact more strongly, and their differences become smaller. At the highest flow rates the levels of these two states become nearly the same, as shown in figure 17–3.

The maximum difference in heights, representing the maximum order in the system, is difficult to predict. In plant systems it represents a maximum

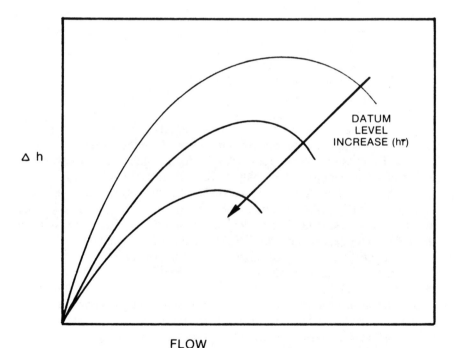

Δ h

FLOW

Figure 17–3. Structure Represented by Δh as a Function of Flow in a Nonequilibrium System

structure or information content maintained by an energy flow from high available energy, solar energy, to transpiration and evaporative thermal losses and respiration.

At highest flows, as in a climax ecological community, the energy goes into maintenance instead of growth. In reference to Boulding's [5] ideas regarding a spaceship economy, the energy goes to maintaining the capital stock.

Another consideration is the question of the effect of the reference or environmental state on the structure. If the environmental temperature increases, does the energy flow required to reach a maximum structure increase or decrease? The model indicates that at a higher reference level the structure maximum will decrease. The maximum structure will occur at a lower energy flow. For example, if the enviromental temperature is increased beyond a certain point, a plant will decrease its productivity.

In the model, the larger holes at the top of the receiver tank represent the observation that the high-potential-energy states are less stable and will tend to decay to equilibrium states more rapidly. If the order measure I is multiplied by Planck's constant h to give

$$I' = hI = \frac{A}{kT/h} \, ,$$

a ratio of stored molecular-bond energy to a rate constant representing the rate of decay to equilibrium is then obtained.

This I' could be considered as a measure of the time the energy is stored in the system as it flows through. This is similar to taking the total mass M of a system and dividing by the flux of material through, with the ratio representing the average time a given mass spends in the system.

The I' measure also agrees with ecological thinking; for example, Margalef's comment [6] that a complex ecosystem rich in information needs a lower amount of energy to maintain such a structure. In a highly organized system, a high I' is maintained by a small energy flow kT/h compared with its available-energy content. The model corresponds to ecological principles to this extent but does not say why a highly organized system would evolve or whether the organization is optimized in a given system.

A photochemical example might be considered. Suppose CO_2, N_2, and H_2O are in a container at equilibrium with a datum reservoir. With little or no energy flow through the system, it will remain near equilibrium. Now suppose the system is irradiated with photons. Photochemical reactions will occur that will increase the available energy by the occupation of more improbable states. These nonequilibrium states will decay back to equilibrium states, but the more stable states will persist longer, leading to reactions among the more stable states. Selection for these more stable states can then occur. The higher the stability of these states, the larger is the I' information residence time and also the residence time of the mass. Margalef [6] discusses this in terms of biomass residence time, but it is probably better visualized as an information residence time in this analogy.

A principle from this observation would state that the steady state that is finally reached is one in which the system is maximally far from equilibrium, and the energy flow is adjusted through the rise in the thermal states until this maximum-nonequilibrium condition is reached. In the hydraulic analog, the Δh between the excited modes and the thermal modes could be maximized by controlling the energy flow, as shown in figure 17–4.

A organism controls both the outflows and inflows of energy. A plant transpires to increase the through-flow when the input flow is too high. It also moves its leaves relative to the sun to reduce the input flow. Photorespiration, as well as photosynthesis, increases with higher light intensity, thus increasing the flow through the system. This important point is often neglected in discussions of plant growth. Photorespiration accompanying photosynthesis provides a function whereby the structure-building flow can be increased.

In the hydraulic analog, reduction of the output alone will not be effective in reducing the input. If the output hole is decreased in size, representing a higher impedance to flow to the environment, the system will move to a higher kinetic-temperature level or higher thermal-energy state, but not a decreased flow. A rise in the reference level will do the same thing.

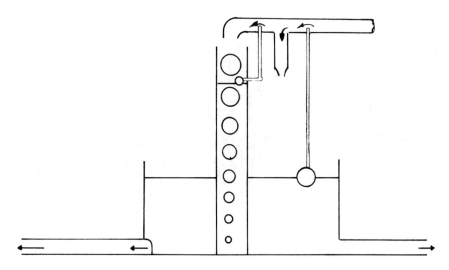

Figure 17–4. Illustration of a Model for Structure Maximization by Input-Flow Control

Saturating the environment of a plant with CO_2 or raising the environmental temperature will change the plant's thermal state in a similar manner.

Now consider the control of the input flow. One possibility is to keep the total input flow constant but divide it between the excited- and the thermal-state containers. Any diversion to the larger tank (thermal state) in this model will move the system toward a lower I or h. Shade control is not of this type. The shade control diverts flow outside the tank. A system with a flow-diversion control would attempt, if maximizing I, to get all energy flow into the excited states.

An energy-circuit description of this system is shown in figure 17–5. In this discussion A_1 represents the energy storage in the chemical-bond form, A_2 the thermal storage, and A_3 the energy level of the environment. This energy diagram can be written in system form as

$$A_1 = J - K_1(A_1 - A_2),$$
$$A_2 = K_1(A_1 - A_2) - K_2(A_2 - A_3).$$

In the diagram the flow between A_1 and A_2 is proportional to the level in A_1, or $K_1 = CA_1$. (This represents the holes increasing in size as A_1 increases). The environment is represented by A_3, so that

$$A_3 = K_3(A_2 - A_3) - M.$$

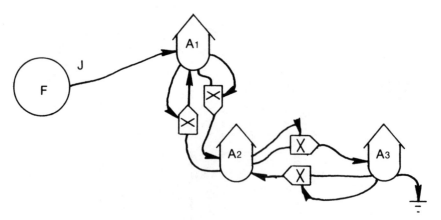

Note: Odum's [13] notation is used in this diagram.

Figure 17–5. Energy Circuit Diagram of the Simple Hydraulic-Bioenergy Analog

Then

$$A_1 = J - CA_1(A_1 - A_2),$$

$$A_2 = CA_1(A_1 - A_2) - K_3(A_2 - A_3),$$

$$A_3 = K_3(A_2 - A_3) - M.$$

In the fluid analog K_3 and K_2 depend on the hole sizes. If the environment is fixed and not affected by the energy flow, then $A_3 = 0$ and

$$M = K_3(A_2 - A_3).$$

An analog simulation of these equations shows the characteristics of figure 17–3. As J increases, a redistribution of energy must occur. As this J increases, higher energy levels become accessible, and more chemical combinations are possible. There is a limit, however, since dissociation will occur at high-energy electronic states. With electronic states involved, reaction rates proceed rapidly, and stable compounds are difficult to maintain.

The limit of high temperatures gives an upper limit where all states are in equilibrium with the source of the energy at high flow rates. Sustained order requires an intermediate flow rate. Although this conclusion is logical, it is not established because of the tentative nature of the argument about the equilibrium at high temperatures. It appears to be true for photosynthetic processes but is modified by the interaction of solar-energy flows with the availability of water and CO_2 required to provide replacement materials lost through respiration.

A crucial part of any analysis of biological systems involves the determination of the "structural entropy." Thermodynamic measurements of the properties of biological species indicate higher entropy values than expected from structural analysis [17,21]. The difference has not yet been resolved. It is generally agreed that biological molecular arrangements are highly organized (improbable configurations). The value of this organization is realized only by other biological organisms. This includes high efficiency of assimilation of organic matter (prey) by predators and utilization of organic fertilizers for plant growth (the development of high-information structures from low-information molecules). Gorski [1] even describes an organism in terms that indicate that it begins with a given organizational capacity, which is used up during the assimilation and growth process.

Biological processes in waste-water systems can use structure to advantage. Combustion processes, on the other hand, use only the energy content of waste products. The implication for future engineering work is that such processes as fermentation and digestion will provide more efficient waste-recycling opportunities.

Primary Productivity: Biomass-Production Systems

The use of primary productivity of plants as an effective energy resource requires decisions about the quality as well as the quantity of the biomass produced. In discussing primary productivity, one must be careful to differentiate reported biomass-production figures. For our purposes we will examine first the material production expressed as organic matter in $kg/m^2/year$. Productivity in these terms will be the organic matter produced per unit ground area per time interval. If a rate is desired, division by the length of the growing season will give a measure of the intensivity of the productivity. During growth spurts, even higher rates would be obtainable. In annual or semiannual crop harvesting, the productivities are reported as maximum annual organic productivities [18]. The comparison is simpler than for long-standing harvest of trees, where maximum yearly productivity varies considerably with age. The organic productivity can be used as the basis for rough calculation of important parameters of fuel-energy content, material production, oxygen production, and assimilative capacity, as discussed later.

One is rarely interested in only the organic productivity. In food production, the protein, nutritional, and calorific value are more important. In construction materials the volume of the dry matter is more important than the weight of organic matter. In an organic fuel, knowledge of the biochemical structure—whether cellulose, lignin, glucose, and so on—and the water content is required. The energy required to dry the crop for storage is also important. If the water content of the material before combustion is high, the heating value will be reduced. For a chemical feedstock many other factors must be considered.

For rough calculations from reported productivity statistics, the following approximations are generally applicable.

Dry-Weight-to-Fresh-Weight Ratio

This is perhaps the most variable parameter, since the dry weight may be from 5 percent of the fresh weight in water plants to 80 percent in desert plants and woody materials. The observed large area of submerged water weeds in eutrophic water is illusory in terms of productivity, since the water content is 80 percent by weight. Dry weights are usually reported as weights obtained after heating samples in an oven at 104–105°C until the weight remains constant. The dried material, if left in air, will pick up water so that 5–20 percent of its measured weight is water.

Organic Matter from Dry Weight

A useful factor is 95 percent for most materials, although values may range down to 20 percent for diatoms containing mostly silica. Aquatic plants generally have low ratios, on the order of 50–75 percent.

Carbon-to-Organic-Matter Ratios

Organic molecules may contain many different ratios of carbon, hydrogen, oxygen, nitrogen, and so forth. The best estimate for most naturally occurring materials is about 50 percent. This may be a measure of CO_2 assimilative capacity, but it has little meaning for chemical estimates. For estimating combustion value of natural materials, a value of 4–5 kcal/gm is a good approximation. Note that this is greater than the heat of combustion of glucose (3.73 kcal/gm), often used in determining energy efficiency of photosynthesis.

Photosynthetic Quotient

The form and composition of the organic molecules determine the CO_2 assimilative ratio compared with the oxygen production. The general chemical-balance equation

$$CO_2 + H_2O \rightleftharpoons (CHO) + O_2$$

indicates an equal number of molecules of O_2 produced per molecule of CO_2 assimilated, or a photosynthetic quotient

$$PQ \equiv \frac{N_{O_2}}{N_{CO_2}} = 1.$$

Actual measured quotients average about 1.2, increasing to 1.4 for high-productivity agriculture with adequate water and nutrient availability, and decreasing in dry-land agriculture or desert plants. This implies that high-productivity plants in general have higher oxygen-to-carbon ratios and hence

Table 17–1
Productivities of Different Ecosystems on Fertile Sites

Organic Productivity, Kg/m²/year	Percentage Range	Climate	Ecosystem
0.1	50	Desert arid	Desert
0.2	50	Ocean	Ocean phytoplankton
0.2	50	Temperate	Lake phytoplankton
0.3	50	Coastal	Coastal phytoplankton
0.6	50	Temperate	Polluted lake phytoplankton
0.6	20	Temperate	Freshwater submerged macrophytes
1.2	25	Temperate	Deciduous forest
1.7	25	Tropical	Freshwater submerged macrophytes
2.0	25	Temperate	Terrestrial herbs
2.2	15	Temperate	Agricultural annuals
2.8	25	Temperate	Coniferous forest
2.9	15	Temperate	Marine submerged macrophytes
3.0	20	Temperate	Agricultural perennials
3.0	20	Temperate	Salt marsh
3.0	20	Tropical	Agricultural annuals
3.5	15	Tropical	Marine submerged macrophytes
3.8	20	Temperate	Reedswamp
4.0	15	Subtropical	Cultivated algae
5.0	20	Tropical	Rain forest
7.5	15	Tropical	Agricultural perennials and reedswamp

Sources: D.F. Westlake, "Comparison of Plant Productivity," *Biological Review*, vol. 38 (Cambrdge: Cambridge University Press, 1963), p. 385; H. Lieth and R.H. Whittaker, eds., *Primary Productivity of the World* (Chapel Hill: University of North Carolina Press, 1973).

lower combustion values. A quotient of 1.2 implies that the carbon-to-oxygen ratio in the organic matter is 0.278 rather than 0.375, if a quotient of 1 is assumed. This also indicates that the oxygen evolution is higher in high-productivity agriculture and water-associated plants than in dry-land agriculture. Irrigation systems therefore produce more oxygen than one might expect from the simple productivity increase.

Accurate measurements of productivity are difficult to make and analyze. Measures including roots at the peak of biomass are especially difficult. Such factors as respiration (10–60 percent of gross production), symbiosis or sharing of root systems, and insect or herbivore losses can cause large uncertainty in the measurements. Table 17–1 summarizes the type and range of productivity reported under optimum conditions for different ecosystems.

A comment on algae culture and its position on this list should be noted. Algae under optimum conditions have not shown higher productivity than reed swamps, intensive sugarcane culture, or rain forests in terms of overall productivity. There are, however, several advantageous aspects of algae culture that must be considered.

1. Water uptake per unit productivity is smaller for algae, giving them an advantage in water-limited environments.
2. Control of the atmosphere (CO_2, temperature); nutrients; and light is more easily obtained with small-size phytoplankton.
3. Continuous-growing–harvesting systems can be used where advanced chemical control can be applied.
4. Cyanophyte algae (blue-green) have the ability to fix nitrogen from the air and hence to enrich the water in which they are grown. These algae provide nitrogen fixation required to maintain rice production and fish culture in Southeast Asian agriculture.
5. The protein content of algae can be enhanced through selective culture to increase the food value of the crop.

Note that the organic productivity of natural communities is usually adapted to a maximum for a region. Introduction of intensive agriculture in replacing these ecosystems does not generally increase the productivity. The energy requirements to maintain intensive agricultural systems put them at a further deficiency in accounting for efficiency. Counterexamples, where irrigation systems or added nutrients are utilized, can be found. If the objective is maximum biomass productivity under natural conditions, then communities as evolved are usually well adapted.

Specific high-productivity species are noted in table 17–2. Even in these species the efficiency of conversion over the whole year does not generally exceed 2 percent of the total solar radiation. Those interested in commercial applications should consult experts with knowledge of local situations and the present state of hybrid development as well as control practices. The adaptability of these species can be a determining factor in their use in other than their natural environment.

Note that productivity does not measure the value of an ecosystem. The efficacy does not depend on the total organic matter produced. It does aid in evaluating the assimilative capacity of different ecosystems and in pointing out the importance of water-associated vegetation such as marshes, swamps, and littoral and tropical regions.

Photosynthesis Processes

This section will show thermodynamic and kinetic limitations on the photosynthetic process. A simple model corresponding to a logistic system will be considered. The model is used to demonstrate how the efficiency depends on solar radiation and CO_2 availability. In particular, the inherent decrease in the efficiency of primary productivity with increased energy flow is outlined.

Table 17–2
Specific High-Productivity Species

Organic Productivity, Kg/m²/year	Community Species
	Marine
3.2	Laminaria longicuris (seaweed)
0.3	Phytoplankton
3.2	Spartina alteriflora (salt marsh)
	Freshwater
4.5	Scenedesmus sp algae (sewage)
5.8	CO_2-enriched algae growth
4.6	Scirpus lacustris bulrush
7.0	Cyperus papyrus
2.9	Typha hybrid
4.4	Eichhornia crassipes angiosperm (floating)
	Forest
2.0	Common Alder (Deciduous)
2.3	Scots Pine (Coniferous)
4.5	Grand Fir (Coniferous)
5.9	Tropical Rain Forest
	Grasses
8.5	Digitaria decumbers tropical
3.0	Alfalfa
	Cultivated Plants
1.8	Barley
3.4	Corn
3.5	Rice
2.8	Sugar beet
8.7	Sugar cane
2.5	Wheat

Sources: D.F. Westlake, "Comparison of Plant Productivity," *Biological Review*, vol. 38 (Cambridge: Cambridge University Press, 1963), p. 385; I. Zelitch, *Photosynthesis, Photorespiration and Plant Productivity* (New York: Academic Press, 1971); H. Lieth and R.H. Whittaker, eds., *Primary Productivity of the World* (Chapel Hill: University of North Carolina Press, 1973); R.S. Loomis, W.A. Williams, and A.F. Hall, "Agricultural Productivity" *Ann. Rev. Plant Physiol.* 22(1971): 431.

The relationship between maximization of productivity and efficiency is illustrated, indicating that if one wants to increase productivity, one does not seek to maximize efficiency. It will be assumed that sufficient water is available so that it is not limiting on growth. A more complicated model would include water as an input parameter.

Primary production as a biological process is often presented as a mechanism whereby carbon in inorganic form is fixed to produce organic compounds. This conversion is usually not from elemental carbon, but from carbon dioxide for terrestrial systems or dissolved carbonates for aquatic systems. The conversion is mainly through the photosynthetic process. As a rough estimate, each gram of carbon fixed is incorporated in 10 grams of biomass (wet weight).

The most common formulation of the photosynthesis process is represented by a net chemical-balance equation where glucose is formed:

$$6 \ CO_2 + 12 \ H_2O \xrightarrow[\text{green plant}]{\text{light}} 6 \ O_2 + C_6H_{12}O_6 + 6 \ H_2O.$$

The required inputs are CO_2 and water (H_2O). It is important to note that the oxygen evolved as output is from the water molecule, not the CO_2 molecule. Provision of hydrogen as a reducing-agent input, rather than water, could lead to production of a carbohydrate with no oxygen generation. The hydrogen donor for fixing CO_2 to a carbohydrate molecule is hydrogen sulfide in certain photosynthetic bacterial reactions where the chemical balance is

$$CO_2 + 2 \ H_2S \xrightarrow{\text{light}} 2S + \text{carbohydrate} + H_2O.$$

Note that this reaction without light input is the basis of the deep-sea biological systems near volcanic rifts.

For photosynthetic processes, two separate measures are used to express the molecular-conversion efficiency.

$$\text{Light-conversion efficiency: } E_\lambda \equiv \frac{\text{chemical energy in organic bonds}}{\text{light energy absorbed}}.$$

$$\text{Quantum yield: } \phi \equiv \frac{\text{moles of carbohydrate converted}}{\text{moles of photons absorbed}}$$

If the chemical bond energy is all in carbohydrates, the two are related as

$E_\lambda = H/(N \ hc/\lambda)\phi.$

H = heat of combustion of one mole of CH_2O (4.7×10^{12} ergs).

N = Avogadro's number (6.023×10^{23}).

h = Planck's constant (6.06×10^{-27} erg·sec)

c = velocity of light (3×10^{10} cm/sec).

λ = wavelength of light (cm).

The denominator represents the energy content of a mole of photons, often termed an *Einstein*.

If glucose is formed, the hydrogen ion H^+ addition to the CO_2 molecule requires 678 kcal/mole of glucose. Experimental photosynthesis work has shown that the reduction of CO_2 in the photosynthesis process requires 8–10 light quanta per molecule of CO_2, or 2,020 kcal/mole of glucose. This gives a conversion efficiency for photons in the active spectrum of

$$E = \frac{678}{2,020} \times 100 = 33 \text{ percent.}$$

The energy absorbed may or may not be incorporated in the photosynthesis process. Measurement of the quantum-yield variation with light wavelength shows characteristics similar to those shown in figure 17–6 [8,9]. This yield is highest in the 0.4–0.7-micron wavelength region of the spectrum. The average yield in this region is about 0.084. The fraction of the energy in a typical solar spectrum in this wavelength band is about 0.45 (see figure 14–1). This gives a potential of $0.084 \times 0.45 = 0.038$ of the solar energy converted to photosynthetic-energy bonds. Low-light-level con-

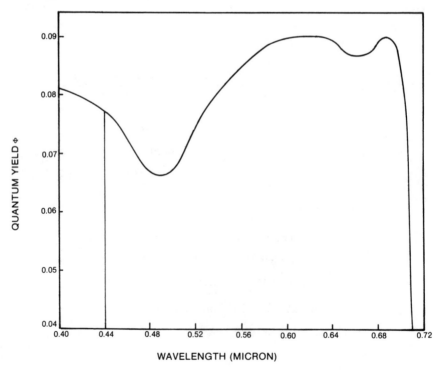

WAVELENGTH (MICRON)

Figure 17–6. Typical Quantum Variation with Light Wavelength for Algae

version efficiencies of 14 percent and overall 5-percent efficiencies can be obtained in the laboratory. The next section expands on the limitations of kinetics and thermodynamics on the efficiencies.

The mass-flux dependence on the potentials of light intensity and chemical availability will first be established through a model with reference to experimentally observed photosynthetic activity. Reference to recent summaries [8,16] of the present state of knowledge of the photosynthetic process should be consulted for specific details and an understanding of the complexities of biological process. In this section an extremely simplified model based on Michaelis-Merten enzyme kinetics and a cycling receptor system will be posed. Once the flux-potential relations are established, they will be incorporated into an energy-flux model steady-state thermodynamic system to illustrate the efficiency-energy-flux characteristics. Conclusions will be drawn from this analysis indicating (1) the limitations of increasing photosynthetic productivity and'(2) the inherent difficulties in extending the high-efficiency results of photosynthetic experiments carried out at low light intensities to high-productivity agriculture.

Let us start with a qualitative model and then attempt to put the statements into quantitative form for analysis (see figure 17–7). (1) Light

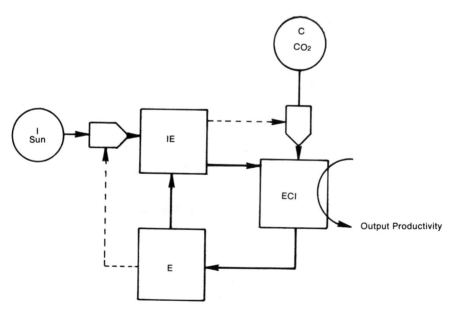

Note: Control lines are indicated by dotted lines.

Figure 17–7. Flow Diagram of Photosynthesis as a Simple Cycling-Receptor System with Control of Uptakes of CO_2 and Sunlight by Substrate Availability

with energy in the range of the activation energy of chlorophyll molecules is accepted and transformed into biomass through an acceptor system. (2) The activated acceptor combines with CO_2 and water to produce the synthesized biomass. (3) Along with the photosynthesis process, a photorespiration process is occurring in which activated substrates are utilized for metabolic processes producing CO_2 and providing energy for increased activated acceptor formation. These may be represented very roughly by linear-rate equations, as follows:

$$E + I \rightleftharpoons IE,$$

$$IE + C \rightleftharpoons ECI,$$

where E = concentration of uptake sites free to take up light energy.

I = light intensity or number of photons in the activation range times their activation energy delivered.

C = CO_2 availability or concentration of CO_2.

IE = light-activated molecules.

ECI = activated photosynthetic-product molecules.

The linear rate relations will be subject to a constraint on the total substrates present in the plant. This will be noted as

$$E + IE + ECI = Z. \qquad (17.1)$$

The productivity is taken as proportional to the activated substrate ECI.

$$\text{Productivity} = K \cdot ECI \qquad (17.2)$$

A diagram indicating the flows and controls is shown in figure 17–7. Kinetic equations for the cycling-receptor system are:

$$\frac{d(IE)}{dt} = K_1 E \cdot I - K_2 IE \qquad (17.3)$$

$$\frac{d(ECI)}{dt} = K_3 IE \cdot C - K_4 ECI \qquad (17.4)$$

If the system is in steady state such that the substrate compartment totals are constant then

$$\frac{dZ}{dt} = 0 = \frac{dE}{dt} + \frac{d(IE)}{dt} + \frac{d(ECI)}{dt} \qquad (17.5)$$

and dE/dt can be calculated from the equations 17.3 to 17.5. An overall balance would include the input of water and the output of oxygen as well. We are interested in the output as a function of the light and CO_2 inputs. If a climax state has been reached then no net photosynthesis is present and the respiration equals the photosynthesis rate. The energy goes into maintenance of the structure.

From examination of the rate equations, and considering that we are interested in that adjustment of the activation energies that will maximize the output, then ECI will be maximum when $d(ECI)/dt = 0$. This implies that there is a mechanism in the cycling-receptor system such that the interactions will provide a steady-flow system. In reality, of course, the growth process will increase the total substrate Z as well, but for a steady-state production we are interested in how the output productivity depends on the input factors of light and CO_2 concentration.

This zero rate of change with time of the substrates is an approximation in keeping with Weiss' view of a biological structure [17]. In this view the cycling of substrates varies rapidly with time but the time-average populations of each of the compartments remains relatively constant. Similarly local equilibrium in irreversible thermodynamic theory assumes local equilibrium for defining properties, which in any individual section are changing with time. The rate of change of the total compartments is slow relative to the production of biomass since most of the growth occurs in structures (that is, stems, roots, seeds, branches, and so on), which are large compared to the active photosynthetic apparatus of the leaves. It is this relative constancy with which we are dealing in this section.

Assuming a steady state system, then equations 17.3 to 17.5 become

$$E \cdot I = \frac{K_2}{K_1} IE \qquad (17.6)$$

$$IE \cdot C = \frac{K_4}{K_3} ECI . \qquad (17.7)$$

These together with equation 17.1 can be solved in terms of Z, C, I for the photosynthesis factor.

$$ECI = \frac{Z \cdot I \cdot C}{\left(\dfrac{K_4}{K_3} + c \right) I + \dfrac{K_4}{K_3} \dfrac{K_2}{K_1}} \qquad (17.8)$$

If the photosynthesis is proportional to the substrate ECI then

$$J_p = K \cdot ECI . \qquad (17.9)$$

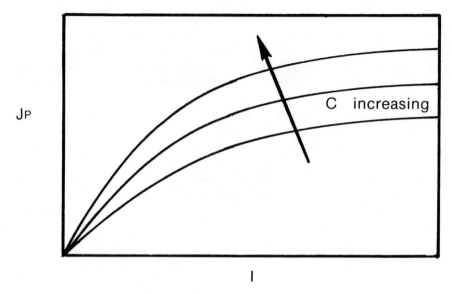

JP

C increasing

I

Figure 17–8. Photosynthetic Productivity as a Function of Light Intensity for Variable CO_2 Concentrations, as Predicted by the Cycling-Receptor Model

At high light intensities, the limit of net photosynthesis approaches

$$J_N = \frac{KZ \cdot C}{\dfrac{K_4}{K_3} + C} \tag{17.10}$$

Figure 17–8 shows the productivity as a function of light intensity for variable CO_2 concentrations predicted by this cycling receptor model. These characteristic curves can be compared with typical experimental observations shown in figure 17–9. The model has limitations at high light intensities, but the general performance is demonstrated.

Examination of figure 17–9 indicates that increasing CO_2 concentrations at high light intensities can significantly increase productivity. This is typical of moderate-productivity plants. Algae cultures can achieve high productivities with artificially supplied CO_2 in closed environments. Much work has been done to develop air circulation that will increase the availability of CO_2 at leaf surfaces. In most crop situations it is usually water or nutrients that limit growth, rather than CO_2. The availability of uptake sites associated with stomatal opening is highly dependent on water presence at these sites. The

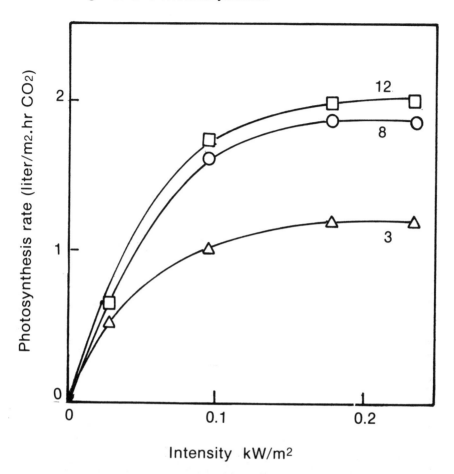

Note: Productivity is determined by the uptake rate of CO_2. CO_2 concentration values are to be multiplied by 10^{-4} to give actual concentrations in mole fractions [9, 11].

Figure 17–9. Typical Experimental Variation of Photosynthetic Productivity with Light Intensity for Different CO_2 Atmospheric Concentrations

closure of these stomatal openings in water-deficit situations reduces CO_2 uptake more than CO_2 depletion in the crop vicinity does.

Normal CO_2 concentration in the atmosphere is approximately 3×10^{-4} moles/mole. With increased fossil-fuel combustion, this concentration in the atmosphere has been observed to rise. This rise would be expected to increase the photosynthesis rate. However, there is a limit to this increase in photosynthesis rate with higher CO_2 concentration. Experimental studies indicate that increasing the CO_2 concentration above 10×10^{-4} does not increase productivity.

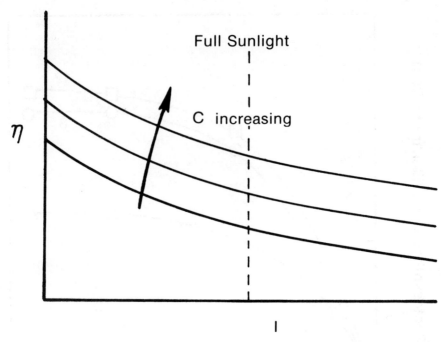

Note: CO_2 concentration changes noted by C.

Figure 17–10. Photosynthetic Efficiency as a Function of Light Intensity from the Cycling-Receptor Model

For discussion purposes, let us define a photosynthetic efficiency as the productivity divided by the light intensity. Using equation 17.8 this efficiency can be written as

$$\eta \equiv \frac{J_p}{I} = \frac{Z \cdot C}{\left(\dfrac{K_4}{K_3} + C\right) I + \dfrac{K_4 \, K_2}{K_3 \, K_1}} . \qquad (17.11)$$

The dependence of the efficiency on the light intensity using this model is shown in figure 17–10. This can be compared with an experimental efficiency variation with intensity, given in figure 17–11. Examination indicates that the actual process efficiency at low light intensity is not dependent on CO_2 concentration. The general behavior with light intensity is indicative of the importance of rate processes in photosynthesis. This is in agreement with the discussion of energy conversion in chapter 7. It is important to note that the high efficiencies of photosynthetic processes reported at low light levels are not applicable to normal light levels.

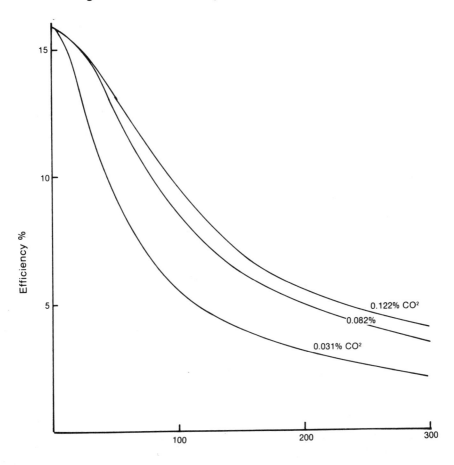

SOLAR INTENSITY

Note: Experimentally determined by measurement of CO_2 uptake, assuming glucose formation [8, 9].

Figure 17–11. Energy Efficiency of a Typical Photosynthetic Process as a Function of Light Intensity for Variable Environmental Concentrations of CO_2

The efficiency can also be found as a function of the productivity by solving equation 17.8 for I in terms of J_p to give

$$I = \frac{\dfrac{K_4}{K_3}\dfrac{K_2}{K_1}J_p}{KZC - \left(\dfrac{K_4}{K_3} + C\right)J_p} . \qquad (17.12)$$

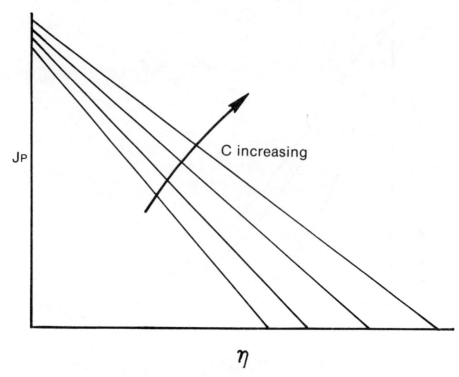

Figure 17–12. Photosynthetic Productivity Variation with Efficiency for a
Cycling-Receptor Model

Substituting into equation 17.11 gives the efficiency as a function of
productivity:

$$\eta = \frac{KZC - \left(\dfrac{K_4}{K_3} + C\right) J_p}{\dfrac{K_4}{K_3}\dfrac{K_2}{K_1}} \qquad (17.13)$$

The productivity as a function of efficiency can then be plotted as shown in
figure 17–12. Comparison of this behavior with the thermodynamic discussion of chapter 7 indicates that such a photosynthetic system is represented by a source of infinite impedance. This is characteristic of a plant that
does not change its absorptivity with light intensity or output productivity.

It should be noted that the experimental results shown were achieved
under controlled conditions at constant temperature. If the temperature is not
controlled by removing thermal energy, the plant temperature will increase,
giving a performance like that shown in figure 17–13. The productivity

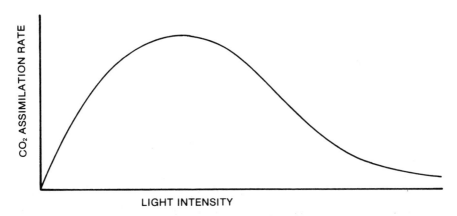

Figure 17–13. Experimental Variation of Productivity with Light Intensity in Experiments Where the Temperature Is Not Controlled

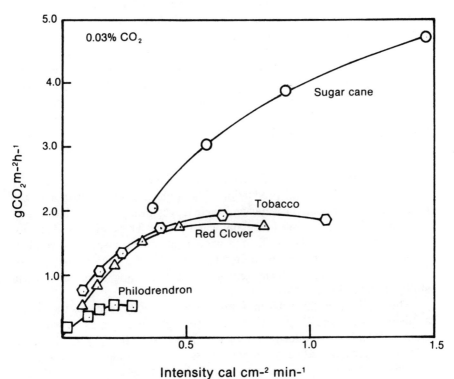

Note: The high-productivity sugarcane is able to obtain high photosynthetic rates at high intensity through the 4-carbon cycle pathway [9, 10].

Figure 17–14. Experimental Photosynthetic Productivity as a Function of Light Intensity for a Range of Productivity Plants

decrease at high light intensity is in agreement with the discussion earlier in this chapter.

Important contributions to the understanding of biological systems in terms of combined cycling receptor and thermodynamic systems have been made by Prigogine at the Universite Libre de Bruxelles [22,25] and Eigen at the Max-Planck Institut für Biophysikalische Chemie [23]. Their work should be consulted for a better understanding of the role of thermodynamics in biology.

In connnection with this efficiency measure at low light intensities, low-productivity crops often show efficiencies equivalent to those of high-productivity crops at low light intensity, as shown in figure 17–14. High-productivity crops are those that are able to increase productivity at higher light intensity, either by reducing photorespiration or by utilizing different enzyme pathways.

The discovery of the four-carbon enzyme pathway for photosynthesis shown in figure 17–15 ranks among the first-rate discoveries in plant biology [12]. Comparison of this four-carbon cycle with the previously known Calvin cycle shown in figure 17–16 indicates the important CO_2 preprocessing mechanism. This preprocessing reduces the chemical potential of CO_2 in the leaf receptors so that assimilation of CO_2 continues at very low environmental CO_2 concentrations. If the temperature is controlled in these four-carbon plants, then the chemical rate of enzyme recycling limits productivity [13]. The species of primary importance with four-carbon uptake systems are corn, sugarcane, bermuda grass, and sorghum. Water loss in these four-carbon plants may be one-fifth that of Calvin-cycle plants at higher light intensity [14].

The reduction in water requirements is one of the principal research needs for increasing agricultural production. The highest-productivity plants are located with their roots in water and their leaves in air. This includes reed grass and sugarcane, as well as many water plants. These high-productivity plants require high water inputs to provide augmented transpiration cooling. Although many of these are four-carbon plants, the four-carbon photosynthesis species have advantages in dry-land farming because of their ability to control water loss.

The requirement for water in primary productivity far exceeds the photosynthetic needs. The efficiency of water use, defined as grams of water per gram of dry matter produced, may vary from 200 to 10,000 [1]. As a general rule, the water required per yield of biomass is lower for high-yield plants. In dry climates the higher ratios are encountered because of the higher transpiration requirements.

Water-deficit effects on yield have been studied by many researchers [15,16,20]. Water deficits typically reduce photosynthesis through three factors:

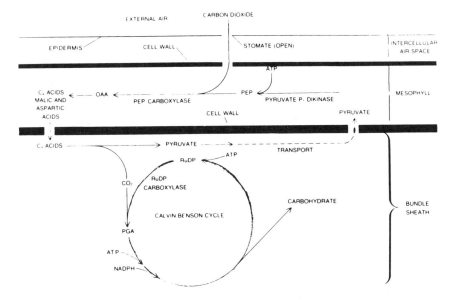

Source: Olle Bjorkman and Joseph Berry, "High Efficiency Photosynthesis," *Scientific American* 229, no. 4 (1973): 82. Copyright 1973 by Scientific American, Inc. All right reserved. Reprinted with permission.

Note: This allows photosynthesis to continue at high light intensity and in reduced CO_2 environments

Figure 17–15. Four-Carbon Photosynthetic Pathway by Which High Productivity Plants Are Able to Preprocess CO_2

1. stomate closing, which reduces CO_2 assimulation;
2. decreased enzyme activity;
3. dehydration reduction of CO_2 diffusion through cell-wall membranes.

Transpiration reduction methods are a common practice in orchards in the southwestern United States through disking or cultivation to maintain exposed soil. Specialized crop production is enhanced at the expense of assimilation capacity.

Revegetation of coal strip-mine areas in dry regions presents other water problems. Since transpiration removes water from vegetation more rapidly than does evaporation from bare soil, a moisture buildup may occur when the vegetation is removed. This buildup enhances the chances for revegetation early in a renovation project. It may, however, lead to later depletion of soil moisture if the new varieties of plants have transpiration rates too high for the soil environment. Initial success in revegetation may eventually then fail because of moisture depletion. On the other hand, the extra moisture storage

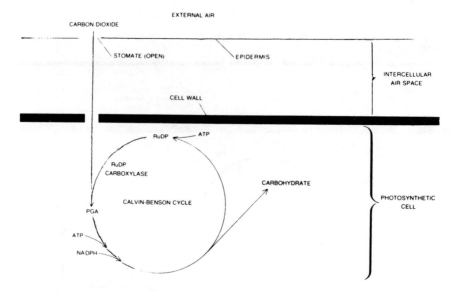

EXTERNAL AIR

CARBON DIOXIDE

STOMATE (OPEN)   EPIDERMIS

INTERCELLULAR
AIR SPACE

CELL WALL

RuDP ← ATP

RuDP
CARBOXYLASE

CARBOHYDRATE

CALVIN-BENSON CYCLE   PHOTOSYNTHETIC
CELL

PGA

ATP

NADPH

Source: From Olle Bjorkman and Joseph Berry, "High Efficiency Photosynthesis," *Scientific American* 229, no. 4(1973): 83. Copyright 1973 by Scientific American, Inc. All rights reserved. Reprinted with permission.

Figure 17–16. Three-Carbon Photosynthetic Pathway (Calvin Cycle) Having Limited Ability to Fix CO_2

in bare soils may allow time for deep-rooted vegetation to reach permanent moisture levels that otherwise might remain unused.

In order to expand the productivity of plants, the most effective method is still the use of irrigation. Genetic developments have recently contributed to enhanced agricultural productivity through the introduction of new plants that utilize less water per unit of organic material produced. This promises to be an important research area in both the near- and long-term future.

Environmental Photosynthesis Effects

Air normally contains approximately 0.03 percent CO_2, and lowering this value decreases photosynthetic production, as shown in the preceding section. As CO_2 is soluble in water, the oceans act as a buffer, absorbing CO_2 in bicarbonates as the CO_2 in the air increases and then releasing it when the air CO_2 decreases. This balance capacity seems to be limited, as observed by the atmospheric increase in CO_2 in the last hundred years as a result of increased fossil-fuel combustion.

The environmental capacity to assimilate CO_2 by the oceans and photosynthetic processes are reduced by many technological processes. Coal strip mining, paving, erosion, toxic-waste disposal, air pollution (SO_2, acid rain, and so forth) and herbicide use all contribute. Hutchinson [24] has pointed out that the cultivation of crops that are photosynthetically active for a short time compared with native vegetation can increase CO_2 in the atmosphere. He also notes that the reduction in the organic content of most soils in intensive agriculture has had a major influence on the amount of CO_2 in the atmosphere. He has calculated that domestic agriculture has contributed more to atmospheric CO_2 than all the industrial contributions from combustion of fossil fuels.

Photosynthesis processes currently act as a control on the surface temperature of the earth by CO_2 and O_2 regulation. Carbon dioxide absorbs solar radiation in the atmosphere as well as radiation emitted by the earth's surface. The net result of an increase in atmospheric CO_2 is to trap more radiation and increase the atmospheric temperature.

In lakes in the United States affected by development of high-density housing or poorly controlled effluent flows, a common result is eutrophication. The nutrient-rich effluents accelerate photosynthetic processes, with concurrent respiration increases. The result is a buildup of nutrients and biomass in a lake. In addition, eutrophication can lead to dramatic increases in growth of algae, which not only are undesirable from the standpoint of lake use, but also lead to the development of an unstable ecosystem dominated by a few specialized species. Although eutrophication is a natural process, the excess-nutrient accumulation is a resource that has other uses. Application of energy for nutrient removal in conventional processes is usually a waste. Irrigation of crops with eutrophic water is one way to use this resource more effectively.

The production of commercial fertilizers discussed previously is a very energy intensive process. The recycling of nutrient-rich water is therefore a double benefit. The advantages of land waste disposal of sewage are now being tested in many parts of the world. Crop irrigation with treated sewage is a viable alternative to biochemical wastewater treatment with return of the water to water courses. There are some difficulties in using land or crop disposal methods for situations in which heavy-metal salts or other toxic substances may enter the effluent.

Estimates of the needs for land area and the processing energy required for fuel production from waste producers and new agricultural production severely limit the potential for biomass fuels. Most analyses suggest that any serious effort to produce biomass fuels on a large scale must rely on ocean areas. The environmental implications of large-scale ocean production of biomass fuels are largely unknown. The potential for both benefit and loss is enormous. The production and assimilation functions of the environment must be well understood for success.

References

1. Gorski, F. *Plant Growth and Entropy Production.* Krakow, Poland: Zaklad Fizjologii Roslin Pan, 1966.
2. Jaynes, E.T. "Information Theory and Statistical Mechanics." *Phys. Rev.* 106(1957):620; 108(1957):171.
3. Callen, H.B. *Thermodynamics.* New York: Wiley, 1960.
4. Morowitz, H.J. *Energy Flow in Biology.* New York: Academic Press, 1968.
5. Boulding, K.E. "The Economics of the Coming Space-Ship Earth." In *Sixth Resources for the Future Forum,* ed. H. Jarrett. Baltimore, Md.: Johns Hopkins University Press, 1966.
6. Margalef, D.R. *Perspectives in Ecological Theory.* Chicago: University of Chicago Press, 1968.
7. Westlake, D.F. "Comparison of Plant Productivity." *Biological Review,* vol. 38. Cambridge: Cambridge University Press, 1963, p. 385.
8. Rosenberg, J.L. *Photosynthesis.* New York: Holt, Rinehart and Winston, 1965.
9. Rabinowitch, E., and Govindjee, *Photosynthesis.* New York: Wiley, 1969.
10. Gaastra, D. "Photosynthesis of Crop Plants as Influenced by Light, Carbon Dioxide, Temperature and Stomatal Diffusion Resistance." *Med. Landbonwhogeschool te Wageningen, Nederland* 59(1959):1.
11. Zelitch, I. *Photosynthesis, Photorespiration and Plant Productivity.* New York: Academic Press, 1971.
12. Hatch, M.D., and Slack, G.R. "Photosynthetic CO_2 Fixation Pathways." *Ann. Rev. Plant Physiol.* 21(1970):141.
13. Odum, H.T. *Environment, Power and Society.* New York: Wiley, 1970.
14. Bjorkman, O., and Berry, J. "High Efficiency Photosynthesis." *Sci. Amer. 228,* no. 4(1973):80.
15. Lieth, H., and Whittaker, R.H., eds. *Primary Productivity of the World.* Chapel Hill: University of North Carolina Press, 1973.
16. Calvin, M. "Photosynthesis as a Resource for Energy and Material." *Amer. Scient.* 64(1976):270.
17. Weiss, P. *The System of Life.* Mt. Kisco, N.Y.: Futura, 1973.
18. Chang, J-H. "Potential Photosynthesis and Crop Productivity." *Ann. Assoc. Amer. Geog.* 60(1970):92.
19. Loomis, R.S.; Williams, W.A.; and Hall, A.F. "Agricultural Productivity." *Ann. Rev. Plant Physiol.* 22(1971):431.
20. Slayter, R.D. "Efficiency of Water Utilization by Arid Zone Vegetation." *Ann. Arid Zone* 3(1964):1.

21. Morowitz, H.J. *Foundations of Bioenergetics*. New York: Academic Press, 1978.
22. Prigogine, I., and Glansdorf, P. *Thermodynamic Theory of Structure, Stability and Fluctuations*. London: Wiley-Interscience, 1971.
23. Eigen, M. "Self-organization of Matter and the Evolution of Biological Macromolecules." *Naturwessenschaften* 58(1971):465.
24. Hutchinson, G.E. "The Biochemistry of the Terrestrial Atmosphere." In *The Earth as a Planet*, ed. G.P. Kuiper. Chicago: University of Chicago Press, 1954, p. 371.
25. Nicolis, G., and Prigogine, I. *Self-organization in Nonequilibrium Systems*. New York: Wiley, 1977.

Appendix 17A:
Diversity and
the Engineer

The idea of *diversity* is a familiar one to ecologists, social scientists, and artists. In ecology, measures of diversity have been used to draw conclusions about the stability and persistance of ecosystems [1–3]. In social science, the value of diversity of cultures for community health and creative behavior is often discussed. Many science-fiction stories deal with the negative aspects of a monoculture manifesting little variety.

It is clear that diversity is an important element in the development of interesting futures. The engineer, however, has little experience in comprehending or analyzing the implications of his designs for diversity.

Diversity promises to become a more important concept in the future, as the limits of resources require new designs. In an engineering context, the design of furnaces that can use a variety of different fuels will become more important. In this situation diversity of resources is an important parameter. The engineer may deal with this by choosing a design that is adaptable to different fuels.

In the chemical and plastics industry, design for diversity of resources will include process equipment that can be adapted to use of multiple resources. An example is the adaptation of plastics processing for using methyl-based resources as well as ethylene-based resources for polymer synthesis.

In the automotive industry the design of automobile components using a single resource like aluminum would represent a lack of understanding of the importance of diversity of supply. Substitutability in this area is important not only from the resource point of view, but also from the standpoint of the need in some cases to deal with supplier labor disputes without curtailing production. In the energy field, the effects of long coal strikes on industry are well documented. The European dependence on OPEC oil supplies is the prime example of a lack of understanding of the importance of supply diversity.

In many cases diversity cannot be achieved without trading off optimization. Multifuel engines or furnaces are often less efficient than those designed for a specific fuel.

Environmental diversity is an even less familiar design parameter for an engineer. The engineer deals primarily with development of new technology to provide goods and services to people. The environment enters largely as a constraint, not as an optimization parameter.

There may be environmental-quality restrictions on the water or air needed by an industrial-processing plant. These may dictate moving the operations to another location. They may also require reprocessing before use.

The penalty for failure to meet output or input constraints may be directly included in optimization procedures if quantitative factors can be assigned. Some aspects of environmental quality resist reduction to quantitative terms. These deal with uniqueness factors and diversity of ecosystems. Recent examples are the Tellico Dam–snail darter controversy and the Alaskan pipeline–Artic tundra situation.

The value of ecological diversity can be demonstrated, but not in direct economic terms. The specific value of endangered species is uncertain. Their value lies not only in the diversity they provide, but also in their genetic singularity. Any stable ecosystem has feedback control systems operating to provide persistence. Removal or addition of an important species can lead to a radically different ecosystem. Examples include the introduction of rabbits to Australia and the lack of predator species to control rats in cities. The use of nonspecific pesticides to control insect pests, killing predatory insects as well, is another example of the impact of changes in diversity.

Perhaps the most serious concern for species diversity was demonstrated during the corn blight in the 1970s in the United States. The development of species of corn that were blight resistant was greatly hindered because most corn varieties were derived from only two major varieties. A resistant variety was found, and the result was successful not only in that the blight was controlled, but also, and more important, in that the search was begun for new varieties that could be used for plant development in the future.

This leads to the most important feature of diversity, which is, of course, diversity in the experience and thought of people involved in creative-design activities. Synectics [8] and other creative-design methodologies have emphasized the importance of analogy and the ability to relate living systems to physical systems. Economic theory and biology have often exchanged or shared ideas in the development of new theories of system behavior. Engineering has often drawn from nature ideas for computer development, structural mechanics, and chemical processing.

Diversity in ecological science is introduced as a measure of both the number of different species present and the distribution of individuals among those species. McArthur [1] followed by many others [2,3] has attempted to provide a numerical measure of diversity. A familiar measure is the information content of an ecosystem. Based on ideas from information-communication theory, this is formulated on the basis that the information content of an ecosystem relates to its stability and its ability to adapt to environmental changes.

The information theory index

$$H = -\Sigma \frac{N_i}{N} \ln \frac{N_i}{N}$$

is introduced, where N_i represents the number of individuals of species i in a sampled ecosystem of N individuals. In this interpretation, the more species present, and the more uneven the distribution, the higher the information content of the ecosystem.

Various attempts to correlate diversity indexes to stability have so far proved inconclusive. The early literature abounds with environmental justification for improving diversity to increase the stability of an ecosystem. Environmental arguments for preserving endangered species have been based on the idea that the more diverse an ecosystem, the greater its stability. This concept is very appealing but has so far resisted substantiation. Its resistance to validation has involved the fact that in a diverse environment a minor species may play a crucial role. This minor species role does not appear as a major factor, however, in the usual diversity measures.

The most general conclusion that can be drawn from the literature on diversity-stability is that environments that have remained stable over the longest time have produced the highest-diversity ecosystems. The relationship of diversity to stability remains an important ecosystem-science problem [4–6]. The crucial element in its development involves the identification of the importance of interactions between species rather than the number of interactions. Ecosystem science should continue to provide insights for engineers as they attempt to relate technological change to environmental factors [7].

References

1. MacArthur, R.H. "Fluctuations of Animal Populations, and a Measure of Community Stability." *Ecology* 36(1955):533.
2. Margalef, D.R. "Information Theory in Ecology." *Gen. Syst.* 3(1958):36.
3. Pielou, E.C. "Species Diversity and Pattern-Diversity in the Study of Ecological Succession." *J. Theoret. Biol.* 10(1966):370.
4. Goodman, D. "The Theory of Diversity-Stability Relationships in Ecology." *Quart. Rev. Biol.* 50(1975):237.
5. May, R.M. *Stability and Complexity in Model Ecosystems*. Princeton, N.J.: Princeton University Press, 1973.
6. Ehrenfeld, D.W. "The Conservation of Non-Resources." *Amer. Scientist* 64(1976):648.
7. Odum, E.P. "The Strategy of Ecosystem Development." *Science* 164(1970):262.
8. Gordon, W.J.J. *Synectics*. New York: Collier Books, 1961.

Index

About the Author

Robert H. Edgerton has been on the faculties of Dartmouth College and Oakland University. He received the Ph.D. in mechanical engineering from Cornell University. Dr. Edgerton has published technical work in the areas of energy conservation, solar energy, resource management, biomedical engineering, and environmental engineering. His professional activities have included serving as president of the Michigan Energy Conservation Group, as a board director for several environmental and solar-energy groups, and as a member of Michigan State advisory committees. He was the recipient of the first Michigan Society of Professional Engineers award as engineering educator of the year. Student groups under his supervision have won major awards in national design competitions with an innovative wind machine and a methanol-fueled vehicle. His present research is in the application of thermodynamics to solar-energy conversion processes.